Methods in Molecular Biology™

Series Editor
John M. Walker
School of Life Sciences
University of Hertfordshire
Hatfield, Hertfordshire, AL10 9AB, UK

For further volumes:
http://www.springer.com/series/7651

Nucleic Acid Detection

Methods and Protocols

Edited by

Dmitry M. Kolpashchikov and Yulia V. Gerasimova

Department of Chemistry, University of Central Florida, Orlando, FL, USA

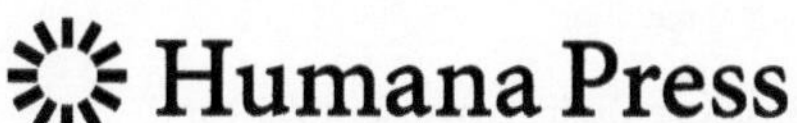

Editors
Dmitry M. Kolpashchikov
Department of Chemistry
University of Central Florida
Orlando, FL, USA

Yulia V. Gerasimova
Department of Chemistry
University of Central Florida
Orlando, FL, USA

ISSN 1064-3745 ISSN 1940-6029 (electronic)
ISBN 978-1-62703-534-7 ISBN 978-1-62703-535-4 (eBook)
DOI 10.1007/978-1-62703-535-4
Springer New York Heidelberg Dordrecht London

Library of Congress Control Number: 2013943589

Printed on acid-free paper

Humana Press is a brand of Springer
Springer is part of Springer Science+Business Media (www.springer.com)

Preface

Nucleic acid detection has an outstanding potential in molecular diagnostics of cancer, infectious diseases, and genetic disorders. It is undoubtedly successful in environmental monitoring, food control, genetic linkage analysis, as well as in forensic casework. Some of the early sequence-specific detection techniques were introduced in the early 1960s and have been evolving and diversifying since then. Among most useful modern technologies the following are of general importance: gel and capillary electrophoresis (CE), polymerase chain reaction (PCR), quantitative real-time PCR (qPCR), DNA microarrays, fluorescent in situ hybridization (FISH), and Southern blot. This volume is a collection of techniques and emerging approaches for the detection of both DNA and RNA. The presented protocols reflect some of the trends in improving nucleic acid detection methods.

Chapter 1 is relevant to forensic DNA analysis and describes an efficient strategy for the recovery of trace amounts of DNA from touched objects followed by its amplification for the analysis of short tandem repeats. Chapter 2 describes a technique for tissue sample preparation that keeps the morphology intact, thus further RNA analysis can reveal the cancerous components from non-cancerous.

Chapters 3 and 4 describe new qPCR chemistries. One common advantage of the methods is the possibility of using the same fluorescent reporter for many targeted sequences, which may significantly reduce the cost of qPCR reagents. Chapters 5 and 6 describe other qPCR compatible cost-efficient fluorescent probes that have an additional advantage of improved selectivity and specificity. Overall, there is a trend of using *universal* fluorescent reporters adaptable for the detection of any DNA or RNA target in real-time formats, which promises to make probe-based qPCR more affordable in the near future.

As alternatives to PCR, isothermal DNA amplification techniques are becoming popular. They have the advantage of eliminating the need for PCR thermal cycler thus representing less-demanding and affordable alternatives to PCR in point-of-care (POC) diagnostics. This volume includes examples of helicase-dependent amplification (Chapter 7) and loop-mediated isothermal amplification (LAMP) (Chapters 8–10).

An alternative to DNA amplification is signal amplification approach, which has received significant attention in the last decade. These types of assays do not amplify the amount of target DNA. Instead, the presence of an analyte activates enzymes that amplify the signal by multiple processing of a (fluorogenic) substrate. Invader assay and 5′ fluorogenic exonuclease assay (TaqMan) are examples of such approaches. Yet another signal amplification technique is detailed in Chapter 11 of this book.

Most common methods for nucleic acid detection rely on fluorescent outputs. However, the demand of POC diagnostics calls for the assays with visually recognized signals. Such assays do not require any instrumentation and minimize the processing time and the user expertise required. Techniques based on peroxidase-like DNA enzyme (Chapters 12 and 13) and gold nanoparticles (Chapter 14) are included in this volume. Other alternatives to fluorescence are electrochemiluminescence (Chapter 15), interferometric reflectance imaging (Chapter 16), and electrochemical detection using graphene oxide (Chapter 17).

FISH has been employed for the chromosome analysis since the early 1980s but still undergoes improvements (Chapter 19) and faces new applications (Chapter 2). A FISH-related technology uses short peptide nucleic acid (PNA) strands to unwind a local dsDNA fragment followed by sequence-specific analysis of the opened fragment (Chapter 18). This very promising approach has the advantages of sequence specificity, low detection limits, and mild hybridization conditions.

In recent years significant attention was devoted to the detection of micro RNAs (miRNAs) as they are considered to be important cancer biomarkers (Chapters 2, 20–22). Another extensively explored application is RNA imaging in live cells. This field is driven both by the great success of green fluorescent protein-based protein imaging and growing understanding of the great diversity and importance of intracellular RNA. Chapters 23 and 24 describe new hybridization probes that might be useful for the purposes of intracellular RNA monitoring.

Besides the traditional issues of detection limits, selectivity, and reliability, the one common trend in the new nucleic acid detection methods is to make them suitable for POC diagnostics, which means, reduced assay time, cost efficiency, and simple and straightforward formats. The other trend is detection of micro RNA and RNA in living cells. While one volume cannot accommodate all the existing developments in the field of nucleic acid detection we hope that the presented methods will be useful for those who are interested in DNA or RNA analysis.

Orlando, FL, USA

Dmitry M. Kolpashchikov
Yuila V. Gerasimova

Contents

Part VII Detection of microRNA

Part VIII RNA Imaging in Live Cells

Contributors

SUNMIN AHN • *Department of Biomedical Engineering, Boston University, Boston, MA, USA*
WANYUAN AO • *Great Basin Corporation, Salt Lake City, UT, USA*
HIROKA AONUMA • *Department of Tropical Medicine, The Jikei University School of Medicine, Tokyo, Japan*
NEETIKA ARORA • *Australian Infectious Diseases Research Centre, School of Chemistry and Molecular Biosciences, University of Queensland, St. Lucia, QLD, Australia*
EDWARD R. ATWILL • *Department of Population Health and Reproduction, University of California Davis, Davis, CA, USA*
ATHANASE BADOLO • *University of Ouagadougou, Ouagadougou, Burkina Faso; Centre National de Recherche et de Formation sur le Paludisme, Ouagadougou, Burkina Faso*
MEREDITH B. BAKER • *Division of Cardiology, Emory University School of Medicine, Atlanta, GA, USA*
JACK BALLANTYNE • *National Center for Forensic Science, Orlando, FL, USA; Department of Chemistry, University of Central Florida, Orlando, FL, USA*
GANG BAO • *Department of Biomedical Engineering, Georgia Institute of Technology and Emory University, Atlanta, GA, USA*
ROSS T. BARNARD • *Australian Infectious Diseases Research Centre, School of Chemistry and Molecular Biosciences, University of Queensland, St. Lucia, QLD, Australia*
GRAEME BARNETT • *Qponics Limited, Brisbane, QLD, Australia*
MAXIM V. BEREZOVSKI • *University of Ottawa, Ottawa, ON, Canada*
HONG-YUAN CHEN • *State Key Laboratory of Analytical Chemistry for Life Science, School of Chemistry and Chemical Engineering, Nanjing University, Nanjing, China*
SHOUYI CHEN • *Department of Microbiology, Guangzhou Center for Disease Control and Prevention, Guangzhou, China*
EVAN M. CORNETT • *Burnett School of Biomedical Sciences, College of Medicine, University of Central Florida, Orlando, FL, USA*
SIMON R. CORRIE • *Australian Institute for Bioengineering and Nanotechnology, University of Queensland, St. Lucia, QLD, Australia*
VENKATARAMAN DHARUMAN • *Molecular Electronics Lab, Department of Bioelectronics and Biosensors, Alagappa University, Karaikudi, India*
XIAOYAN DONG • *Beijing FivePlus Molecular Medicine Institute, Beijing, China*
ZHIYUAN FANG • *Key Laboratory of Regenerative Biology, South China Institute for Stem Cell Biology and Regenerative Medicine, Guangzhou Institutes of Biomedicine and Health, Chinese Academy of Sciences, Guangzhou, China*
DAVID S. FREEDMAN • *Department of Electrical and Computer Engineering, Boston University, Boston, MA, USA*

YULIA V. GERASIMOVA • *Department of Chemistry, University of Central Florida, Orlando, FL, USA*
ERIN K. HANSON • *National Center for Forensic Science, Orlando, FL, USA*
ANDREAS HERRMANN • *Molecular Biophysics, Department of Biology, Humboldt University, Berlin, Germany*
AKIRA ITO • *Department of Parasitology, Asahikawa Medical University, Asahikawa, Japan*
KUMARASAMY JAYAKUMAR • *Molecular Electronics Lab, Department of Bioelectronics and Biosensors, Alagappa University, Karaikudi, India*
ROBERT JENISON • *Great Basin Corporation, Salt Lake City, UT, USA*
JEFF J. JOHNSON • *Harper Cancer Research Institute, University of Notre Dame, Notre Dame, IN, USA; Department of Chemistry and BiochemistryUniversity of Notre Dame, Notre Dame, IN, USA*
HIROTAKA KANUKA • *Department of Tropical Medicine, The Jikei University School of Medicine, Tokyo, Japan*
TATYANA V. KARAMYSHEVA • *Institute of Cytology and Genetics, Siberian Branch, Russian Academy of Sciences, Novosibirsk, Russian Federation, Novosibirsk, Russia*
NASRIN KHAN • *Department of Chemistry, University of Ottawa, Ottawa, ON, Canada*
ANDREA KNOLL • *Organic and Bioorganic Chemistry, Department of Chemistry, Humboldt University, Berlin, Germany*
DMITRY M. KOLPASHCHIKOV • *Department of Chemistry, University of Central Florida, Orlando, FL, USA*
SUSANN KUMMER • *Max Planck Institute Chemistry, Gottingen, Germany*
RICHARD LAI • *Australian Infectious Diseases Research Centre, School of Chemistry and Molecular Biosciences, University of Queensland, St. Lucia, QLD, Australia*
PAVEL P. LAKTIONOV • *Institute of Chemical Biology and Fundamental Medicine, Siberian Branch, Russian Academy of Sciences, Novosibirsk, Russian Federation*
JUNTAO LI • *Department of Microbiology, Guangzhou Center for Disease Control and Prevention, Guangzhou, China*
XUNDE LI • *Department of Population Health and Reproduction, University of California Davis, Davis, CA, USA*
FANG LIANG • *Australian Infectious Diseases Research Centre, School of Chemistry and Molecular Biosciences, University of Queensland, St. Lucia, QLD, Australia*
PUCHANG LIE • *Key Laboratory of Regenerative Biology, South China Institute for Stem Cell Biology and Regenerative Medicine, Guangzhou Institutes of Biomedicine and Health, Chinese Academy of Sciences, Guangzhou, China*
YINJIAO MA • *Department of Pharmacology, Jinling Hospital, Nanjing University School of Medicine, Nanjing, China*
ELISA MOKANY • *SpeeDx Pty Ltd., Eveleigh, NSW, Australia*
EVGENIY S. MOROZKIN • *Institute of Chemical Biology and Fundamental Medicine, Siberian Branch, Russian Academy of Sciences, Novosibirsk, Russian Federation*
SHIZUKA NAKAYAMA • *Department of Chemistry and Biochemistry, University of Maryland, College Park, MD, USA*
JIA LING NEO • *Department of Chemistry, Defence Medical and Environmental Research Institute, DSO National Laboratories, Singapore, Singapore*
AGATHE NKOUAWA • *Department of Parasitology, Asahikawa Medical University, Asahikawa, Japan*

Kiyoshi Okado • *Department of Tropical Medicine, The Jikei University School of Medicine, Tokyo, Japan*
Akimitsu Okamoto • *The Research Center for Advanced Science and Technology (RCAST), The University of Tokyo, Tokyo, Japan*
Darnley Pearson • *Australian Infectious Diseases Research Centre, School of Chemistry and Molecular Biosciences, University of Queensland, St. Lucia, QLD, Australia*
Guang-sheng Qian • *State Key Laboratory of Analytical Chemistry for Life Science, School of Chemistry and Chemical Engineering, Nanjing University, Nanjing, China*
Rajendiran Rajesh • *Department of Chemistry, Pondicherry University, Pondicherry, India*
Nikolay B. Rubtsov • *Institute of Cytology and Genetics, Siberian Branch, Russian Academy of Sciences, Novosibirsk, Russian Federation, Novosibirsk, Russia*
Yasuhito Sako • *Department of Parasitology, Asahikawa Medical University, Asahikawa, Japan*
Charles D. Searles • *Division of Cardiology, Emory University School of Medicine and Atlanta Veterans Administration Medical Center, Atlanta, GA, USA*
Oliver Seitz • *Organic and Bioorganic Chemistry, Department of Chemistry, Humboldt University, Berlin, Germany*
Zonggao Shi • *Harper Cancer Research Institute, University of Notre Dame, Notre Dame, IN, USA*
Herman O. Sintim • *Department of Chemistry and Biochemistry, University of Maryland, College Park, MD, USA*
Theo Sloots • *Queensland Paediatric Infectious Diseases Laboratory, Queensland Children's Medical Research Institute, the University of Queensland, Herston, QLD, Australia*
Irina Smolina • *Department of Biomedical Engineering, Boston University, Boston, MA, USA*
M. Sharon Stack • *Harper Cancer Research Institute, University of Notre Dame, Notre Dame, IN, USA; Department of Chemistry and Biochemistry, University of Notre Dame, Notre Dame, IN, USA*
Wenhong Tian • *College of Life Science, Jilin University, Changchun, Jilin, China*
Alison V. Todd • *SpeeDx Pty Ltd., Eveleigh, NSW, Australia*
M. Selim Ünlü • *Department of Biomedical Engineering, Boston University, Boston, MA, USA; Department of Electrical and Computer Engineering, Boston University, Boston, MA, USA*
Mahesh Uttamchandani • *Department of Chemistry, Defence Medical and Environmental Research Institute, DSO National Laboratories, Singapore, Singapore; Department of Chemistry, National University of Singapore, Singapore, Singapore; Department of Biological Sciences, National University of Singapore, Singapore, Singapore*
Rangarajan Venkatesan • *Department of Chemistry, Pondicherry University, Pondicherry, India*
Valentin V. Vlassov • *Institute of Chemical Biology and Fundamental Medicine, Siberian Branch, Russian Academy of Sciences, Novosibirsk, Russian Federation*
David Whiley • *Queensland Paediatric Infectious Diseases Laboratory, Queensland Children's Medical Research Institute, the University of Queensland, Herston, QLD, Australia*

MEI-SHENG WU • *State Key Laboratory of Analytical Chemistry for Life Science, School of Chemistry and Chemical Engineering, Nanjing University, Nanjing, China*
XIAOBING WU • *Beijing FivePlus Molecular Medicine Institute, Beijing, China*
ZHIJIAN WU • *Unit on Ocular Gene Therapy, National Eye Institute, NIH, Bethesda, MD, USA*
ZHUO XIAO • *Key Laboratory of Regenerative Biology, South China Institute for Stem Cell Biology and Regenerative Medicine, Guangzhou Institutes of Biomedicine and Health, Chinese Academy of Sciences, Guangzhou, China*
JING-JUAN XU • *State Key Laboratory of Analytical Chemistry for Life Science, School of Chemistry and Chemical Engineering, Nanjing University, Nanjing, China*
TETSUYA YANAGIDA • *Department of Parasitology, Asahikawa Medical University, Asahikawa, Japan*
DAVID CHE CHENG YEH • *Australian Infectious Diseases Research Centre, School of Chemistry and Molecular Biosciences, University of Queensland, St. Lucia, QLD, Australia*
LINGWEN ZENG • *Key Laboratory of Regenerative Biology, South China Institute for Stem Cell Biology and Regenerative Medicine, Guangzhou Institutes of Biomedicine and Health, Chinese Academy of Sciences, Guangzhou, China*
KANG LIANG ZHANG • *Australian Infectious Diseases Research Centre, School of Chemistry and Molecular Biosciences, University of Queensland, St. Lucia, QLD, Australia*
XIRUI ZHANG • *Department of Biomedical Engineering, Boston University, Boston, MA, USA*
GUOHUA ZHOU • *Department of Pharmacology, Jinling Hospital, Nanjing University School of Medicine, Nanjing, China*
BINGJIE ZOU • *Department of Pharmacology, Jinling Hospital, Nanjing University School of Medicine, Nanjing, China*

Part I

Advances in Sample Preparation

Chapter 1

"Getting Blood from a Stone": Ultrasensitive Forensic DNA Profiling of Microscopic Bio-Particles Recovered from "Touch DNA" Evidence

Erin K. Hanson and Jack Ballantyne

Abstract

In forensic casework analysis it is sometimes necessary to obtain genetic profiles from increasingly smaller amounts of biological material left behind by persons involved in criminal offenses. The ability to obtain profiles from trace biological evidence is routinely demonstrated with the so-called touch DNA evidence (generally perceived to be the result of DNA obtained from shed skin cells transferred from donor to an object or a person during physical contact). The current method of recovery of trace DNA employs cotton swabs or adhesive tape to sample an area of interest. While of practical utility, such a "blind-swabbing" approach will necessarily co-sample cellular material from the different individuals whose cells are present on the item, even if the individuals' cells are located in geographically distinct locations on the item. Thus some of the DNA mixtures encountered in such touch DNA samples are artificially created by the swabbing itself. Therefore, a specialized approach for the isolation of single or few cells from "touch DNA evidence" is necessary in order to improve the analysis and interpretation of profiles recovered from these samples. Here, we describe an optimized and efficient removal strategy for the collection of cellular microparticles present in "touch DNA" samples, as well as enhanced amplification strategies to permit the recovery of short tandem repeat profiles of the donor(s) of the recovered microparticles.

Key words "Touch DNA" evidence, Trace biological evidence, Cell isolation and recovery, DNA profiling, Short tandem repeat (STR) analysis, Forensic science

1 Introduction

"Touch DNA" evidence is generally perceived to be the result of DNA obtained from shed skin cells transferred from donor to an object or a person during physical contact (i.e., Locard's exchange principle: "every contact leaves a trace") [1–6]. Therefore, it is often assumed that if DNA is recovered from an object that is typically handled, for example a knife, then the DNA originated from shed skin cells from the handler. However, in the absence of any supporting scientific data, this could be a precarious assumption.

Dmitry M. Kolpashchikov and Yulia V. Gerasimova (eds.), *Nucleic Acid Detection: Methods and Protocols*, Methods in Molecular Biology, vol. 1039, DOI 10.1007/978-1-62703-535-4_1,

Numerous published reports demonstrate the ability to obtain DNA profiles from a variety of objects that have been in direct contact with a person including items such as paper and documents [7, 8], keyboards [9], bedding and fabrics [5], shoe insoles [1], firearms and fired cartridges cases [10], drinking containers, pens, and briefcase handles [11]. These studies involved the collection of biological material through swabbing, tape lifts, or removal of portions of the material itself and do not include an assessment of the type of biological material present. Additionally, these studies typically involved simulated conditions where there was prior knowledge of direct contact with skin and these same assumptions cannot be applied to forensic casework analysis where the precise circumstances and events of a crime are not typically known. While in many cases it may in fact be the direct result of skin contact, the presence, for example, of trace amounts of saliva from incidental (not surface to surface) contact cannot be excluded [12]. While it is not the responsibility of forensic casework analysts to interpret the circumstances or events of a crime, there is a responsibility to report scientific findings, not assumptions, in service of the law. It could be considered negligent to report that DNA is from shed skin cells if there is no scientific basis for such a conclusion and could ultimately have serious repercussions for justice if such misrepresentations, even if inadvertently implied, are presented in court.

Very few studies involving the analysis of touch DNA samples have included an examination of the nature of the biological material present in these samples [13, 14]. In the study conducted by Kita et al., microscopic examinations of human neck skin using histological and immunological staining provided a representation of the characteristics of outer epidermal skin layers (flattened cells with condensed nuclei which have lost their typical shape) [14]. Using antibodies to single-stranded DNA they were able to identify fragmented DNA in stripped nuclei on the skin surface and uppermost layers [14]. However, when a fingerprint on a glass slide was examined, stripped nuclei were rarely detected and trace DNA was prevalent [14]. Therefore, this study demonstrates that assumptions made about the surface of some areas of skin may not directly apply to the material that is transferred to objects through touch or contact. Additionally, swabs of human skin surface were examined and the presence of degraded DNA was observed after the electrophoretic analysis of extracted DNA [14]. However, this study ignored the possibility that the degraded DNA observed was actually due to the presence of low-molecular-weight bacterial DNA from the microbiota known to be present on the surface of human skin. Allesandrini et al. performed experiments involving fingerprints on glass slides and found a considerable level of interindividual variability in the number of nucleated cells (0–9 depending on the donor) and "stripped nuclei" (0–8 depending on the donor) [13]. While these studies provide an indication that potential skin cellular material is present in "touch DNA" samples, they do not provide any methodologies to conclusively identify a skin source of origin. Additionally, the authors did not demonstrate that the recovered DNA profiles came directly from the identified cellular material. It is possible that the source of DNA did not originate from the shed skin cells but from "naked DNA" from other body secretions such as sweat or sebaceous fluid. It is, therefore, essential that critical research be done to characterize the biological material recovered from touched objects and to conclusively demonstrate a direct link between the biological material and recovery of DNA profiles from it. Here, we describe an optimized and efficient removal strategy for the collection of cellular microparticles present in "touch DNA" samples, as well as enhanced amplification strategies to permit the recovery of short tandem repeat (STR) profiles of the donor(s) of the recovered microparticles.

A critical factor in the analysis of touch DNA evidence is the successful recovery of the trace biological material present. As mentioned previously, the collection of touch DNA evidence typically involves swabbing of the suspected area with a cotton swab. However, with certain types of objects swabbing may fail to recover biological material contained in grooves and crevices. When biological material is recovered with a swab, successful analysis of samples recovered in this manner will, depending upon whether used wet or dry, be influenced by the absorptivity and adsorptivity of the swab used and also the efficiency of release of the biological material. The potential for loss of sample is increased during the physical manipulations required to remove the absorbed or adsorbed material from the collection swab. Additionally, the use of generalized swabbing techniques often results in the recovery of a non-resolvable or challenging admixed DNA profile as a result of the failure to conduct a prior inspection of biological material that is present to determine if separate recovery of the particles could be achieved [6, 11, 15]. If a mixture was present due to multiple-contributor handling of the object, the contact between the object and second donor may result in a significant loss of biological material from the first donor. If the small amount of material remaining from the original donor was not selectively isolated and is instead recovered in a generalized sampling of biological material, standard extraction and analysis techniques may fail to recover that minor profile.

In order to avoid the above-described challenges of "blind swabbing" strategies, our experimental schema for the recovery of bio-particles from "touch DNA" involves the following (adapted from [16] and personal communications with McCrone Associates, Inc. Westmont, IL): (1) transfer of the biological material from touched objects and surfaces to a low-retention adhesive material, (2) microscopic examination of the recovered biological material, and (3) isolation of single or few bio-particles or cells (identified as cells at this stage only if a nucleus can be identified) using a water-soluble adhesive. The use of the water-soluble adhesive permits the transfer of the recovered particles to a lysis buffer or directly into a PCR amplification reaction since the adhesive will dissolve, thus releasing the particles into solution. The entire transfer process can be visualized under the microscope ensuring successful transfer of the recovered bio-particles for further analysis. All subsequent reactions (e.g., STR profiling) are performed in the same reaction tube, thereby eliminating potential sample loss from additional manipulations or sample transfers. Several reports claiming to be able to recover DNA profiles from single cells have been published but none have adduced convincing data that full autosomal STR profiles can be consistently obtained [17, 18]. The sensitivity of routinely used commercially available STR amplification kits using standard manufacturer's conditions will be insufficient for the analysis of trace biological evidence unless the microparticles recovered contain a sufficient number of cells. However, next-generation STR multiplex systems are available that have an increased robustness (ability to overcome inhibitors) and sensitivity (≤100 pg) [19, 20]. We have utilized one of these STR multiplex systems, AmpFlSTR® Identifiler® Plus PCR amplification kit (Applied Biosystems by Life Technologies), in combination with increased amplification cycle number in order to achieve improved sensitivity and permit STR profiles to be recovered from bio-particles recovered from "touch DNA" samples.

While the developed protocols permit the recovery and analysis of bio-particles from "touch DNA" samples, they are still in their infancy and will require additional work prior to routine use in forensic casework. A majority of the cells found in the outer epidermal layer of skin are dead or dying keratinocytes (i.e., corneocytes) [21]. During terminal differentiation of these cells, there is a degradation of the cell nucleus and possible nuclear material [21]. Therefore, the cellular material recovered from touched objects may contain nonnucleated cells. Without the presence of a nucleus, it will be difficult to determine if the recovered material is in fact cellular biological material using a microscopic examination. Therefore, at this time it is difficult to determine which and how many bio-particles should then be selected in order to successfully recover DNA profiles from "touch DNA" evidence. Therefore, additional work will be required in order to develop a more comprehensive characterization of the biological cellular material present in touched objects. Additionally, while genetic profiles from trace biological evidence can be obtained, the true nature of touch DNA evidence has remained elusive, generally perceived to be the result of DNA obtained from shed skin cells yet never confirmed with scientific certitude. This is largely due to the perception that it is not possible to ascertain the tissue source of origin of the biological material in touch DNA evidence. The uncertainty with regard to the source of trace biological material is now being exploited in some criminal proceedings in an attempt to diminish the significance of trace biological evidence. Thus far, research has failed to provide operational crime laboratories with feasible methods to identify the tissue source of origin of touch DNA samples.

We are hopeful that the methods developed here (and by others also working on "touch DNA" evidence) will help the forensic community further characterize the biological material in touch or contact DNA evidence and subsequently lead to a molecular identification of the presence of skin cells. The approach described in this chapter involves the following steps: (1) collection of suspected bio-particles onto adhesive material (WF Gel-Film® (Gel-Pak®)); (2) optional cell staining; (3) removal of bio-particles from Gel-Film® surface with water-soluble adhesive; (4) nucleic acid isolation (i.e., direct lysis); (5) enhanced STR amplification of isolated DNA; and (6) detection of amplified products using capillary electrophoresis. Such an approach is expected to yield significantly more probative information from trace evidentiary items.

2 Materials

2.1 General Supplies and Equipment

1. Sterile microcentrifuge tubes: 0.5 or 1.5 mL.
2. Tweezers (cleaned thoroughly between samples).
3. KimWipes®.
4. 70 % ethanol.
5. Deionized water.
6. Glass microscope slides.
7. Disposable transfer pipets.
8. Pipets: 0.2–2, 2–20, 10–100, 100–1,000 μL.
9. Sterile, aerosol-resistant pipet tips.
10. Vortex.
11. Mini-centrifuge.
12. Stereomicroscope (*see* **Note 1**).
13. Thermal cycler (*see* **Note 2**).
14. 3130 Genetic Analyzer (Applied Biosystems (AB) by Life Technologies) (*see* **Note 3**).
15. GeneMapper® software (AB by Life Technologies).

2.2 Recovery of Bio-Particles from Touched Objects

1. WF Gel-Film® (Gel-Pak®), ×8 retention level (*see* **Note 4**).
2. Trypan blue stain (0.4 %) (*see* **Note 5**).

2.3 Sample Preparation

1. Double-sided tape.
2. Glass blocks (3″ × 1.5″ × ½″) (*see* **Note 6**).
3. Tungsten needle with holder.
4. 3 M™ water-soluble wave solder tape 5414 transparent.
5. Sterile 0.2 mL PCR amplification tubes.

2.4 Nucleic Acid Isolation: Direct Lysis

1. Direct lysis buffer (*forensic*GEM™ Saliva, ZyGEM) (*see* **Note 7**).
2. Sterile water.

2.5 Autosomal Short Tandem Repeat Profiling

1. AmpFlSTR® Identifiler® Plus PCR amplification kit (AB by Life Technologies) (*see* **Note 8**).
2. Sterile water.

2.6 Product Detection: Capillary Electrophoresis (See Note 9)

1. Hi-Di™ formamide (AB by Life Technologies).
2. GeneScan™ 500 LIZ™ size standard (AB by Life Technologies).
3. 96-well plates.
4. Plate septa, 96-well (AB by Life Technologies).
5. Performance-optimized polymer, POP-7 (AB by Life Technologies).

3 Methods

3.1 Recovery of Bio-Particles from Touched Objects (See Note 10)

3.1.1 Without Initial Glass Slide Support

1. Cut the Gel-film® to a desired size, ensuring that the size is appropriate for subsequent attachment to the glass microscope slide support.
2. Prior to removing the top protective plastic film layer from the Gel-film® surface, make a shallow cut along the Gel-film® to create an area for handling separate from the sample collection area.
3. Remove the top protective plastic layer from the sample collection portion of the Gel-film®. From this point on, only handle the Gel-film® in the designated handling area in order to avoid contamination of the sample collection area.
4. Place the Gel-film® surface in contact with the touched object of interest. Apply a small amount of pressure to ensure efficient collection of bio-particles. Multiple samplings from the same area of the touched item can be collected on the same piece of Gel-film® if desired (Fig. 1). Avoid using a rubbing motion during collection as this may disturb bio-particles on the Gel-film®.
5. Once collection is complete, the Gel-film® should then be attached to a glass microscope slide support. Holding the Gel-film® only in the designated handling area, remove the back

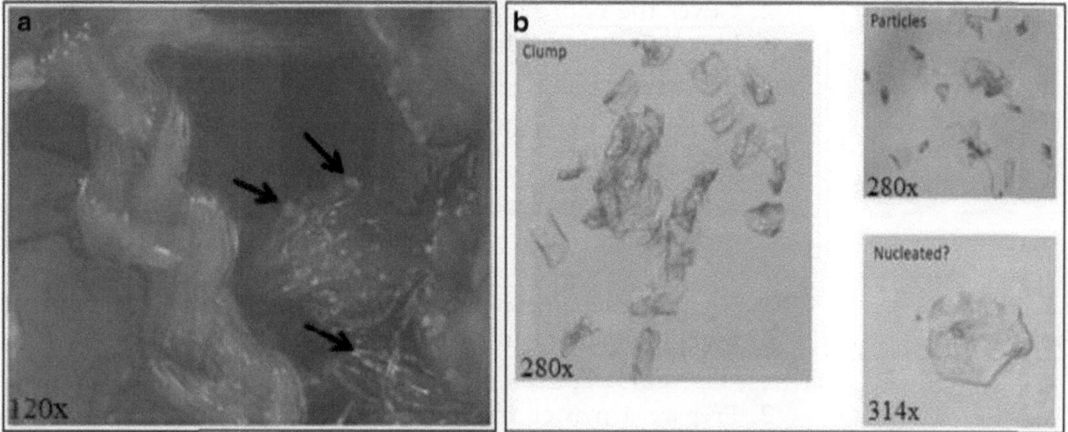

Fig. 1 Bio-particles on worn clothing item. (**a**) Worn clothing item examined using the stereomicroscope (120× magnification). *Arrows* indicate location of possible bio-particles. (**b**) Recovered bio-particles after transfer to Gel-film®. Bio-particles can be observed as "clumps" (*left panel*) or individual "particles" (*top right panel*). Occasionally, the presence of a possible nucleus can be observed (*bottom right panel*)

protective covering to expose the adhesive backing. Carefully place the Gel-film® onto a clean glass microscope slide and press down on the edges and corners with sterile tweezers to secure the Gel-film® on the glass slide.

3.1.2 With Initial Glass Slide Support

1. Cut the Gel-film® to the desired size, ensuring that the size is appropriate for subsequent attachment to the glass microscope slide support.
2. Remove the back protective covering of the Gel-film® to expose the adhesive backing. Place the Gel-film® onto a clean glass microscope slide and press to secure the Gel-film® on the glass slide. The Gel-film® contains a top protective layer, so the entire surface of the Gel-film® can be pressed down onto the slide.
3. Once the Gel-film® is secure on the glass slide support, the top protective layer can be removed using sterile tweezers.
4. Place the Gel-film® surface in contact with the touched object of interest. Apply a small amount of pressure to ensure efficient collection of bio-particles. If too much pressure is applied, the glass slide could break. Multiple samplings from the same area of the touched item can be collected on the same piece of Gel-film® if desired (Fig. 1). Avoid using a rubbing motion during collection as this may disturb bio-particles on the Gel-film®.

3.1.3 Optional Cell Staining (See **Note 5***)*

1. Place the sample (Gel-film® containing recovered bio-particles from touched object on the glass slide support) onto a slide-staining rack over a sink.

2. Cover the entire sample surface of the Gel-film® with trypan blue stain using a disposable transfer pipet.
3. Incubate at room temperature for 1–2 min.
4. Remove excess stain. Gentle flooding with sterile water can be used if needed.
5. Allow the samples to air-dry before proceeding to sample preparation.

3.2 Sample Preparation

1. Remove the appropriate number of 0.2 mL PCR tubes from their container and place them on a rack. Label the tubes appropriately.
2. Prepare a master mix of lysis buffer (*forensic*GEM™ Saliva): 6.9 μL of sterile water, 2.1 μL of 10× buffer—BLUE, and 1 μL of *forensic*GEM™ (volumes listed are per sample). Vortex the master mix well and briefly centrifuge in a mini-centrifuge.
3. Pipet 10 μL of the prepared lysis buffer in the bottom of the 0.2 mL PCR tube. Cap each tube loosely or tightly if desired.
4. Place a piece of double-sided tape onto a clean glass microscope slide or directly on the glass block. This will be used to hold the 0.2 mL PCR tube in place during sample preparation. This slide should be prepared fresh for each collection.
5. Place a piece of double-sided tape (or two side-by-side depending on the width of the tape) onto a clean glass microscope slide. Place a piece of the 3 M™ water-soluble wave solder tape on top of the double-sided tape. This slide can be stored for future use. Place the first 0.2 mL PCR tube flat on its side and place on the double-sided tape. Use gentle force to secure the tube on the double-sided tape. Set this tube aside until the sample is collected.
6. Place the 3 M™ water-soluble wave solder tape slide on a glass block and view under the microscope (low magnification) (Fig. 2). Using the tip of the tungsten needle, lightly scrape the surface of the tape in order to collect a small amount ("ball") of the adhesive material on the tip of the tungsten needle. The size of the "ball" can be increased if a larger number of bio-particles will be collected or multiple collections can be placed into the same collection tube.
7. Carefully remove the tungsten needle with adhesive from under the microscope. The needle can be placed in a rack (tip side up) if needed.
8. Place the sample (Gel-film® containing recovered bio-particles from touched object on the glass slide support) on a glass block and place on microscope stage. Adjust the focus and magnification until the bio-particles can be easily viewed (*see* **Note 11**). Determine which bio-particles will be collected (*see* **Note 12**).

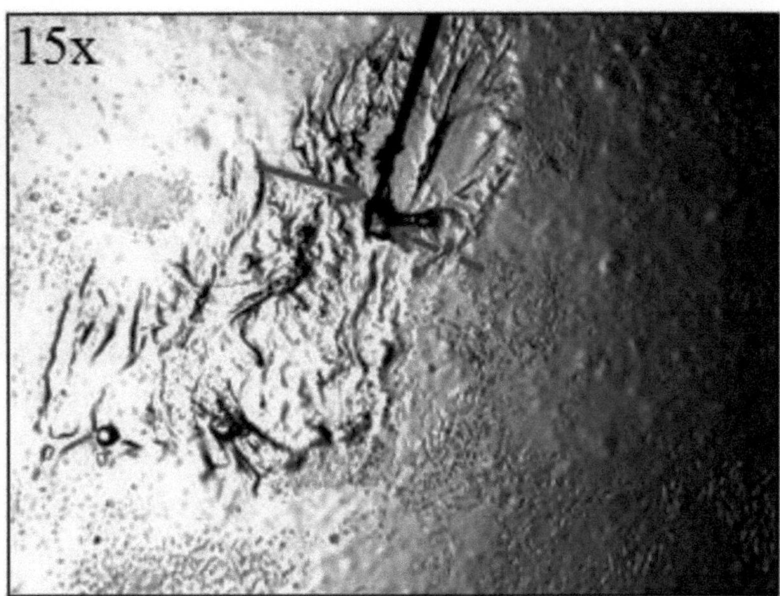

Fig. 2 Collection of 3M™ water-soluble adhesive on the tip of tungsten needle. The tip of the tungsten needle (*solid arrow*) is used to collect a small "ball" of adhesive (*dashed arrow*)

9. Retrieve the tungsten needle with adhesive and bring over the surface of the Gel-film® where the targeted bio-particles are located (*see* **Note 13**).
10. Place the tungsten needle with adhesive over the bio-particles of interest and press down so that the adhesive is in contact with the bio-particles. Lift the needle up to ensure that the bio-particle has been removed from the Gel-film® surface. Repeat this process until the desired number of bio-particles has been collected (*see* **Note 14**).
11. Once the desired number of bio-particles has been collected, keep the tungsten needed in hand and replace the sample with the prepared 0.2 mL PCR tube on the microscope stage. Lower the magnification on the microscope so that the bottom of the 0.2 mL tube containing the lysis buffer is in focus.
12. Carefully insert the tungsten needle into the 0.2 mL PCR tube, avoiding contact with the tube walls, until the needle is placed into the lysis buffer (Fig. 3). Hold the needle in the lysis buffer until the adhesive dissolves and the bio-particles are released into solution (*see* **Note 15**).
13. Remove the 0.2 mL PCR tube from the double-sided tape and place upright. Close the lid to the tube tightly. If necessary, briefly centrifuge the tube in a mini-centrifuge to collect all liquid at the bottom of the tube.

Fig. 3 Transfer of collected bio-particles into reaction tube. Bio-particles, collected using the water-soluble adhesive on the tip of the tungsten needle, are placed directly into the lysis buffer (10 μL) at the bottom of a 0.2 mL PCR tube. The needle is held in the lysis buffer until the adhesive dissolves and the cells are released into solution. The *red arrow* indicates the location of a bio-particle released into solution

14. Clean the tungsten needle between samples.
15. Repeat **steps 6–12** until all samples have been collected.

3.3 Nucleic Acid Isolation (Direct Lysis)

1. The direct lysis reaction should be performed immediately after sample collection.
2. Place samples in a thermal cycler and use the following lysis program: 75 °C for 15 min, and 95 °C for 5 min.
3. Briefly centrifuge samples, if needed, to collect all volume in the bottom of the tube.

3.4 Autosomal Short Tandem Repeat Profiling

1. Remove two additional 0.2 mL PCR tubes from their container and place them on a rack for the amplification positive and negative control. Label the tubes appropriately.
2. Add 10 μL of the AmpFlSTR control DNA 9947a (0.10 ng/μL) to the appropriate 0.2 mL tube ("positive control"). Add 10 μL of sterile water to the appropriate 0.2 mL tube ("negative control") (*see* **Note 16**).
3. Prepare the amplification master mix in a microcentrifuge tube (0.5 or 1.5 mL) according to the manufacturer's recommended protocol (volumes per sample): 10 μL AmpFlSTR® Identifiler® Plus Master Mix, and 5 μL AmpFlSTR® Identifiler®

Plus Primer set (*see* **Note 17**). Vortex the master mix and centrifuge briefly.

4. Add 15 μL of the prepared master mix to each 0.2 mL sample tube for a 25 μL total reaction volume (*see* **Note 18**).
5. Place samples in the thermal cycler (*see* **Note 2**) and amplify the samples using the following cycling conditions: 95 °C for 11 min; 36 cycles (*see* **Note 19**) of 94 °C for 20 s and 59 °C for 3 min; 60 °C for 10 min; and 4 °C hold.

3.5 Product Detection (Capillary Electrophoresis) (See Note 3)

1. Add 1 μL of the amplified sample or allelic ladder to 9.7 μL of Hi-Di™ formamide and 0.3 μL of GeneScan™ 500 LIZ™ size standard in a 96-well plate (*see* **Note 20**). Add a 96-well plate septa to the plate. For blank wells, use 10 μL of Hi-Di™ formamide.
2. Use electrophoresis conditions specific by the manufacturer in the amplification kit manual or internally validated conditions (*see* **Note 21**).
3. Analyze raw data with the GeneMapper® Software (Fig. 4).

4 Notes

1. The stereomicroscope used in our experiments is the Leica M205C which reportedly has the highest resolution available (1,280×) for such a microscope and a high zoom ratio (20.5:1). It is also equipped with a digital camera and image-processing system (required for documentation of collected bio-particles). Alternative stereomicroscopes that provide high magnification (up to 350×) can also be used.
2. An Applied Biosystems® GeneAmp® PCR system 9700 thermal cycler (AB by Life Technologies) was used for all experiments. Alternative thermal cyclers can be used. The use of thermal cyclers with different ramp speeds (e.g., fast thermal cyclers) may require validation of the lysis or STR amplification protocols.
3. A 3130 Genetic Analyzer (AB by Life Technologies) was used for product detection. Alternative genetic analyzer instruments can be used (e.g., 310, 3130xL, 3500, and 3500xL). However, the reagents and consumables as well as electrophoretic conditions may vary by instrument.
4. The WF Gel-film® is available in different retention levels (×0, ×4, ×8). We have successfully recovered bio-particles from touched objects using all three retention levels and therefore any of the retention levels can be used. However, in order to increase efficiency of bio-particle recovery we currently use the WF Gel-film x8. Gel-film® is also available in precut sizes or

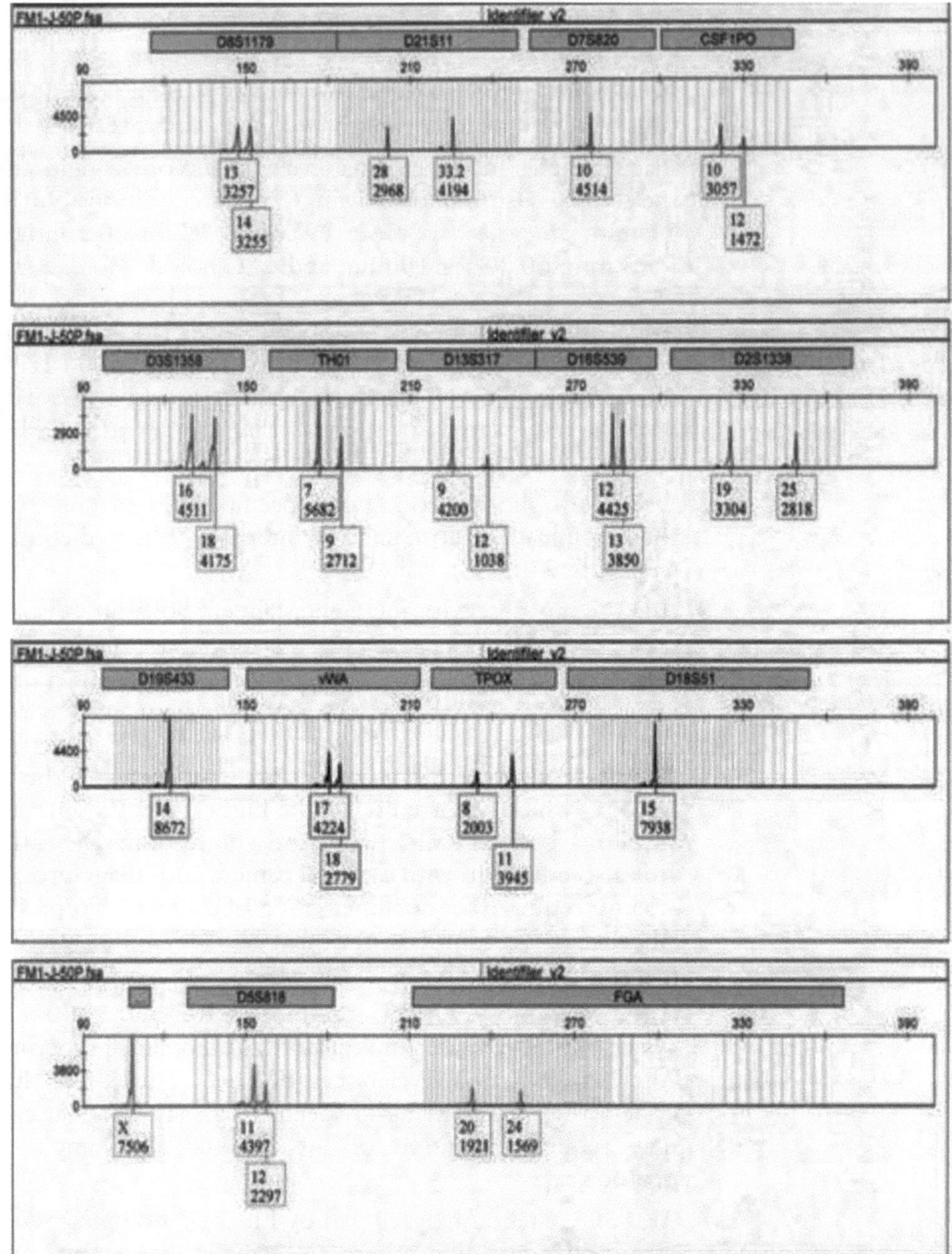

Fig. 4 Autosomal STR profile obtained from 50 "bio-particles" recovered from a jacket sleeve. Bio-particles from a worn jacket sleeve were recovered using Gel-film®. An autosomal STR profile (AmpFlSTR® Identifiler®) was obtained from a 50-bio-particle sample using direct lysis and an increased cycle number amplification (36 cycles). The accuracy of this profile was determined by comparison to a donor reference sample. Fifteen autosomal STR loci and amelogenin (sex determination) are co-amplified in a single reaction, separated by capillary electrophoresis and displayed as an electropherogram. Allele numbers are designated below each peak at each locus. The *x*-axis represents fragment size (bp) and the *y*-axis represents signal intensities (*RFU* relative fluorescence units; provided under each corresponding allele number). Each locus is labeled with a fluorescent dye: *top channel*—6-FAM™ (*blue*); *second channel*—VIC® (*green*); *third channel*—NED™ (*yellow*; shown in *black*); *bottom channel*—PET® (*red*)

can be requested in custom sizes and shapes. Small strips of Gel-film®, suitable for placement on a standard glass microscope slide, are used for bio-particle collection. Therefore, any size of Gel-film® can originally be used.

5. Staining is not required but is recommended in order to better visualize possible nuclei. Our protocols have been developed using a trypan blue stain. Other stains can be used instead of trypan blue. However, this would require the use of different reagents and staining protocols. Additionally, any potential PCR inhibitory effects from other stains would have to be evaluated prior to routine use.
6. Glass blocks can be used to place microscope slides containing Gel-Pak® samples, 3M water-soluble adhesive, or 0.2 mL PCR tubes for easier manipulation. They are not required.
7. There are numerous other direct lysis buffers available. Our protocols have been optimized for the *forensic*GEM™ Saliva kit (ZyGEM) and therefore any alternative lysis buffer would require additional validation.
8. There are numerous other STR amplification kits. Our protocols have been optimized for the AmpFlSTR® Identifiler® Plus PCR amplification kit (AB by Life Technologies). The use of any alternative STR amplification kit would require additional validation.
9. All materials listed are based on the use of the 3130 Genetic Analyzer. These reagents and consumables may vary depending on each laboratory or type of instrument used.
10. Some items may require greater Gel-film® flexibility in order to achieve better contact between the item and Gel-film® surface. If more flexibility is needed, follow the "without glass slide support" protocol (*see* Subheading 3.1.1).
11. Typically higher magnification (150× or greater) is needed to view the bio-particles with sufficient resolution in order to identify potential nuclei. However, the magnification level can vary by sample and by user.
12. At this time, the nature of "bio-particles" from touch DNA evidence is still being evaluated and requires additional characterization in order to determine the quantity of bio-particles needed for analysis (*see* Subheading 1). Currently, we are able to obtain DNA profiles when ~50 bio-particles are collected. If nucleated cells can be identified, fewer cells may be needed for analysis. At this time, it is recommended to collect multiple samplings from each touched object sample with increasing number of particles collected (e.g., 25, 50, 100, >100).
13. Be careful not to touch the tungsten needle to other parts of the Gel-film® when getting the needle into position for collection. It is often easier to bring the needle into the correct

position when viewed using a lower magnification. Once the bio-particles of interest have been identified, the magnification can be lowered to view the tungsten needle. With the needle in position over the general area of interest, the magnification can be increased to bring the bio-particles into focus.

14. The collection of multiple bio-particles needs to be viewed under the microscope to ensure that all targeted particles are collected. After collection of numerous particles, it is possible that there will not be sufficient free surface area on the adhesive for more particles to be collected efficiently. Continued attempts to use the same adhesive may result in loss of previously collected bio-particles (e.g., they will be redeposited onto the Gel-film® surface which will be visible under the microscope). If this occurs, sometimes the needle can be rotated to expose another side of the "ball" where additional particles can be collected. Alternatively, another sampling can be collected and transferred to the same collection tube in order to recover the desired number of bio-particles.
15. View insertion of the tungsten needle into the PCR tube under the microscope. If it is difficult to determine if the adhesive has completely dissolved, slowly pull the tungsten needle back out of solution. If adhesive is still present, place the needle back into solution and recheck. Repeat this process until the adhesive is completely dissolved in solution. Since such a small amount of the adhesive is typically used, it does not take long for the adhesive to dissolve completely.
16. The manufacturer recommends the use of low TE buffer (10 mM Tris–HCl, 0.1 mM EDTA, pH 8.0). This is not supplied in the kit and would need to be prepared by the user if used instead of sterile water.
17. Prepare the master mix in excess (typically 1–2 samples) to account for volume loss during pipetting.
18. The amplification is performed in the sample 0.2 mL reaction tube to eliminate potential sample loss from additional transfers.
19. The manufacturer's recommended cycle number is 28 (standard protocol) or 29 (for added sensitivity when <125 pg DNA inputs are used). However, the use of increased cycle number is recommended for improved sensitivity since very low template DNA amounts are expected with the bio-particle samples. Selection of an appropriate amplification cycle number may be laboratory dependent and requires internal validation.
20. The manufacturer's recommended protocol includes the use of 8.7 μL of Hi-Di™ formamide and 0.3 μL of GeneScan™ 500 LIZ™, with 9 μL added to each well and 1 μL of amplified product or allelic ladder for a total reaction volume of 10 μL.

21. The electrophoresis conditions used in our protocol are as follows: 16-s injection time, 1.2 kV injection voltage, 15 kV run voltage, 60 °C, 1,200-s run time, and Dye Set G5 (6-FAM™, VIC®, NED™, PET®, LIZ®).

References

1. Bright JA, Petricevic SF (2004) Recovery of trace DNA and its application to DNA profiling of shoe insoles. Forensic Sci Int 145:7–12
2. Goray M, Eken E, Mitchell RJ, van Oorschot RA (2010) Secondary DNA transfer of biological substances under varying test conditions. Forensic Sci Int Genet 4:62–67
3. Hall D, Fairley M (2004) A single approach to the recovery of DNA and firearm discharge residue evidence. Sci Justice 44:15–19
4. Pang BC, Cheung BK (2007) Double swab technique for collecting touched evidence. Leg Med (Tokyo) 9:181–184
5. Petricevic SF, Bright JA, Cockerton SL (2006) DNA profiling of trace DNA recovered from bedding. Forensic Sci Int 159:21–26
6. Wickenheiser RA (2002) Trace DNA: a review, discussion of theory, and application of the transfer of trace quantities of DNA through skin contact. J Forensic Sci 47:442–450
7. Balogh MK, Burger J, Bender K, Schneider PM (2003) STR genotyping and mtDNA sequencing of latent fingerprint on paper. Forensic Sci Int 137:188–195
8. Barbaro A, Cormaci P, Teatino A, La Marca A, Barbaro A (2004) Anonymous letters? DNA and fingerprints technologies combined to solve a case. Forensic Sci Int 146:S133–S134
9. Castello A, Alvarez M, Verdu F (2004) DNA from a computer keyboard. Forensic Sci Commun 6. http://www.fbi.gov/about-us/lab/forensic-science-communications/fsc/july2004/case/2004_03_case01.htm
10. Horsman-Hall KM, Orihuela Y, Karczynski SL, Davis AL, Ban JD, Greenspoon SA (2009) Development of STR profiles from firearms and fired cartridge cases. Forensic Sci Int Genet 3:242–250
11. van Oorschot RA, Jones MK (1997) DNA fingerprints from fingerprints. Nature 387:767
12. Port N, Bowyer V, Graham E, Batuwangala MS, Rutty GN (2005) How long does it take a static speaking individual to contaminate the immediate environment? Forensic Sci Med Pathol 2:157–163
13. Alessandrini F, Cecati M, Pesaresi M, Turchi C, Carle F, Tagliabracci A (2003) Fingerprints as evidence for a genetic profile: morphological study on fingerprints and analysis of exogenous and individual factors affecting DNA typing. J Forensic Sci 48:586–592
14. Kita T, Yamaguchi H, Yokoyama M, Tanaka T, Tanaka N (2008) Morphological study of fragmented DNA on touched objects. Forensic Sci Int Genet 3:32–36
15. Raymond JJ, van Oorschot RA, Gunn PR, Walsh SJ, Roux C (2009) Trace evidence characteristics of DNA: a preliminary investigation of the persistence of DNA at crime scenes. Forensic Sci Int Genet 4:26–33
16. Kelley-Primozic K (2008) Sperm recovery. Evidence Technol Mag 6:18–19
17. Di MD, Giuffre G, Staiti N, Simone A, Le Donne M, Saravo L (2004) Single sperm cell isolation by laser microdissection. Forensic Sci Int 146:S151–S153
18. Hubert R, Weber JL, Schmitt K, Zhang L, Arnheim N (1992) A new source of polymorphic DNA markers for sperm typing: analysis of microsatellite repeats in single cells. Am J Hum Genet 51:985–991
19. Ensenberger MG, Thompson J, Hill B, Homick K, Kearney V, Mayntz-Press KA et al (2010) Developmental validation of the PowerPlex 16 HS System: an improved 16-locus fluorescent STR multiplex. Forensic Sci Int Genet 4:257–264
20. Wang DY, Chang CW, Lagace RE, Calandro LM, Hennessy LK (2012) Developmental validation of the AmpFlSTR(R) Identifiler(R) Plus PCR Amplification Kit: an established multiplex assay with improved performance. J Forensic Sci 57:453–465
21. Darmon M, Blumenberg M (eds) (1993) Molecular biology of the skin—the Keratinocyte. Academic, San Diego, CA

Chapter 2

Detecting MicroRNA in Human Cancer Tissues with Fluorescence In Situ Hybridization

Zonggao Shi, Jeff J. Johnson, and M. Sharon Stack

Abstract

The technique of nucleic acid in situ hybridization is an effective method for identifying the existence and abundance of nucleic acids in tissue sections or cytological preparations. Such a method has the advantage of keeping morphological relationships intact while identifying changes at the molecular level. As a noncoding regulatory RNA, microRNA has been found to intricately control many physiological and pathological conditions. We provide here a representative fluorescence in situ hybridization protocol for microRNA detection, and note commonly used alternatives, and some troubleshooting points. The method described is based on formalin-fixed paraffin-embedded oral cancer tissues but should be broadly applicable to similarly processed tissues of other types of cancer.

Key words Squamous cell carcinoma, microRNA, Fluorescence in situ hybridization (FISH), Formalin-fixed paraffin-embedded (FFPE) tissue

1 Introduction

Since first discovered in the nematode *C. elegans* in 1993 [1], microRNA as a category of small regulatory RNA has been associated with numerous physiological and pathological processes in various species, especially in human. Because many microRNAs have been found to play important roles in cancer initiation and progression, there has been an ever-increasing amount of interest in the detection methodology of microRNAs [2].

As a small piece of RNA, all traditional methods for RNA detection, such as Northern blotting, quantitative polymerase chain reaction (qPCR), and microarray, could be and have been attempted for detecting the existence of microRNA or quantifying them in biological samples. However, due to the fact that mature microRNA is only about 22 nucleotides in length, many microRNAs differ from each other by only a few nucleotides. This poses a challenge in terms of specificity and sensitivity in developing a detection method. Some significant modifications have been made to

Dmitry M. Kolpashchikov and Yulia V. Gerasimova (eds.), *Nucleic Acid Detection: Methods and Protocols*, Methods in Molecular Biology, vol. 1039, DOI 10.1007/978-1-62703-535-4_2, © Springer Science+Business Media New York 2013

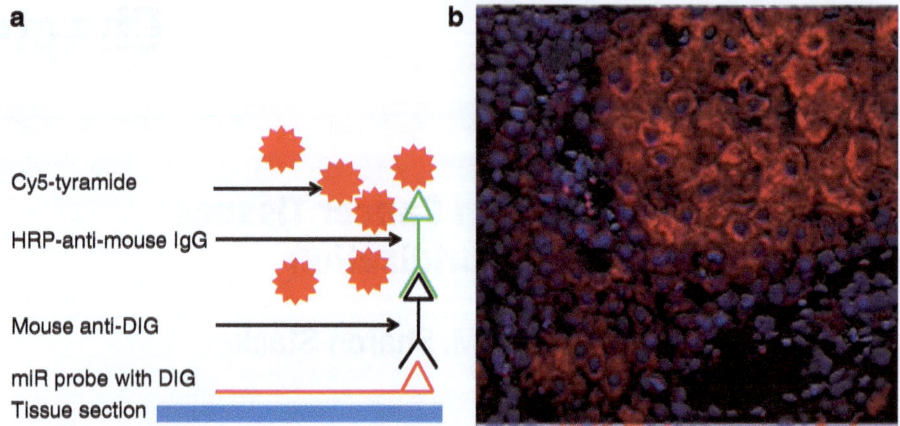

Fig. 1 Key steps in microRNA fluorescence in situ hybridization and typical result. (**a**) Diagram shows a processed microRNA containing tissue section that is first hybridized with DIG-labeled LNA-incorporated probe, followed by mouse anti-DIG, and HRP-conjugated anti-mouse IgG, prior to visualization of the signal by the addition of Cy5-conjugated tyramide (modified from Shi et al. Journal of Oncology, 2012, 903581). (**b**) Shown here is a positive result (*red* staining) from a section of formalin-fixed paraffin-embedded oral cancer tissue that was hybridized with mir-146a probe. *Blue* parts are cell nuclei stained by DAPI. Original magnification 400×

address the uniqueness of microRNA and newer methods have been under development [3].

Cancer tissues are inherently heterogeneous. Oftentimes, carcinoma cell populations are interwoven with nonmalignant stromal cells. So to further delineate the role of microRNA in cancer pathology, it is important to maintain the morphological relationships within the tissue and to differentiate the cancerous components from noncancerous while measuring the expression of microRNAs. In situ hybridization (ISH) is one of the methods to achieve this and has been applied for many years. As there are no antibody-based detection methods to gauge microRNA directly, ISH for microRNA in tissue sections has received a good amount of attention [4].

Our protocol for fluorescence in situ hybridization (FISH) technique to detect microRNA starts with the proper processing of tissues and then proceeds with the deparaffinization, pretreatment of the section, and then hybridization with a labeled probe. Post-hybridization immunohistochemistry (IHC) is used to visualize the positive signals with fluorescence microscopy. A diagram of the key scheme is presented in Fig. 1a. Some parts of the content have been previously published by us in an open access journal paper [5].

2 Materials

All solutions should be prepared with RNase-free Milli-Q-grade water at room temperature, unless otherwise specified.

2.1 Tissue Processing

1. RNase AWAY™ reagent (Life Technologies Corporation).
2. 10 % neutral buffered formalin (NBF): 10 % formalin, phosphate buffer, pH 7.0. To make 1 L of NBF, use 100 mL of formalin and 900 mL of water, 6.5 g of anhydrous disodium hydrogen phosphate (Na_2HPO_4), and 4.0 g of monohydrate sodium dihydrogen phosphate ($NaH_2PO_4 \cdot H_2O$). There is no need to adjust the pH. This solution is commercially available (Richard-Allan Scientific, Kalamazoo, MI).
3. Histological grade ethanol, xylene, and paraffin (melting point at 55–57 °C).
4. 4 % paraformaldehyde (PFA).
5. 100 mM Glycine.
6. Tissue cassettes, automatic tissue processor, tissue embedding center, and rotary microtome.
7. Oven or incubator that could be maintained at any temperature from ambient to 65 °C.

2.2 Pretreatment and Hybridization

1. 10× PBS: 0.1 M PBS, pH 7.4. Add 10.9 g of Na_2HPO_4 (anhydrous), 3.2 g of NaH_2PO_4 (anhydrous), and 90.0 g of NaCl to 900 mL water in a glass beaker. Mix and adjust pH to 7.4. Working solution of PBS is used at 1×. Dilute the stock 1:10 with RNase-free Milli-Q-grade water before use and adjust pH if necessary.
2. PBS-T: 0.1 % Tween-20, 1× PBS.
3. TE buffer: 10 mM Tris–HCl, pH 8.0, 1 mM EDTA. To make 100 mL of TE buffer, add 1 mL of 1 M Tris–HCl, pH 8.0, and 0.2 mL of 0.5 M EDTA to a beaker and fill up with Milli-Q water to 100 mL (*see* **Note 1**).
4. Proteinase K solution: 20 μg/mL Proteinase K in TE buffer.
5. Hydrochloric acid (HCl) solution: 24 mM HCl in ethanol. Add 0.2 mL of concentrated (12 M) hydrochloric acid to 100 mL of ethanol.
6. 25 μM Locked nucleic acid (LNA)-based probes (Exiqon, Vedbaek, Denmark) with the trademark of miRCURY LNA™ detection probe (*see* **Note 2**). Probe sequence for miR-146a is AACCCATGGAATTCAGTTCTCA and it is labeled with digoxigenin (DIG) at the 5′ end. The human miR-146a target sequence is UGAGAACUGAAUUCCAUGGGUU. Negative control probe sequence is GTGTAACACGTCTATACGCCCA and it is 5′-DIG labeled as well.
7. Sakura Finetek Tissue-Tek Slide Staining Set.
8. Plastic coverslip.
9. Moisture chamber.

10. 20× SSC: 3 M Sodium chloride (NaCl), 0.3 M sodium citrate. Dissolve 175.3 g of NaCl and 88.2 g of sodium citrate in 800 mL of Milli-Q-grade water; adjust the pH to 7.0 with a few drops of 1 M HCl, and adjust the volume to 1 L with additional water. Sterilize by autoclaving.
11. 50× Denhart's solution: 1 % Ficoll 400 (w/v), 1 % polyvinylpyrrolidone (PVP) (w/v), 1 % bovine serum albumin (BSA) (w/v). Add 10 g of Ficoll 400, 10 g of PVP, and 10 g of BSA to 900 mL of distilled water, and then fill up to 1 L. Filter the solution prior to storage through a 0.2-μm filter and store at −20 °C. Warm up to room temperature before use.
12. Pre-hybridization solution: 50 % deionized formamide, 2× SSC, 1× Denhart's, 0.02 % sodium dodecyl sulfate (SDS), 0.5 mg/mL yeast tRNA, 0.5 mg/mL salmon sperm DNA.
13. Hybridization solution: 50 % Deionized formamide, 2× SSC, 1× Denhart's, 10 % dextran sulfate, 0.5 mg/mL yeast tRNA, 0.5 mg/mL salmon sperm DNA (*see* **Note 3**).
14. Soaking solution: 50 % formamide, 1× SSC.
15. Detergent solution: 0.02 % SDS, 1×SSC.

2.3 Immunofluorescence

1. Blocking solution: 10 % normal goat serum in PBS.
2. Horseradish peroxidase (HRP)-conjugated anti-mouse IgG (Santa Cruz Biotechnologies, Inc., Santa Cruz, CA).
3. Mouse anti-digoxigenin (Roche Applied Science, Indianapolis, IN).
4. Tyramide Signal Amplification (Cy5-TSA) kits (Perkin Elmer, Waltham, MA, catalog # NEL-705A) (*see* **Note 4**). The kit contains HRP–streptavidin, blocking reagent, amplification diluent, and Cyanine 5-conjugated tyramide.
5. Working solution of Cy5-TSA. To prepare working solution dilute Cy5-TSA from the kit (*see* **item 4** of Subheading 2.3) 1:50 with 1× amplification diluent from the kit.
6. ProLong Gold Antifade mounting medium with 4′,6-diamidino-2-phenylindole (DAPI) (Life Technologies, CA).
7. Fluorescence microscope that is equipped with proper filters for DAPI and Cy5 dyes.

3 Methods

Use caution to maintain an RNase-free environment by wearing gloves and mask during all the procedures wherein RNA has been exposed until the hybridization step is finished (*see* **Note 5**).

3.1 Processing of Tissues

1. Fixation: Trim surgically removed tissue slice so as not to exceed 0.4 cm × 2.5 cm × 2 cm (in thickness × length × width) to be accommodated by the size of tissue cassette and to ensure the efficiency of downstream processing steps. Immediately fix the tissue in 10 % NBF for 24–72 h (*see* **Note 6**).
2. Dehydration: Secure tissue slices in a cassette; wrap up tissues with sponge pieces if tissue size is too small to prevent slices from floating away. Immerse tissue-containing cassettes in the following solutions in order: 70 % ethanol for 30 min, 80 % ethanol for 30 min, 90 % ethanol for 30 min, 100 % ethanol for 45 min, and 100 % ethanol for 45 min again.
3. Clearing: Immerse tissue cassettes in xylene for 45 min and another round of xylene for 45 additional minutes.
4. Infiltration: Immerse tissue cassettes in molten paraffin (65 °C) for 45 min and repeat with new molten paraffin for 45 min.
5. Embedding: Put tissue into paraffin block mold with proper orientation using embedding center (Tissue-Tek TEC, Sakura), fill up with molten paraffin, and let solidify on cold plate.
6. Sectioning: Use microtome (Leica RM2125) for sectioning. Set thickness at 5 μm and mount sections on positively charged slides (*see* **Note 7**). Drain and store the slides at room temperature.

3.2 Deparaffinization of Sections

1. Baking: Heat the sections in a metal rack at 65 °C for 1 h.
2. Dewaxing: Put slides on a histological staining rack and immerse sections in xylene for 10 min; repeat this step in another container of xylene.
3. Rehydration to water: Pass through in serial ethanol solutions (100, 90, 80 %), 1 min each, and into 1× PBS for 5 min.

3.3 Pretreatment of Sections

1. Wash sections in 1× PBS for 5 min, and repeat.
2. Put a section in the horizontal position in an open-lid moisture chamber; such a position is needed whenever a small amount of liquid is applied to the sections. Apply 300 μL of 24 mM HCl solution to each section and incubate for 10 min at room temperature (RT) to quench endogenous peroxidase (*see* **Note 8**).
3. Wash slides in a container with 1× PBS for 5 min twice.
4. Apply proteinase K solution, 500 μL each; incubate at 37 °C for 10 min.
5. Wash slides in a container with 1× PBS for 5 min twice.
6. Post-fixation: Apply 500 μL of 4 % PFA to each section in the horizontal position; fix for 10 min.
7. Wash slides in a container with 1× PBS for 5 min twice.

8. Apply 300 μL of 100 mM glycine, and incubate for 10 min to neutralize any residual PFA.
9. Wash slides in a container with 1× PBS for 5 min twice.
10. Wash slides in a container with 2× SSC wash for 5 min.
11. Apply 100 μL of pre-hybridization solution, cover with plastic coverslip, and incubate in moisture chamber filled with Kimwipes, which were soaked in soaking solution, at 50 °C for 2 h (*see* **Note 9**).

3.4 Hybridization and Stringent Wash

1. After pre-hybridization time is up, gently discard the solution by tipping over the slide to a waste can. Wipe clean slide area outside tissue section.
2. Without any wash, add 100 μL of hybridization solution containing probe at 25 nM (1 μL of 25 μM DIG-labeled probe to 1,000 μL) to sections. Cover with plastic coverslip, and incubate overnight (18 h) at 50 °C in moisture chamber (*see* **Note 10**).
3. Start stringent washes. Discard hybridization solution and plastic coverslip by immersing the slides individually into a container of 2× SCC. Use a water bath shaker for gentle shaking, 30 rpm.
4. Wash in 2× SSC at 37 °C, for 15 min.
5. Wash at 2× SCC at 50 °C for 15 min, with gentle shaking.
6. Wash in 1× SSC at 37 °C for 15 min.
7. Wash in 1× SSC at 50 °C for 15 min, with gentle shaking.
8. Wash in detergent solution at 37 °C for 15 min.
9. Wash in detergent solution at 50 °C for 15 min, with gentle shaking.
10. Wash in PBS-T at RT for 5 min; repeat three times.

3.5 Post-hybridization Immunofluorescence

1. Serum blocking: Add 300 μL of blocking solution per section and incubate at RT for 1 h.
2. Without washing, add 300 μL of mouse anti-DIG diluted 1:250 in blocking and incubate at RT for 0.5 h.
3. Wash in PBS-T for 5 min; repeat three more times (*see* **Note 11**).
4. Add 300 μL of HRP-conjugated goat anti-mouse IgG 1:500 dilution in 1× PBS and incubate at RT for 0.5 h.
5. Wash in PBS-T for 5 min; repeat three more times.
6. Add working solution of Cy5-TSA, 200 μL per slide; incubate at RT for 10 min.
7. Wash in PBS-T for 5 min; repeat three more times.
8. Wash in 1× PBS for 5 min.

9. Air-dry for 10 min and then mount glass coverslip using ProLong Gold. Let solidify overnight and seal with nail polish before observation under a fluorescent microscope.
10. Document the results with a camera-equipped fluorescent microscope. Typical results: Technical negative control slides should produce no signal on Cy5 channel while positive staining locates in the cytoplasm of selective cell populations. DAPI from the mounting medium stains all cell nuclei (Fig. 1b).

4 Notes

1. Proteinase K solution is stable over a broad pH range (4.0–12.5), but the optimum is pH 8.0. Proteinase treatment is very critical in unmasking the nucleic acid targets. Concentration and incubation time with different tissues may vary. Pepsin or collagenase may be used as alternatives.
2. A probe is a labeled nucleic acid whose nucleotide sequence is complementary to target nucleic acid, which dictates the specificity of hybridization. While other choices are possible, LNA-based probe is the default type of probe for microRNA detection and has been widely used because of its supremacy in specificity and sensitivity [4]. As for tracer labeling, biotin-labeled probe is another choice; however certain human tissues are rich with endogenous biotin, which could lead to significant background staining. Further, blocking of such endogenous biotin is not easily achievable. In contrast, DIG as a hapten is from a plant source; there is no similar analog in animal tissues, so anti-DIG antibodies have much less background binding.
3. Pre-hybridization and hybridization solutions usually have the same components except that the latter will be added with probes before use. Factors affecting hybridization include temperature, pH, concentration of monovalent cations, and presence of organic solvents such as formamide. The use of formamide significantly reduces the melting temperature of nucleic acid hybrids, therefore making it possible for hybridization to happen at a much lower temperature than melting temperature (T_m). The best condition for each probe and the probe concentration may vary and should be optimized.
4. Tyramide as a substrate of HRP has greatly enhanced the sensitivity of immunohistochemistry that uses HRP-labeled antibody [6, 7]. But because of its high sensitivity, the suppression of endogenous peroxidase sometimes becomes challenging.
5. An important consideration is to use diethylpyrocarbonate (DEPC)-treated or otherwise made RNase-free water in all

solution preparation and to exercise caution to maintain an RNase-free work environment during all procedures. Avoid RNase contamination by wearing disposable mask and gloves during the entire reagent preparation and other material preparation processes since skin contact and salivary fluid shed from talking are rich sources of RNases. Treat with RNase AWAY™ reagent all surfaces that may come in contact with RNA-containing specimen. RNase-free environment is not necessary after hybridization. Some protocols use RNase treatment as a negative control for RNA hybridization, but we do not recommend it.

6. Formalin fixation time was often a controversial topic. Considering the fact that this fixative penetrates tissue at a speed of 0.5 mm/h and protein cross-linking happens more slowly, a minimum of 24 h is required for a typical tissue slice; but overfixation may cause difficulty in unmasking the targets for ISH and IHC [8, 9]. Frozen sections cut by cryostat could be fixed for 30 min in 10 % NBF and used for FISH as well.
7. The thickness of paraffin sections affects the results of ISH; thicker sections would provide more targets for probes to bind but may compromise the revelation of a clear morphological structure. The use of commercially available positively charged microscope slides gives improved cellular adhesion. This is to prevent that the lengthy incubation and extensive washing steps in FISH procedures may detach tissue sections from slide surfaces. Poly-L-lysine- or amino silane-coated glass slides work equally well for this purpose and could be self-prepared. Proper baking before the dewaxing is also very important in retaining the section on slides.

   Processing tissues for formalin-fixed paraffin-embedded (FFPE) materials is a routine part of a standard histology lab. The majority have been automated by using tissue processor. Our model of this equipment is Leica TP1020. Making perfect sections from paraffin blocks is often challenging to beginners, so it is better to have this part performed by a well-trained histotechnologist. The quality of microRNA preservation in FFPE material has been a matter of concern since previously mRNA has been shown to be poorly preserved in such materials. Improper fixation or processing would certainly degrade all biomolecules within tissue specimens. However, recently some labs have verified that microRNA survives even better than mRNA in FFPE material, probably because of its short size and association with complicated protein complex [10, 11]. In addition to successful microRNA FISH on FFPE specimens, we have been able to isolate decent quality RNA for microRNA quantitative PCR via laser capture microdissection on FFPE materials.

8. Alternatively, use 3 % hydrogen peroxide (H_2O_2) in methanol incubation for 10 min. However, 24 mM HCl solution for suppressing endogenous peroxidase may have the additional benefit for improving signal-to-noise ratio because of partial hydrolysis of target sequence and extraction of proteins.
9. A pre-hybridization is always necessary to prevent nonspecific binding of probes, which would lead to background noise in staining.
10. The DIG-labeled LNA-incorporated mir-146a probe from Exiqon has a T_m of 85 °C for RNA target. Hybridization temperature is a critical factor affecting specificity and sensitivity of FISH; it has been recommended to be 30–45 °C lower than T_m with 50 % formamide present in hybridization solution and optimized with other factors upon specific probes [12]. Also, the choice of LNA-based probes in microRNA ISH has become a norm because LNA-modified oligos have a better mismatch discrimination [4, 13].
11. Substituting 0.1 % Triton X-100 for 0.1 % Tween-20 in the PBS-T buffer for wash may help decrease the background.

References

1. Lee RC, Feinbaum RL, Ambros V (1993) The C. elegans heterochronic gene lin-4 encodes small RNAs with antisense complementarity to lin-14. Cell 75:843–854
2. Andorfer CA, Necela BM, Thompson EA, Perez EA (2011) MicroRNA signatures: clinical biomarkers for the diagnosis and treatment of breast cancer. Trends Mol Med 17: 313–319
3. Qavi AJ, Kindt JT, Bailey RC (2010) Sizing up the future of microRNA analysis. Anal Bioanal Chem 398:2535–2549
4. Nielsen BS (2012) MicroRNA in situ hybridization. Methods Mol Biol 822:67–84
5. Shi Z, Johnson JJ, Stack MS (2012) Fluorescence in situ hybridization for microRNA detection in archived oral cancer tissues. J Oncol 2012:903581
6. McKay JA, Murray GI, Keith WN, McLeod HL (1997) Amplification of fluorescent in situ hybridisation signals in formalin fixed paraffin wax embedded sections of colon tumour using biotinylated tyramide. Mol Pathol 50:322–325
7. Bobrow MN, Moen PT Jr (2001) Tyramide signal amplification (TSA) systems for the enhancement of ISH signals in cytogenetics. Curr Protoc Cytom Chapter 8, Unit 8 9
8. Helander KG (1994) Kinetic studies of formaldehyde binding in tissue. Biotech Histochem 69:177–179
9. Fox CH, Johnson FB, Whiting J, Roller PP (1985) Formaldehyde fixation. J Histochem Cytochem 33:845–853
10. Glud M, Klausen M, Gniadecki R, Rossing M, Hastrup N, Nielsen FC et al (2009) MicroRNA expression in melanocytic nevi: the usefulness of formalin-fixed, paraffin-embedded material for miRNA microarray profiling. J Invest Dermatol 129:1219–1224
11. Liu A, Tetzlaff MT, Vanbelle P, Elder D, Feldman M, Tobias JW et al (2009) MicroRNA expression profiling outperforms mRNA expression profiling in formalin-fixed paraffin-embedded tissues. Int J Clin Exp Pathol 2:519–527
12. Nuovo GJ (2010) In situ detection of microRNAs in paraffin embedded, formalin fixed tissues and the co-localization of their putative targets. Methods 52:307–315
13. Kloosterman WP, Wienholds E, de Bruijn E, Kauppinen S, Plasterk RH (2006) *In situ* detection of miRNAs in animal embryos using LNA-modified oligonucleotide probes. Nat Methods 3:27–29

Part II

New Real-Time and Instantaneous Assays

Chapter 3

MNAzyme qPCR: A Superior Tool for Multiplex qPCR

Elisa Mokany and Alison V. Todd

Abstract

Multicomponent nucleic acid enzymes (MNAzymes) are nucleic acid enzymes composed of multiple oligonucleotide partzymes that only associate to form catalytic complexes in the presence of a target nucleic acid. Once assembled, MNAzymes cleave a separate substrate (probe) between fluorophore and quencher labels to produce a fluorescent signal indicative of the presence of the target. MNAzymes are particularly useful as tools for monitoring the accumulation of amplicons during real-time quantitative PCR (qPCR). The partzyme pairs have sensor domains that are complementary to adjacent regions in the amplicons such that their partial catalytic core domains form a complete active MNAzyme core. The probe-binding domain of the partzymes can be complementary to any one of a series of well-characterized universal probes. Since there is no need to synthesize and optimize new target-specific probes for each new target, MNAzyme qPCR provides a flexible alternative which allows target-specific interrogation with a generic readout. A series of universal probes have been designed that perform with high reliability, yielding consistent and reproducible results for any target, making the development of multiplex qPCR assays faster, cheaper, and simpler. This chapter describes a 5plex MNAzyme RT-qPCR method which simultaneously quantifies five mRNA transcripts with high efficiency and specificity using five unique universal probes, each labeled with a different fluorophore.

Key words Real-time PCR, Quantitative PCR, qPCR, MNAzymes, Multiplex, DNAzymes, 5plex, Quintuplex, Endogenous controls

1 Introduction

1.1 Real-Time Quantitative PCR

Real-time quantitative PCR (qPCR) has revolutionized nucleic acid research due to its unrivalled sensitivity and speed. Manufacturers of qPCR instruments are now providing machines that allow multiplex detection of up to five targets within a single reaction. Multiplex analysis provides many benefits including minimization of assay costs by measurement of multiple targets in a single reaction, the ability to gain maximum amount of information from small samples, the option to include more than one reference control within the same reaction as the target, and the inclusion of positive and negative controls within the same reaction

Dmitry M. Kolpashchikov and Yulia V. Gerasimova (eds.), *Nucleic Acid Detection: Methods and Protocols*, Methods in Molecular Biology, vol. 1039, DOI 10.1007/978-1-62703-535-4_3,

as the target. Despite the availability of multiplex-ready qPCR instruments, the great promise of multiplex analysis has not been realized as the majority of qPCR detection methods do not provide simple, robust multiplex assays.

Intercalating dyes such as SYBR® Green 1 and EvaGreen® provide an inexpensive approach to single-target detection (reviewed in [1]), but do not permit multiplexing and lack the specificity that is optimum for clinical applications. Target-binding probes used in technologies such as TaqMan® and Molecular Beacons improve specificity (reviewed in [1]), but new probes must be designed and optimized for each new target, and further optimization is required when multiplexing target-specific probes together. Few researchers attempt to multiplex more than two targets together using TaqMan® probes, and the amount of optimization required to set up multiplex assays reduces downstream cost benefits [2].

1.2 Multicomponent Nucleic Acid Enzymes

Multicomponent nucleic acid enzymes (MNAzyme) detection provides a robust and inexpensive way to fully utilize the multiplex capacity of qPCR instruments. MNAzymes are enzymes composed of multiple, nucleic acid oligonucleotides which only attain catalytic activity in the presence of their target [3]. They were engineered from uni-molecular DNAzymes which are capable of recognizing, and subsequently cleaving, a nucleic acid substrate [4]. Compared to DNAzymes, MNAzymes recognize a target nucleic acid, resulting in formation of catalytic complexes that then cleave a separate nucleic acid substrate (probe). As such, the target nucleic acid is not destroyed in the process and MNAzymes allow uncoupling of target detection from signal generation associated with cleavage of a universal probe. In their simplest format, MNAzymes are composed of two-component oligonucleotide "partzymes" (A and B: Fig. 1). Each partzyme contains a target-binding domain (sensor arm), a partial catalytic core, and a probe-binding domain (probe arm). The sensor arms are complementary to the target nucleic acid (DNA, RNA, or PCR amplicons) and when a partzyme pair aligns adjacently on a target their partial catalytic cores form the complete and active catalytic core of the MNAzyme. A universal reporter probe can then bind and subsequently be cleaved between a fluorophore and quencher to produce a signal that can be detected and monitored in real time (Fig. 1).

1.3 MNAzyme qPCR and Reverse Transcriptase qPCR

This chapter describes the use of MNAzymes for multiplex analysis of up to five targets at once in a single qPCR reaction. MNAzyme qPCR is illustrated in Fig. 1. The MNAzyme qPCR reaction contains the template (i.e., RNA or DNA) used to produce PCR amplicons, PCR primers, partzyme A, partzyme B, the universal reporter probe, and protein enzymes for polymerization and/or reverse transcription. Partzymes A and B are designed so that their

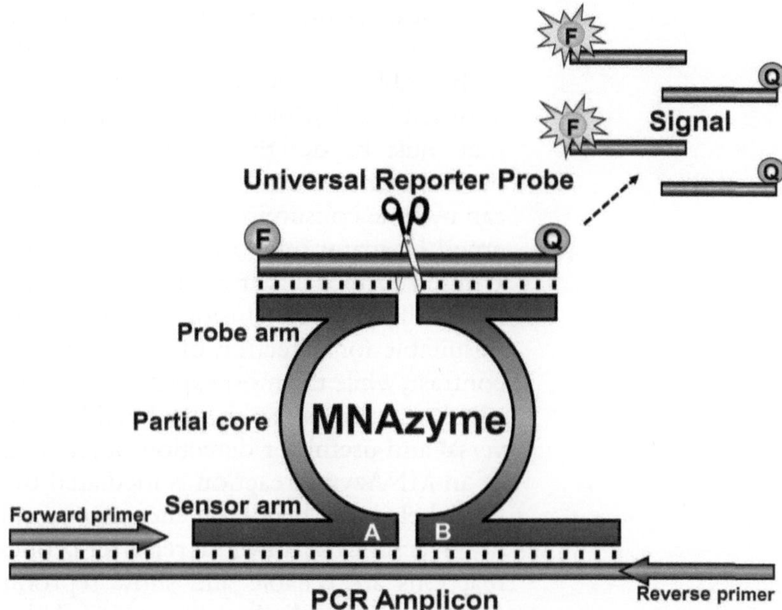

Fig. 1 MNAzyme qPCR: Partzymes A and B, each composed of a target-binding domain (sensor arm), a partial catalytic core (partial core), and probe-binding domain (probe arm), assemble adjacently on PCR amplicons forming an MNAzyme with an active catalytic core which can cleave a universal reporter probe emitting a fluorescent signal that can be monitored in real time

target-binding sensor arms (*see* Fig. 1) are complementary to the target. The fluorescently labeled reporter probe also binds to the partzymes, and once the active catalytic core is formed, the probe is then cleaved producing a signal that is indicative of successful amplification of the target gene or transcript. MNAzyme qPCR reactions have been shown to be specific and sensitive, with the capacity to measure target concentration over a broad dynamic range, extending over 5 orders of magnitude [3]. A suite of universal reporter probes with known performance provide the ability to rapidly and easily set up multiplex reactions for diverse applications. This method can also detect low copy numbers of target and performs flawlessly on a plethora of real-time instrumentation and in most standard commercial reagents.

1.4 Advantages of MNAzyme qPCR

The specificity of MNAzyme qPCR reactions is extremely high because signal generation requires four separate target-binding events to occur in a precise orientation, namely, the hybridization of two PCR primers and two partzymes (Fig. 1). In contrast, other probe-based chemistries, such as TaqMan® and Molecular beacons, have only three determinants of specificity derived from the binding of two PCR primers and one target-specific probe. This additional level of specificity in MNAzyme qPCR is particularly valuable in clinical settings [5]. MNAzyme technology employs universal reporter probes, making it the only qPCR detection technology with highly specific target interrogation and a generic readout. TaqMan® and Molecular beacons both use gene-specific probes that must be designed and optimized for each new assay. The development of sensitive assays which use target-binding probes can be time consuming since both probe and assay design are governed by many rules and the process is not completely infallible. Unused probes either related to design failures or which are left over following conclusion of the study of a specific target are unsuitable for detection of any other target and are discarded. In contrast, while the most expensive oligonucleotide in an MNAzyme qPCR reaction is also the probe, this component is completely universal and useful for detection of any target. The target-specificity of an MNAzyme reaction is mediated by the partzymes which are not labeled and hence are cheap to manufacture.

The series of universal reporter probes available for MNAzyme reactions are reliable and show reproducible, consistent performance when applied to any target. This is exemplified by at least one universal probe, which has been demonstrated to successfully detect thousands of different targets by qPCR. The use of a universal probe with known performance eliminates the cost involved in probe failures and reduces the time required for assay development as a new probe is not needed for each new target. As any probe can be used for any target, there is an overall decrease in the cost of goods as there is no waste of the probe when an assay is discontinued. The readily adaptable nature of MNAzymes also reduces the complexity of assay design and development. The majority of the MNAzyme assays described in this chapter went through just a single round of design with the resulting single-target assays and did not require subsequent optimization. The robust nature of MNAzymes has thus allowed the development of universal MNAzyme qPCR reaction conditions allowing MNAzymes to be substituted into preexisting qPCR reactions with minimal optimization. In general, only highly specialized assays, such as those targeting SNPs or variant genotypes, require optimization.

The ease of design provided by the universal MNAzyme probes extends to multiplex applications. The probes have been designed and tested to ensure that there is no cross-reactivity with partzymes designed to bind to the other probes. The probes have also been tested to ensure that they provide consistent reliable performance when mixed together in any multiplex reaction. New targets can be added to a reaction without the need to test for cross-reactivity between each assay or the requirement to develop a new gene-specific fluorescently labeled probe.

The simplicity of multiplexing with MNAzymes is outlined in Fig. 2. Each new target requires only the design of a new set of primers and new target-sensing arms of partzymes, while the rest

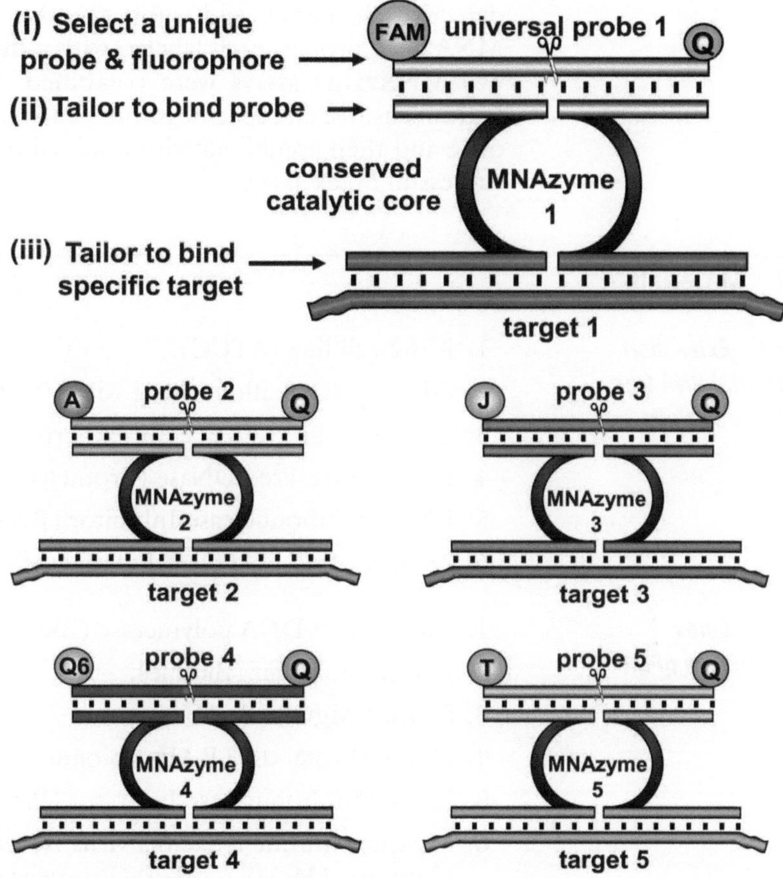

Fig. 2 Multiplexing with MNAzymes: When combining two or more target sequences (*i*) select a unique universal reporter probe that is labeled with a unique fluorophore (*see* **Note 4**), and tailor the component partzymes of the MNAzyme to (*ii*) bind to the selected probe and (*iii*) bind to each specific target

of the reaction components are completely universal. The catalytic core region of each partzyme is the same for every target sequence and for each of the universal probes. Reports in the literature provide examples of duplex and triplex qPCR reactions with various chemistries; however demonstrations of four or more targets multiplexed together in qPCR are rare as this currently requires considerable optimization [6]. MNAzymes have already been used successfully in numerous 4- and 5plex reactions and have also been successfully assayed in 6plex reactions on a new real-time instrument that can read up to six channels simultaneously (unpublished data).

1.5 5plex MNAzyme RT-qPCR

In this chapter we demonstrate an MNAzyme reverse transcriptase qPCR (RT-qPCR) reaction where five transcripts were reverse transcribed, amplified, and simultaneously detected by MNAzymes in a one-step reaction. MNAzymes were designed to specifically detect each target and subsequently cleave one of the five MNAzyme probes, each labeled with a different fluorophore. The five MNAzyme assays were combined with ease and produced highly sensitive and specific results. Each individual assay was tested once and then combined with minimal optimization to produce a successful 5plex assay.

2 Materials

2.1 Extraction of Total RNA from Cultured Cells

1. K562 cell line (ATCC).
2. QIAamp RNA Blood Mini Kit (50) (Qiagen).
3. 10× MULTI-CORE™ Buffer (Promega).
4. RQ1 RNase-Free DNase (Promega).
5. RNasin® Ribonuclease Inhibitor (Recombinant) (Promega).
6. 25 mM $MgCl_2$ (Applied Biosystems).

2.2 5plex MNAzyme qPCR

1. Immolase™ DNA polymerase (Bioline).
2. 10× Immobuffer (Bioline).
3. 50 mM $MgCl_2$ (Bioline).
4. 100 mM total dNTP Mix (Bioline).
5. RNasin® Ribonuclease Inhibitor (Promega).
6. Moloney murine leukemia virus Reverse Transcriptase (RNase H minus) (M-MLV (H−)) (Promega).
7. Nuclease-free water (NF water).
8. Oligonucleotides (*see* Tables 1 and 2).
9. Total RNA (*see* Subheading 1).

Table 1
Probe sequences and labeling

Type	Name	5′ Labeled fluorophore	3′ Labeled quencher[a]	Sequence (5′–3′)[b]
Probe 1	Sub2-A350	Alexa Fluor® 350	BHQ1	AAGGTTTCCTCguCCCTGGGCA
Probe 2	Sub3-Q670	Quasar® 670	BHQ2	CAGCACAACCguCACCAACCG
Probe 3	Sub4-TxRd	Texas Red®-X	BHQ2	CATGGCGCACguTGGGAGAAGTA
Probe 4	Sub6-FAM	6-FAM™	BHQ1	ATCACGCCTCguTCCTCCCAG
Probe 5	Sub7-JOE	JOE	BHQ1	TTAACATGGCACguTGGCTGTGATA

[a]BHQ1 = Black hole quencher 1 and BHQ2 = Black hole quencher 2
[b]Uppercase represents DNA and lowercase represents RNA

Table 2
Primer and partzyme sequences for each target

Type	Name[a]	Sequence (5′–3′)[b]
Forward primer 1	5RPLP0	CATTCTATCATCAACGGGTA
Reverse primer 1	3RPLP0	CAAAGGCAGATGGATCAG
Partzyme A 1	RPLP0_A/2-P	**CAAACGAGTCCTGGCCTTGTCT** ACAACGA*GAGGAAACCTT*
Partzyme B 1	RPLP0_B/2-P	*TGCCCAGGGA*GGCTAGCT **GTGGAGACGGATTACACCTTC**
Forward primer 2	5ACTB	CATTGCCGACAGGATGCAGA
Reverse primer 2	3ACTB	GAGCCGCCGATCCACACG
Partzyme A 2	ACTB_A/3-P	**AGATCAAGATCATTGCTCC** ACAACGA*GGTTGTGCTG*
Partzyme B 2	ACTB_B/3-P	*CGGTTGGTGA*GGCTAGCT **TCCTGAGCGCAAGTACTC**
Forward primer 3	5KRAS	GCCTGCTGAAAATGACTGAATA
Reverse primer 3	3KRAS	AATGATTCTGAATTAGCTGTATC
Partzyme A 3	KRAS_A/4-P	**TAAACTTGTGGTAGTTGGAG** ACAACGA*GTGCGCCATG*
Partzyme B 3	KRAS_B/4-P	*TACTTCTCCCAA*GGCTAGCT **CTGGTGGCGTAGGCAAGAGTGCC**
Forward primer 4	5HMBS	ACCCACACACAGCCTACTTTC
Reverse primer 4	3HMBS	TACCCACGCGAATCACTCTCA
Partzyme A 4	HMBS_A/6-P	**GCCATGTCTGGTAACGGCAA** ACAACGA*GAGGCGTGAT*
Partzyme B 4	HMBS_B/6-P	*CTGGGAGGAA*GGCTAGCT **TGCGGCTGCAACGGCGGAA**
Forward primer 5	5HPRT1	CTTTGCTGACCTGCTGGATTA
Reverse primer 5	3HPRT1	CCTGTTGACTGGTCATTACAA
Partzyme A 5	HPRT1_A/7-P	**ACTGAATAGAAATAGTGATAGAT** ACAACGA*GTGCCATGTTAA*
Partzyme B 5	HPRT1_B/7-P	*TATCACAGCCAA*GGCTAGCT **CCATTCCTATGACTGTAGATT**

[a]Targets amplified and quantified in this are the transcripts KRAS (NM_033360), RPLP0 (NM_001002), HMBS (NM_000190), ACTB (NM_001101), and HPRT1 (NM_000194.2)

[b]Bases in bold bind to the target sequence, bases underlined represent the partial catalytic cores, and bases in italics bind to the reporter probe

10. 96 well non-skirted clear plate (ABgene®).
11. Ultra clear 8 strip optical caps (ABgene®).
12. MicroAmp cap installing tool (Applied Biosystems).
13. Mx3005P QPCR system (Agilent Technologies).

3 Methods

The following protocol was accomplished in three separate rooms to minimize the possibility of contamination of reactions with previous PCR product. PCR master mixes were set up in one room, RNA extraction and addition of RNA to reactions were performed in a second room, and thermocycling of reactions was performed in a third room (*see* **Note 1**).

3.1 Extraction and Quantification of K562 Total RNA from Cultured Cells

Extract total RNA from cultured K562 cells by following the instructions of the "QIAamp RNA mini protocol" ("QIAamp RNA blood mini handbook" from Qiagen) using the protocol "for pelleted cells" (*see* **Note 2**). Use 600 μL of buffer RLT and 1 mL of buffer RPE for every 5×10^6 cells. Place the eluted RNA immediately onto ice at the end of the protocol. Digest potential contaminating DNA as follows:

1. Make up the following volume of DNA digestion mix for each 50 μL of RNA solution: 7 μL of MULTI-CORE buffer (10× solution), 5 μL of RQ1 RNase-free DNase (1 unit/μL), 2 μL of RNasin® Ribonuclease Inhibitor (40 units/μL), and 6 μL of $MgCl_2$ (25 mM).
2. Add 20 μL of the DNA digestion mix to every 50 μL of RNA solution.
3. Incubate at 37 °C for 15 min (digestion of DNA).
4. Incubate at 75 °C for 5 min (inactivation of DNase) and place on ice.
5. To measure the concentration of total RNA, centrifuge at 10,000 × *g* for 15 s and return to ice. Dilute an aliquot of total RNA, 1 in 100 in NF water, and measure the absorbance of the concentrated stock at $A_{260\mathrm{nm}}$ on a spectrophotometer (following the manufacturer's instructions for use). Calculate the concentration of the RNA solution using the following formula:

$$\text{Concentration } (\mu\text{g/mL}) = 40 \ (1 \text{ OD unit of RNA} = 40\ \mu\text{g/mL}) \times A_{260} \times 100 \ (\text{dilution factor}).$$

6. Where possible dilute RNA to 100 ng/μL in NF water. Aliquot into smaller sample sizes to avoid multiple cycles of freeze–thaw. Store at −80 °C.

3.2 Calibrator Curve from Total RNA from K562 Cells

Calibration curves provide a much better insight into the performance of an assay compared to assessment of a target at a single concentration. The performance of the assay with serial dilutions of template provides information about the sensitivity, specificity, linearity, and efficiency of each reaction.

The following instructions refer to creating a calibration curve from K562 total RNA (extracted in Subheading 3.1) in a background of NF water. All solutions should be aspirated below the meniscus, and the inside of the pipette tip should be rinsed in the solution at least three times when aspirating from one tube and dispensing into another. Briefly vortex and spin down each calibrator solution before use.

Construct a series of RNA solutions with different concentrations by performing a fourfold serial dilution. Make up calibrator 1 containing 100 ng/μL total RNA from K562 cells and then produce calibrators 2–7 as follows:

Calibrator 2 (25 ng/μL): 12.5 μL Calibrator 1, 37.5 μL NF water.

Calibrator 3 (6.25 ng/μL): 12.5 μL Calibrator 2, 37.5 μL NF water.

Calibrator 4 (1.56 ng/μL): 12.5 μL Calibrator 3, 37.5 μL NF water.

Calibrator 5 (0.39 ng/μL): 12.5 μL Calibrator 4, 37.5 μL NF water.

Calibrator 6 (0.098 ng/μL): 12.5 μL Calibrator 5, 37.5 μL NF water.

Calibrator 7 (0.024 ng/μL): 12.5 μL Calibrator 6, 37.5 μL NF water.

Calibrators can either be prepared in advance or freshly made from calibrator 1, prior to addition to reaction mixes. It is recommended that concentrations of RNA less than 100 ng/μL should not be stored at −80 °C for more than 1 month. To prevent degradation of RNA the serial dilutions below should be used within a 1-week period and should not be thawed more than three times.

3.3 Oligonucleotides

Precision in qPCR reactions is obtained by ensuring that oligonucleotides (oligos) are exactly of the concentration required (*see* **Note 3**). The concentration or yield supplied by oligo manufacturing companies is not always accurate. Utilizing the following method for measurement of the concentration of stock solutions of oligos, followed by confirmation that the final working solution is at the correct concentration, will ensure that the right amount of oligo is being used in the final reaction. Use of oligos that have been supplied lyophilized is recommended (*see* **Note 4** for more instructions on the recommended format for different oligos).

1. Calculate the amount of water in μL (*Y*μL) to add to the lyophilized oligo to get a solution of approximately 250 μM. This calculation is performed using the following formula:

$$[X]\,\mu M = (\text{OD supplied} \times 10^6)/(\text{extinction coefficient (OD units}/\mu\text{mol)} \times Y\,\mu L \text{ NF water}).$$

The "OD supplied" and "extinction coefficient (OD units/μmol)" can be found on the spec sheet supplied with the oligo.

2. Briefly centrifuge the tube of lyophilized oligo before opening.
3. Add the amount of NF water to the lyophilized oligo as calculated in **step 1**.
4. Vortex tube, briefly centrifuge, and then leave at room temperature for at least 1 h (*see* **Note 5**).
5. Vortex tube again and then briefly centrifuge.
6. Make a 1 in 200 dilution of the oligo solution and measure the absorbance at A_{260} using a spectrophotometer. If the absorbance reading does not fall between 0.1 and 0.9 remake the dilution so that the absorbance falls between these values (the majority of spectrophotometers are not accurate outside this range).
7. Use the OD at A_{260} to calculate the exact concentration of the stock solution:

$$[X]\,\mu M = (OD\ A_{260} \times \text{dilution factor } (200) \times 1{,}000) / \text{extinction coefficient (OD units}/\mu\text{mol)}.$$

8. Make a 20 μM working solution from the stock oligo solution.
9. Perform a 1 in 20 dilution of the working stock oligo solution and measure the absorbance at A_{260} in a spectrophotometer (using the same guidelines for absorbance readings as stated in **step 6**):

$$[X]\,\mu M = (OD\ A_{260} \times \text{dilution factor } (20) \times 1{,}000) / \text{extinction coefficient (OD units}/\mu\text{mol)}.$$

This value should be between 19 and 21 μM (*see* **Note 5**).

3.4 MNAzyme RT-qPCR

MNAzyme RT-qPCR can be performed on any real-time thermocycler. This protocol describes a 5plex MNAzyme RT-qPCR, which requires a machine that can read five channels. In this example we have used the Mx3005P QPCR system and readers are referred to the manufacturer's user manual for full details (*see* **Note 6**).

The RT-qPCR described herein is composed of five MNAzyme sets, each one specific for one of the following five transcripts, *KRAS, ACTB, RPLP0, HMBS,* and *HPRT1*. Each MNAzyme was designed to bind to an amplicon from one of the five target transcripts and then cleave one of the five unique universal probes: Sub4, Sub3, Sub2, Sub6, and Sub7 (*see* Tables 1 and 2).

1. Thaw reagents at room temperature (*see* **Note 1**) and then keep all reagents on ice. Ensure that fluorescently labeled probes are protected from exposure to light. Briefly vortex (approximately 2 s) and centrifuge reagents and oligos. To make the master mix add the reagents together as described in

Table 3
Multiplex MNAzyme qPCR master mix

Reagent name	Working stock concentration	Final amount in qPCR	Volume per 25 μL reaction (μL)
NF water			To a total of 20 μL
Immobuffer	10×	1×	2.5
$MgCl_2$	50 mM	10 mM	5.0 (*see* **Note 10**)
dNTPs	25 mM each	200 nM each	0.2
RNasin	40 units	10 units	0.25 (*see* **Note 11**)
Immolase	5 units/μL	2 units	0.4
M-MLV (H−)	200 units/μL	20 units	0.1
5plex Probe mix (Table 4)			2.5
5plex Oligo mix (Table 5)			6.0

Table 4
5plex Probe mix (*see* Note 12)

Reagent name	Working stock concentration (μM)	Final amount in qPCR (μM)	Volume per 25 μL reaction (μL)
NF water			To a total of 2.5 μL
Sub2-A350	20	0.6	0.75 (*see* **Note 13**)
Sub3-Q670	20	0.2	0.25
Sub4-TxRd	20	0.4	0.5 (*see* **Note 13**)
Sub6-FAM	20	0.2	0.25
Sub7-JOE	20	0.2	0.25

Tables 3–5. Briefly vortex, then centrifuge each mix upon completion, and keep the mix on ice at all other times.

2. Aliquot 20 μL of cold master mix into a 96-well non-skirted clear plate (*see* **Note 7**). Add 5 μL of water to the appropriate no-template-control (NTC) wells. Aliquot a sufficient number of reactions to allow each calibrator and test sample to be measured in at least two separate wells. Keep the 96-well plate on ice or on a cooling rack at all times. Place strip-caps over the columns with completed NTC wells and seal using

Table 5
5plex Oligonucleotide mix (see *Note 14*)

Details	Reagent name	Working stock (μM)	Final amount in qPCR (μM)	Volume per 25 μL reaction (μL)
NF water				To a total of 6 μL
Reverse primer 1	3RPLP0	20	0.2	0.25
Forward primer 1	5RPLP0	4	0.04	0.25
Partzyme A 1	RPLP0_A/2-P	20	0.2	0.25
Partzyme B 1	RPLP0_B/2-P	20	0.2	0.25
Reverse primer 2	3ACTB	20	0.3	0.375 (*see* **Note 15**)
Forward primer 2	5ACTB	4	0.06	0.375 (*see* **Note 15**)
Partzyme A 2	ACTB_A/3-P	20	0.2	0.25
Partzyme B 2	ACTB_B/3-P	20	0.2	0.25
Reverse primer 3	3KRAS	20	0.2	0.25
Forward primer 3	5KRAS	4	0.04	0.25
Partzyme A 3	KRAS_A/4-P	20	0.2	0.25
Partzyme B 3	KRAS_B/4-P	20	0.2	0.25
Reverse primer 4	3HMBS	20	0.3	0.375 (*see* **Note 15**)
Forward primer 4	5HMBS	4	0.06	0.375 (*see* **Note 15**)
Partzyme A 4	HMBS_A/6-P	20	0.2	0.25
Partzyme B 4	HMBS_B/6-P	20	0.2	0.25
Reverse primer 5	3HPRT1	20	0.4	0.5 (*see* **Note 15**)
Forward primer 5	5HPRT1	4	0.08	0.5 (*see* **Note 15**)
Partzyme A 5	HPRT1_A/7-P	20	0.2	0.25
Partzyme B 5	HPRT1_B/7-P	20	0.2	0.25

MicroAmp Cap Installing Tool. When assigning reactions to the plate, ensure that NTCs are not placed in the same column as reactions that are going to contain calibrators or test samples. This avoids having to reopen the strip-caps to add template as this is likely to aerosolize reaction mixture and cause cross contamination.

3. In a separate room, thaw previously made calibrators 1–7 on ice (created in Subheading 3.2) or make them fresh according to instructions in Subheading 3.2. Before addition of each

calibrator to the plate, briefly mix and then centrifuge. Before aspirating 5 μL of each calibrator, pipette the volume up and down in calibrator solution at least three times and then dispense into appropriate well.

4. Place strip-caps over wells and seal using MicroAmp Cap Installing Tool.
5. Keep the plate covered and on ice until ready for placing on Mx3005P QPCR System.
6. In a separate room (*see* **Note 1**), set up the Mx3005P QPCR System according to the user's manual.
7. Program the following thermocycling profile (*see* **Note 8**):

   30 min at 50 °C, and 10 min at 95 °C.

   10 cycles of 95 °C for 15 s and 64 °C for 30 s (−1 °C per cycle).

   40 cycles of 95 °C for 15 s and 51 °C for 60 s (acquire fluorescence).
8. Ensure that plate is loaded in the correct orientation, close lid, and select "run."
9. After completion of the run, analyze data from Mx3005P QPCR System according to the guidelines from Agilent technologies (*see* **Note 9**). The amplification plots and calibration curves for the five assays multiplexed together are given in Fig. 3.

3.5 Analysis of RT-qPCR 5plex

Amplifying a serial dilution of calibrators allows the generation of a calibrator curve by plotting the Ct against the log of the starting concentration. Calibration curves should span over 4–5 logs and should contain at least four different concentrations of starting template. A calibration curve can be used to determine the concentration of unknown samples and also assess the quality of the qPCR itself. The Pearson correlation coefficient (Fig. 3, R^2) represents the linearity of the data, indicating the closeness of fit of the calibrator curve to a straight line and the variability between replicates of the same calibrator. This value should be above 0.990 if the calibration curve is to be effectively used to estimate the concentration of target in a sample [7]. The PCR efficiency (Fig. 3, Eff) gives an indication of the robustness of the assay. The optimal amplification efficiency is between 90 and 110 % for a calibration curve of four logs or greater [8]. If your efficiencies fall outside this range you will need to optimize your assay, and to do so you should take into consideration primer design and the reaction conditions.

The 5plex MNAzyme RT-qPCR showcased here provides robust, specific, and sensitive assays (*see* **Note 16**). The R^2 ranges from 0.995 to 0.999 and the Eff all fall within the range of

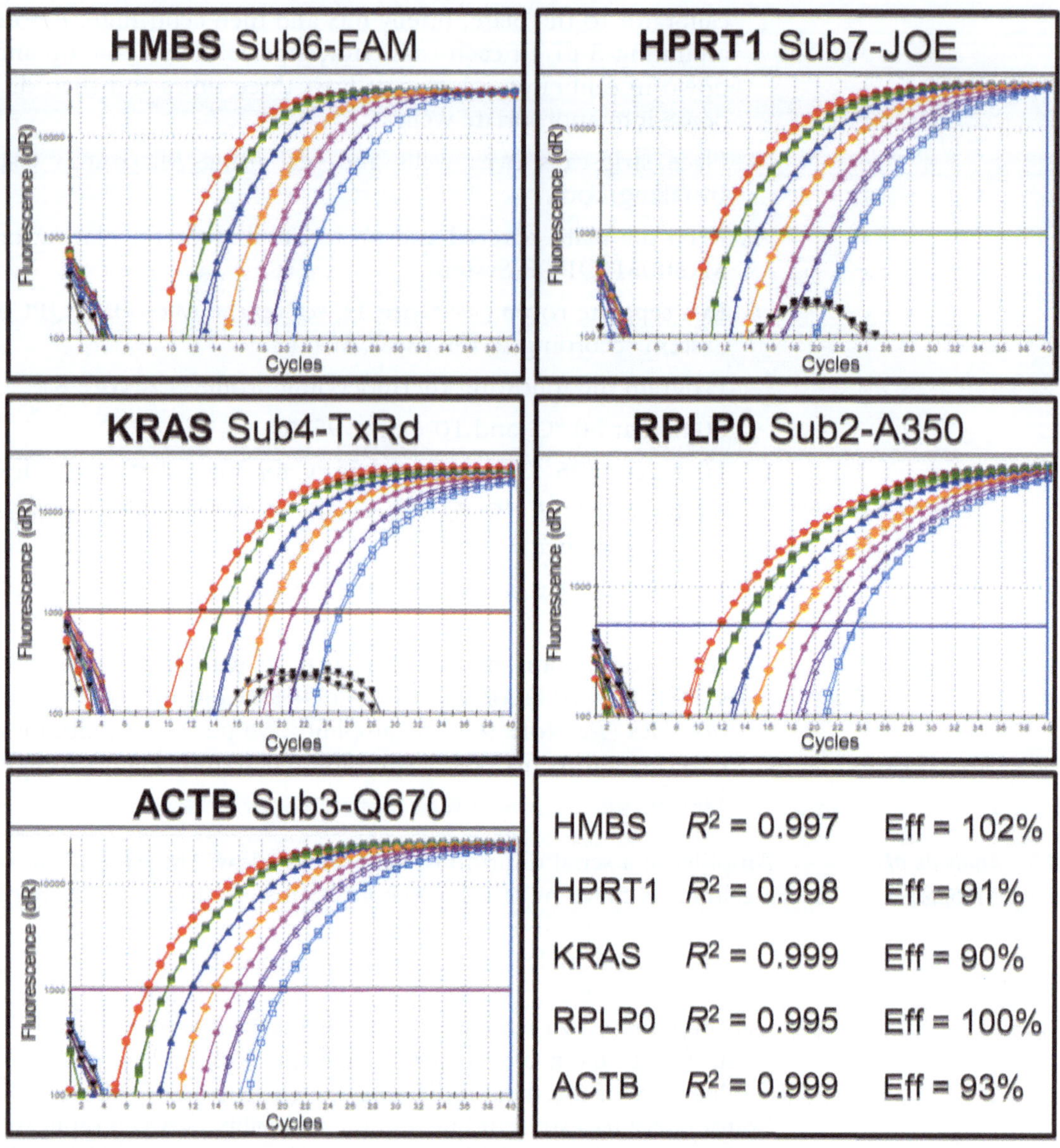

Fig. 3 Amplification plots and calibration curve data for the 5plex MNAzyme single-tube RT-qPCR assay. The final concentration of template in each well was 500, 125, 31, 8, 2, 0.5, and 0.12 ng, from *left* to *right*. R^2 Pearson correlation coefficient, *Eff* Efficiency

90–102 % (Fig. 3). There was no signal generated from the NTCs indicating that the assays are highly specific and that there is no cross-reactivity between oligos for different targets, evident by the single assay signal production for each fluorescent channel.

4 Notes

1. To minimize the risk of contamination, the different steps involved in setup and running of a qPCR experiment should be conducted in physically separated rooms. All rooms should contain everything necessary for their designated tasks, ensuring minimum movement of equipment between the rooms. Oligos and qPCR reagents (Tables 3–5) should be handled in the first room designated the most "clean" room. Template, be it RNA, DNA, or cDNA, should be handled in a second room and thermocyclers should be confined to a third room. Keeping a one-direction flow of work between the rooms will also minimize the risk of contamination. Reactions should be prepared in the clean room, and then moved through to the template-handling room. Once a researcher has moved into the template room, there should be a policy in place that prevents any movement back into the reagent room. Once template is added the reaction plate should be taken directly to the thermocycling room. Once a researcher has been in the thermocycler room they should not go back into either of the other rooms in the same day. It is possible for caps or adhesive covers to be faulty or misapplied and this may result in escape of PCR product in the thermocycler room. It is possible for this contamination to travel in dust and be picked up by researchers and carried into the other rooms. Meticulous laboratory protocol should be followed to avoid contamination. Before starting any procedure all benches and pipettes should be cleaned with a solution such as DNA-ExitusPlus™ (AppliChem) followed by 70 % ethanol. Gloves should be changed frequently. All tubes of reagents should be centrifuged before opening to minimize the production of aerosols. Handling of the target samples should also be minimized. It is advisable to wear laboratory coats, gloves, and hair nets. Shoe covers should be used in the room containing thermocyclers and tacky mats should be installed at the entrance of all rooms to remove dust from shoes. If it is not possible to set up separate rooms due to space restriction, then the laboratory space should be divided into three zones as far apart from each other as possible. It is advisable not to open qPCR plates once the reactions have finished.
2. All steps using the QIAamp RNA Blood Mini Kit are carried out at room temperature. Working quickly will minimize degradation of RNA.
3. In MNAzyme qPCR, this is particularly crucial for the limiting forward primer, which has the lowest concentration. If this concentration is wrong, it can affect the efficiency of the reaction.

4. We order our oligos from Integrated DNA Technologies, Inc (IDT). We have found that desalting provides a sufficient level of purification for partzymes and primers. However, if using another oligo manufacturer you should perform a comparison to ensure that the desalted partzymes work comparably to the ones HPLC purified. Partzymes should have a 3′ phosphate attached to prevent their extension during qPCR. Reporter probes require HPLC purification and are usually considered a custom order by most oligo manufacturers due to the inclusion of ribonucleotides and fluorophore and quencher moieties. Reporter probes can be labeled with any fluorophore and quencher compatible for multiplex qPCR on any instrument. Note that these probes are not considered for use in a 5′ nuclease assay.
5. As stated on the IDT Web site (http://www.idtdna.com/pages/decoded/decoded-articles/core-concepts/decoded/2011/03/16/dna-oligonucleotide-resuspension-and-storage) for resuspending dry oligos "not all oligonucleotides dry identically and some require more time to go into solution than others." We have found that 1 h at room temperature seems to be sufficient for most oligos; if any still have precipitate visible we will often leave at room temperature overnight. If there is a problematic oligo it is often evident at the final step as the working concentration of 20 μM will be <19 or >21 μM. If this does occur repeat the process from **step 5**.
6. Although the standard Mx3005P QPCR System comes with the ability to read five different channels, the standard Cy3 channel did not work well in any 5plex assay we have tested. The manufacturer advised that a custom installation of a channel to read Alexa Fluor® 350 was necessary for successful 5plex assays. Experiments have shown that the Alexa Fluor® 350 channel and/or dye reads very weakly and has proved difficult to work with. If using another instrument, the fluorophores required for 5plex assays may vary, as channel frequencies can vary between manufacturers. Check with manufacturer for ideal fluorophore combinations for multiplexing on their instrument.
7. A new pipette tip should be used for the addition of master mix and RNA calibrators to each well; i.e., do not use one pipette tip to add all the master mix to all the wells on the plate. A multi-displacement pipette, such as the Eppendorf Research® pro, can be used for addition of the master mix. To measure reproducibility at least duplicates of all reactions should be performed.
8. The Mx3005P QPCR system has a manually adjusted gain setting for each fluorescent channel. The manufacturer's instructions tell you the ideal fluorescence plot range for each channel. For each fluorescently labeled probe the gain setting needs to

be investigated. As a starting point set the gain for FAM and JOE to 4×, ALEXA to 8×, and CY5 and TEXRD to 1×. These settings are specific for the reaction mix conditions in use. Altering the reaction mix described here may require adjustment of gain settings.

9. The Mx3005P software has an auto threshold function. To enable meaningful comparisons between multiple runs of the same assay it is advisable to set the threshold at 1,000 for each fluorophore for every run. Note that the Alexa Fluor® 350 fluorophore has a lower fluorescence range and so a threshold of 500 is recommended. The software also has an auto baseline calculation for each well. This baseline can sometimes be miscalculated by the software. This is apparent when the baseline of the normalized plot for that sample deviates extremely from zero compared to other samples on the same plate (the deviation can either be a dip below the level of other samples or a rise above the normalized level of the other samples). This deviation is a result of application of the normalization over the wrong part of the curve. For example, this may occur in a reaction containing a low template concentration where the software has set the baseline using the whole cycle period (1–40). Manual adjustment of the upper limit to an acceptable value, typically cycle 25–30, will improve the plots. The reverse situation can occur for NTCs. The auto baseline calculation will often be set between 2 and 7 cycles, as the software may have detected a change in the slope of the NTC readout. This results in the "normalization" of the NTC so that the plot appears as though amplification has taken place in these reactions. Changing the upper limit, i.e., 7, to cycle 40 will correct the normalization and the plot will then fall below the threshold line.

10. The $MgCl_2$ concentration in an MNAzyme reaction is higher than for standard qPCR. The catalytic function of the MNAzyme is dependent on metal ions and adding more magnesium improves the catalytic rate. This is especially important when combining five targets in a single well; therefore a final concentration of 10 mM $MgCl_2$ greatly improves cleavage activity and thus signal efficiency. For qPCR assays that contain three or less MNAzymes, 8 mM $MgCl_2$ can be used. We have found that this increased concentration of magnesium has not impacted on the specificity of the reaction. It is extremely rare to obtain a signal from the negative controls and NTCs.

11. The addition of RNasin® to the reaction is to protect RNA from test samples and the universal MNAzyme probe from degradation. We find that the RNA bases in the universal probes are quite stable and the addition of RNasin® may not always be needed. It is advisable to add RNasin® if there is an issue with the purity or the stability of total RNA in test samples. This is particularly crucial when using samples in which

the origin or the extraction process is unknown (potentially still containing RNAses).

12. In the method above, the probe mix (Table 4) and primer and partzyme mix (Table 5) have been put into separate tables from the master mix (Table 3); this is to clearly identify the different components of an MNAzyme reaction. If using a different set of targets to the five described above you just need to modify the primer and partzyme table (Table 5). However, when setting up your experimental protocol, all three tables can be combined.
13. Factors that can contribute to the loss of efficiency of an assay in a multiplex form include the labeling of reporter probes with a weaker dye and/or channel, such as Alexa Fluor® 350. Two of the reporter probes (Sub4 and Sub7) in the current suite of probes are cleaved less efficiently by MNAzymes and should be used for higher abundance targets and should not be labeled with Alexa Fluor® 350. New probes are being developed to produce a suite of probes that are cleaved with equal and strong efficiency. In this multiplex the efficiency was improved when concentration of the reporter probe labeled with Alexa Fluor® 350 (Sub2-A350) was tripled and concentration of the weaker probe Sub4 (Sub4-TxRd) was doubled.
14. MNAzyme qPCR makes use of asymmetric PCR primer ratios to ensure that the strand that the MNAzyme binds to is in excess and is not sequestered by double-stranded amplicons. Asymmetric primer ratios are also advantageous for multiplexing since this ensures that PCR reagents/enzymes are not depleted during amplification of the abundant targets, thus allowing more efficient amplification of less abundant targets. A general guide is to have the primers at a ratio of 1:5 (forward to reverse primer, respectively) or greater. Decreasing this ratio makes the reaction too symmetric and may result in reduced efficiency. 200 nM reverse primer and 40 nM forward primer are the universal concentrations to use initially with all assays. These conditions usually produce an efficient and robust assay.
15. Minor optimization is sometimes required for the efficient and robust amplification of all targets in a multiplex reaction. To optimize an MNAzyme multiplex reaction first test each primer set using the universal primer concentrations of 200 nM of the reverse primer and 40 nM of the forward primer. For any assays that do not meet acceptance criteria such as Ct, efficiency, and linearity, increase the concentrations of reverse and forward primer to 300 and 60 nM, respectively. In the 5plex described here, this was required for the targets *ACTB*, *HMBS*, and *HPRT1*. This adjustment to primer concentration improved the assay for *ACTB* and *HMBS*; however *HPRT1* was still inefficient. The reverse and forward primer concentrations for the

HPRT1 assay were further increased to 400 and 80 nM, respectively. This was sufficient to improve the efficiency of the assay. Generally there will be no requirement to increase primer concentrations higher than 400:80 nM. For assays that are outcompeting others, the reverse and forward primer concentrations can be decreased to 100 and 20 nM, respectively. Optimization to one of these four alternate primer concentrations has resulted in routine production of robust multiplex assays using MNAzymes.

16. The Ct values reported here appear low due to the initial ten cycles of touchdown PCR during which there is no acquisition of fluorescence. This must be taken into account when comparing these Ct values with those of other qPCR protocols. The ten cycles of touchdown PCR produce more efficient reactions by increasing the initial specificity of target amplification. While nonspecific PCR products are never detected by MNAzymes, any nonspecific amplification may reduce the amplification efficiency of specific amplicons.

References

1. Bustin SA (2000) Absolute quantification of mRNA using real-time reverse transcription polymerase chain reaction assays. J Mol Endocrinol 25(2):169–193
2. User Bulletin #5, ABI PRISM®, 7700 Sequence Detection System August 10, 1998 (updated 01/2001)
3. Mokany E, Bone SM, Young PE et al (2010) MNAzymes, a versatile new class of nucleic acid enzymes. That can function as biosensors and molecular switches. J Am Chem Soc 132: 1051–1059
4. Santoro S, Joyce G (1997) A general purpose RNA cleaving DNA enzyme. Proc Natl Acad Sci USA 94:4262–4266
5. Klein D (2002) Quantification using real-time PCR technology: applications and limitations. Trends Mol Med 8(6):257–260
6. Kirchner S, Krämer KM, Schulze M et al (2010) Pentaplexed quantitative real-time PCR assay for the simultaneous detection and quantification of botulinum neurotoxin-producing clostridia in food and clinical samples. Appl Environ Microbiol 76(13):4387–4395
7. Real-time PCR fundamentals: hallmarks of an optimized qPCR assay. (http://www.gene-quantification.de/real-time-pcr-guide-bio-rad.pdf). Copyright © 2012 Bio-Rad Laboratories
8. Demidov VV, Broude NE (2004) DNA amplification current technologies and application. Horizon Bioscience, Norfolk, UK

Chapter 4

A New, Multiplex, Quantitative Real-Time Polymerase Chain Reaction System for Nucleic Acid Detection and Quantification

Fang Liang, Neetika Arora, Kang Liang Zhang, David Che Cheng Yeh, Richard Lai, Darnley Pearson, Graeme Barnett, David Whiley, Theo Sloots, Simon R. Corrie, and Ross T. Barnard

Abstract

Quantitative real-time polymerase chain reaction (qPCR) has emerged as a powerful investigative and diagnostic tool with potential to generate accurate and reproducible results. qPCR can be designed to fulfil the four key aspects required for the detection of nucleic acids: simplicity, speed, sensitivity, and specificity. This chapter reports the development of a novel real-time multiplex quantitative PCR technology, dubbed PrimRglo™, with a potential for high-degree multiplexing. It combines the capacity to simultaneously detect many viruses, bacteria, or nucleic acids, in a single reaction tube, with the ability to quantitate viral or bacterial load. The system utilizes oligonucleotide-tagged PCR primers, along with complementary fluorophore-labelled and quencher-labelled oligonucleotides. The analytic sensitivity of PrimRglo technology was compared with the widely used Taqman® and SYBR green detection systems.

Key words Quantitative PCR, Multiplexed PCR, Primer design, PrimRglo, Nucleic acid

1 Introduction

In the field of molecular diagnostics, the capacity to amplify more than one target in a single reaction tube is known as *multiplexing*. The rationale for multiplex reactions is to produce results for multiple nucleic acid targets from a limited quantity of sample in a single tube. Multiplexing is used to amplify a control target using multiple probes within each sample, allowing quantification relative to the internal control. Described here is a novel probe-based method (PrimRglo™) for performing a multiplex real-time quantitative PCR [1]. To demonstrate the reaction, we describe detection and quantification procedures for three types of nucleic acid sequences from *influenza A* and *Neisseria meningitidis* in a single reaction [2, 3].

Dmitry M. Kolpashchikov and Yulia V. Gerasimova (eds.), *Nucleic Acid Detection: Methods and Protocols*, Methods in Molecular Biology, vol. 1039, DOI 10.1007/978-1-62703-535-4_4,

The method is compared with the widely used SYBR green and Taqman qPCR systems. The Taqman® quantitative real-time PCR technique utilizes multiple fluorescent dyes conjugated to hydrolysis probes and allows monitoring the progress of the PCR reaction and comparison of the fluorescence signal from multiple targets during the log-linear phase of amplification.

The PrimRglo™ system described in this chapter has the potential to be expandable to a high degree of multiplexing, not achievable by hydrolysis probe (Taqman) or intercalating fluorophore (SYBR green) methods. The protocol allows quantitative identification of several targets present in one reaction tube. The procedure does not require the design of internal target-specific probes, but utilizes the standard PCR primers modified with adaptor or "tag" sequences. Two additional single-stranded reporter sequences are introduced into the system: one identical (except for a one base-pair mismatch) to the adaptor sequence and labelled with a fluorophore at the 5′ end. The second, complementary to the first, is labelled with a quencher dye at the 3′ end. Signal detection relies upon the hybridization of the fluorescence-labelled oligonucleotide with the quencher oligonucleotide which results in quenching of the fluorescence signal. At the start of the PCR reaction, the chimeric PCR forward primer binds to the quencher oligonucleotide blocking the formation of a reporter complex and increasing the fluorescence signal. As the reaction progresses, the concentration of free forward primer in the reaction is reduced as it is incorporated into newly formed double-stranded PCR product. This permits the interaction between the reporter and quencher oligonucleotides to form more reporter–quencher complexes which results in reduction of the detected fluorescence signal (*see* Fig. 1).

Fluorescently labelled probes and primers specific to a particular nucleotide target sequence are required to carry out a multiplex reaction. Both linear hydrolysis probes (e.g., TaqMan®) or probes with secondary structure (e.g., Molecular Beacon®) can be used for quantification and detection of a particular sequence and they need to be labelled with distinct fluorophores for each target [4]. These systems combine the objectivity of fluorescence detection with the simplicity of the original PCR reaction. Results obtained using fluorescence-based PCR chemistries are now accepted as a gold standard for quantification of viral load in clinical samples [5]. With careful design of primers and probes [5–7] and empirical optimization of primer and probe concentrations [8, 9] the course of the PCR amplification in the multiplex real-time PCR can produce results of comparable analytical sensitivity and specificity as compared to singleplex reactions. The PrimRglo™ method adds to the range of fluorescence-based quantitative methods, and, by virtue of its use of oligonucleotide "tags," is amendable to high-degree multiplexing.

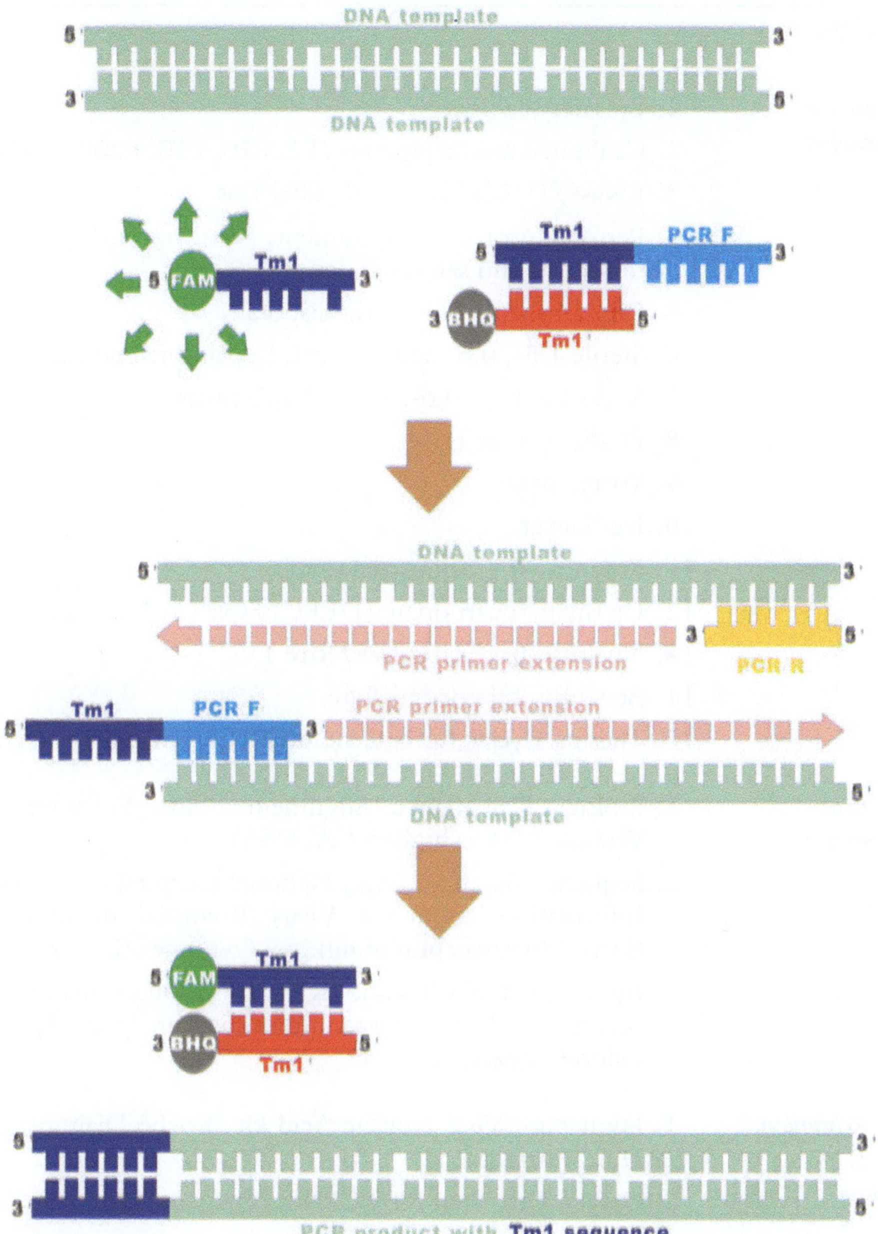

Fig. 1 In this system, signal detection relies on the hybridization between the reporter oligonucleotide, which has a fluorescence reporter attached at its 5′ end (labelled Tm and shown in *purple*), and quencher oligonucleotide, which has quencher attached at its 3′ end (labelled Tm′ in *red*). At the start of the PCR reaction, the chimeric PCR forward primer (consisting of both Tm (*dark blue* color) and target template-specific sequences) interacts with the quencher oligonucleotide. This allows the fluorescence signal to be emitted from the reporter. The interaction between chimeric PCR forward primer and quencher oligonucleotide is favored due to the presence of a single-nucleotide mutation in the reporter probe. During the PCR process, the chimeric PCR forward primers are integrated into the newly formed PCR product. As the concentration of chimeric PCR forward primer is reduced, an interaction between the reporter oligonucleotide and quencher oligonucleotide is established. This allows the quencher to absorb the fluorescence signal emitted from the reporter oligonucleotide, thus decreasing the fluorescence signal with each PCR cycle

2 Materials

2.1 Equipment and Disposables

1. Biosafety cabinet.
2. Calibrated sterile pipettes (P2, P10, P20, P200, P1000).
3. RNase/DNase-free sterile filter tips.
4. Personal protective equipment: Nitrile or latex gloves, laboratory coat, and safety glasses.
5. Clinical sharps bin for tip disposal.
6. Sterile 1.5-, 0.6-, and 0.2-mL Eppendorf microtubes.
7. Racks for 1.5-, 0.6-, and 0.2-mL tubes.
8. Plastic storage boxes.
9. Vortex mixer.
10. Ice bucket.
11. Ethanol 70 % (industrial grade).
12. Diethylpyrocarbonate (DEPC) H_2O (*see* **Note 1**).
13. Nuclease-free water (*see* **Note 1**).
14. Benchtop microcentrifuge.
15. Thermal Cycler, Rotor Gene-6000 (Qiagen, Doncaster, Australia).

2.2 Design of Oligonucleotides

1. Biological Sequence Alignment Editor Software (BioEdit, Version 7.09 or higher, CA, USA).
2. Sequence database (e.g., National Center For Biotechnology Information Influenza Virus Resource database (IVRD) (http://www.ncbi.nlm.nih.gov/genomes/FLU/).
3. Integrated DNA Technologies (IDT) Oligo Analyzer software online (http://www.idtdna.com/analyzer/Applications/OligoAnalyzer/).

2.3 RNA Extraction and Reverse Transcription

1. High Pure Viral Nucleic Acid kit (Roche Diagnostics, Castle Hill, Australia).
2. Qiagen One-Step RT-PCR Kit (Qiagen, Doncaster, Australia).
3. 10 mM Tris–HCl, pH 8.0.
4. Gene-specific forward primer and gene-specific reverse primer (*see* Table 1).
5. RNase AWAY™ Reagent (0.1–1.0 % Alkali hydroxide, 60–100 % distilled water) (Invitrogen, VIC, Australia).
6. –80 °C Freezer.

2.4 Construction of pGEM-T/H1N1/M, pGEM-T/NM-ctrA, and pGEM-T/NM-porA Plasmids

1. QIAquick PCR Purification Kit (Qiagen, Doncaster, Australia).
2. pGEM-T Easy vector (Promega, Sydney, NSW, Australia).
3. X-Gal (5-bromo-4-chloro-3-indolyl-beta-d-galactopyranoside) 40 mg/mL.

Table 1
List of primer and probe sequences involved in influenza A real-time PCR (monoplex)

Name	Sequence	Purpose	5′ modification	3′ modification
TM1-FluA-MF	CTTTAATCTCAATCAATACAAATCCTTCT AACCGAGGTCGAAACGTA	FluA PrimRglo forward primer	NA	NA
FluA-MF	CTT CTA ACC GAG GTC GAA ACG TA	FluA PCR forward primer	NA	NA
FluA-MR	GGT GAC AGG ATT GGT CTT GTC TTT A	FluA PCR reverse primer	NA	NA
FAM-FluA-BHQ	TC AGG CCC CCT CAA AGC CGA G	FluA TaqMan® probe	FAM	BHQ
FAM-TM1	CT TTA ATC TCC ATC AAT ACA AAT C	Reporter forward oligo sequence	FAM	C3
BHQ-TM1′	GA TTT GTA TTG ATT GAG ATT AAA G	Reporter reverse oligo sequence	C3	BHQ1

N.B. The nucleotide underlined indicates the position of mismatch between the reporter and quencher oligonucleotides

4. Isopropyl-beta-d-thiogalactopyranoside (IPTG) 0.1 M.
5. TE buffer: 10 mM Tris–HCl, pH 7.4; 1 mM EDTA, pH 8.0.
6. TE-saturated phenol/chloroform (*see* **Note 2**).
7. TAE buffer: 40 mM Tris–acetate, 2 mM EDTA, pH 8.0.
8. Luria-Bertani (LB) medium.
9. LB plates with antibiotic (*see* **Note 3**).
10. NaCl solution: 0.1 M NaCl, 5 mM Tris–HCl, 5 mM $MgCl_2$, pH 7.0.
11. $CaCl_2$ solution: 0.1 M $CaCl_2$, 5 mM Tris–HCl, 5 mM $MgCl_2$, pH 7.0.
12. SOC medium: 2 % Bacto tryptone, 0.5 % Bacto-yeast extract, 10 mM NaCl, 2.5 mM KCl, 10 mM $MgCl_2$, 10 mM $MgSO_4$, 20 mM glucose.
13. Top 10 Chemically Competent *E. coli* (Invitrogen, VIC, Australia).
14. dNTP mix (2.5 mM in each dNTP).
15. T4 DNA ligase: 4 U/μL of the enzyme solution in storage buffer: 20 mM Tris–HCl (pH 7.5), 1 mM DTT, 50 mM KCl, 0.1 mM EDTA, and 50 % glycerol.
16. 10× Ligation buffer of 400 mM Tris–HCl, 100 mM $MgCl_2$, 100 mM DTT, and 5 mM ATP (pH 7.8 at 25 °C).
17. Constant-temperature incubator.
18. Shaking incubator.
19. QIAprep Spin Miniprep Kit (Qiagen, Doncaster, Australia).

2.5 Quantitative Real-Time PCR

2.5.1 SYBR Real-Time PCR System

1. ABI SYBR green PCR Master Mix (Applied Biosystems, Foster City, CA, USA).
2. PCR Primers for Influenza A matrix protein gene (IDT, VIC, Australia; *see* Table 1).

2.5.2 TaqMan Real-Time PCR System

1. ABI Taqman Universal PCR Master Mix (Applied Biosystems, CA, USA).
2. PCR Primers and reporter oligonucleotides for Influenza A matrix protein gene (IDT, VIC, Australia; *see* Table 1).

2.5.3 PrimRglo Real-Time System

1. ABI Taqman Universal PCR Master Mix (Applied Biosystems, CA, USA).
2. PCR Primers and reporter oligonucleotides for Influenza A matrix protein gene (IDT, VIC, Australia) (*see* Table 1); PCR Primers and reporter oligonucleotides for *N. meningitidis* ctrA/porA genes (IDT, VIC, Australia; *see* Table 2).

Table 2
List of primer and probe sequences involved in NM real-time PCR (duplex)

Name	Sequence	Purpose	5′ modification	3′ modification
NM-porAF	CGGCTCGTTTATCGGCTT	NM porA PCR forward primer	NA	NA
NM-porAR	CGACAAAGGATTCCCTGTTG	NM porA PCR reverse primer	NA	NA
NM-ctrAF	CCGCATTATTCTGCACCA	NM ctrA PCR forward primer	NA	NA
NM-ctrAR	GCCTTTCTTCGATGGGCT	NM ctrA PCR reverse primer	NA	NA
FAM-TM2	TGATTGTA**T**TATGTATTGATAAAG	Reporter forward oligo sequence	FAM	C3
HEX-TM3	GATTGTAAGA**G**TTGATAAAGTGTA	Reporter forward oligo sequence	HEX	C3
BHQ-TM2′	CTT TAT CAA TAC ATA CTA CAA TCA	Reporter reverse oligo sequence	C3	BHQ
BHQ-TM3′	TACACTTTATCAAATCTTACAATC	Reporter reverse oligo sequence	C3	BHQ
TM2-NM-porAF	TGATTGTAGTATGTATTGATAAAGCGG CTCGTTTATCGGCTT	NM porA PrimRglo forward primer	NA	NA
TM3-NM-ctrAF	GATTGTAAGATTTGATAAAGTGTACCG CATTATTCTGCACCA	NM ctrA PrimRglo forward primer	NA	NA

N.B. The nucleotide underlined indicates the position of mismatch between the reporter and quencher oligonucleotides

2.6 Gel Electrophoresis and RNA/DNA Quantification

1. 1.5 % Agarose gel made in 1× TAE buffer containing 0.5 μg/mL ethidium bromide.
2. 100 bp HyperLadder™ IV DNA marker (Bioline (Aust), Alexandria, NSW 1435) ranging from 100 to 1,013 bp (5 μL loaded per lane of agarose gel).
3. 5× loading buffer: 0.25 % w/v bromophenol blue, 0.25 % w/v xylene cyanol, 0.1 M EDTA, 0.2 M EDTA, 30 % glycerol, in H_2O.
4. TAE buffer: 40 mM Tris-base, 20 mM acetic acid, and 1 mM EDTA.
5. UV transilluminator.
6. Nanodrop 1000 UV Spectrophotometers (Thermo fisher scientific, Wilmington, USA).
7. Electrophoresis equipment (power pack and gel electrophoresis tanks).
8. Sequencing facility or a reputable sequencing service provider.
9. Medical waste bins for incineration.
10. Beaker for ethidium bromide waste disposal.
11. Sterile blade.
12. Gloves for gel electrophoresis area.
13. Kimberly–Clark Kimwipes® EXL Delicate Task wipes, 11.4×21.3 cm or lint-free paper.

3 Methods

3.1 Design of Oligonucleotides for the Influenza A Matrix (M) Gene Segment and N. meningitidis ctrA/porA Genes

1. A modified forward primer, plus two additional oligonucleotides, must be used in the PCR reaction for each target gene, in order to monitor the dynamic progress of PCR. One of each pair of additional oligonucleotides is labelled with a fluorophore and the other is labelled with a quencher dye. Design three such sets of extra oligonucleotides, synthesized for the influenza A M gene, *N. meningitidis* ctrA, and *N. meningitidis* porA genes in this protocol (*see* Tables 1 and 2).
2. Use a 24 bp "universal" oligonucleotide sequence as an extension at the 5′ end of each of the PCR forward primers for ctrA, porA, and fluA M gene segments, respectively, resulting in a total length of the primers between 42 and 47 bp (named as Tm in Fig. 1) (*see* **Note 5**).
3. Design reverse sequences complementary to the universal sequence and modify by adding a quencher dye at 3′end and a C3 spacer at the 5′end (named as Tm′ in Fig. 1) (*see* **Note 6**).

4. Also design another set of 24 bp "universal" oligonucleotide sequences with a single bp mismatch and add a fluorescent dye at the 5′ end and a C3 spacer at the 3′end (*see* **Note 6**).
5. Design the PCR primers using the NCBI Influenza Virus Resource Database or the NCBI database as described previously [1, 9–11]. Construct the extra oligonucleotides on the basis of the universal tags on the PCR primers by taking several factors into consideration such as the free energy of self-dimer, hetero-dimer, and melting temperature. This can be achieved using the IDT Oligo Analyzer software online (*see* **Note 7**).

3.2 PCR Template Preparation

3.2.1 Influenza A DNA Preparation

1. Extract RNA from the influenza A sample using High Pure Viral Nucleic Acid kit in accordance with the manufacturer's protocol and elute the sample into 50 μL volume with DEPC water (*see* **Note 8**). Store RNA eluate at −80 °C if not for immediate use (*see* **Note 4**).
2. Set up the NanoDrop™ 1000 machine and calibrate with DEPC-treated water as blank control.
3. Pipette a 1–2 μL sample of RNA onto the end of the receiving fiber of the NanoDrop™ spectrometer. A second fiber-optic cable (the source fibers) is then brought into contact with the liquid sample causing the liquid to bridge the gap between the fiber-optic ends. The instrument is controlled by NanoDrop™ proprietary software run from a personal computer, and the data is logged in an archive file on the computer.
4. Measure and collect data before swabbing away the remaining sample on both fibers before the next analysis (*see* **Note 9**).
5. Amplify the 210 bp fragment of the influenza A matrix gene with Qiagen One-Step RT-PCR Kit according to the standard procedures (*see* **Note 10**).
6. Before conducting the reverse transcription-PCR (RT-PCR), it is recommended that the final concentration for each forward and reverse primer in 25 μL of RT-PCR mixture is 0.8 μmol/L FluA-MF and FluA-MR (*see* Table 1) with a 5 μL volume of RNA eluate as template [10].
7. Perform the RT-PCR procedure under the following thermal cycling conditions: 50 °C for 20 min, 95 °C for 15 min followed by 45 cycles of 95 °C for 15 s and 60 °C for 1 min.
8. Perform gel electrophoresis of the PCR product so generated using 1 % agarose gel stained with ethidium bromide. Visualize PCR product "bands" on a UV transilluminator to verify product size and absence of bands in the "no template" negative control.

3.2.2 *N. meningitidis* DNA Preparation

1. Isolate DNA from *N. meningitidis* bacterial culture supplied by the Sir Albert Sakzewski Virus Research Centre, QLD, Australia, using the High Pure Viral Nucleic Acid kit in accordance with

the standard manufacturer's procedures and elute into a 50 μL volume with elution buffer (*see* **Notes 4** and **8**).

2. Amplify the 131 bp fragment in *N. meningitidis* porA gene and the 156 bp fragment in *N. meningitides* ctrA gene using the same procedure and reaction conditions as described in Subheading 3.2.1.
3. It is recommended that the concentrations for each forward and reverse primer in the final 50 μL of PCR mixture be 0.2 μmol/L porA-MF, 0.4 μmol/L porA-MR, 0.2 μmol/L ctrA-MF, and 0.4 μmol/L ctrA-MR (*see* Table 2). The volume of the template added is 2 μL [11].

3.3 Construction of pGEM-T/H1N1/M, pGEM-T/NM-ctrA, and pGEM-T/NM-porA Plasmids

1. Perform purification of the PCR product using QIAquick PCR Purification kit according to the manufacturer's protocol (*see* **Note 11**). Collect the PCR product into 30 μL of 10 mM Tris buffer, pH 8.0.
2. Set up a ligation reaction and transform "Top 10" chemically competent *E. coli* cells (200 μL) with ligated pGEM-T vector (1–20 ng in 1–15 μL). Spread onto agar plates with X-Gal, IPTG, and ampicillin (*see* **Note 3**). Select white colonies, and screen for the presence of insert by PCR.
3. Purify recombinant plasmids using QIAprep Spin Miniprep Kit according to the standard manufacturer's procedure (*see* **Note 12**).
4. Confirm DNA purity and concentration using the NanoDrop 1000 as described in Subheading 3.2.1.
5. Confirm the existence of the insert by restriction digestion with the enzyme Not I, followed by electrophoresis.
6. Dissolve plasmids in TE buffer to make a 50 ng/μL stock. Store at 4 °C (*see* **Note 4**).

3.4 Quantitative Real-Time PCR

3.4.1 SYBR Green Real-Time System

1. Prepare serial dilutions of pGEM-T/H1N1/M plasmid from 2 to 2×10^{-9} ng/μL with DNase-free water. Place ten 1.5-mL centrifuge tubes in a plastic rack. Aliquot 5 μL of plasmid solution from 50 ng/μL stock into the first tube, add 45 μL of water, and pipette up and down to mix. Follow similar procedures to prepare template solution from 0.2 to 2×10^{-9} ng/μL (*see* **Notes 13** and **14**).
2. Prepare real-time PCR matrix in 0.2-mL tubes as follows:

ABI SYBR green PCR Master Mix (2×) (*see* **Note 19**)	25 μL
TM1-FluA-MF forward primer (final conc. 400 nM)	2 μL
FluA-MR reverse primer (final conc. 400 nM)	2 μL
Water	19 μL

3. Add 2 μL of 0.2 ng/μL (start concentration) template to PCR mix, and pipette or vortex to mix. Centrifuge briefly in a microfuge to push any droplets into the bulk reaction mix.
4. Follow the same procedure to prepare the PCR mix using the other concentrations of templates.
5. Run the PCR reaction under the following condition: 95 °C for 10 min, followed by 40 cycles of 95 °C for 30 s, 45 °C for 30 s, and 72 °C for 30 s, and finish with a final extension of 95 °C for 1 min and 60 °C for 10 min. The fluorescence signal collecting point for each cycle is at the end of 45 °C for 30 s phase.
6. After the amplification process, the standard curve for the reaction will be generated by the Rotor Gene-6000 software. The threshold is determined by the software (*see* **Note 15**).

3.4.2 Taqman Probe Real-Time System

1. Make serial dilutions of pGEM-T/H1N1/M plasmid from 0.2 to 2×10^{-9} ng/μL with DNase-free water as described above.
2. Prepare real-time PCR reaction mix in 0.2-mL tubes as follows: (*see* **Note 16**)

ABI Taqman Universal PCR Master Mix (2×) (*see* **Note 19**)	25 μL
TM1-FluA-MF forward primer (final conc. 400 nM)	2 μL
FluA-MR reverse primer (final conc. 400 nM)	2 μL
FAM-probe-BHQ (final conc. 200 nM)	1 μL
Water	19 μL

3. Add 2 μL of 0.2 ng/μL (starting concentration) template to PCR reaction mix. Pipette or vortex to mix. Centrifuge briefly in a microfuge to push any droplets into the bulk reaction mix.
4. Follow the same procedure to prepare PCR reaction mixes with the range of concentrations of templates.
5. Run PCR reaction under the following condition: 95 °C for 10 min, followed by 40 cycles of 95 °C for 30 s, 45 °C for 30 s, and 72 °C for 30 s, and finish with a final extension of 95 °C for 1 min and 60 °C for 10 min. The fluorescence signal collecting point for each cycle is at the end of 45 °C for 30 s step.
6. After the amplification process, the standard curve for the reaction will be generated by the Rotor Gene-6000 software.

3.4.3 PrimGlo Real-Time System

1. Make serial dilutions of pGEM-T/H1N1/M plasmid from 0.2 to 2×10^{-9} ng/μL with DNase-free water as described above.
2. Prepare real-time PCR matrix in 0.2-mL tubes as follows (*see* **Note 17**):

ABI Taqman Universal PCR Master Mix (2×) (*see* **Note 19**)	25 μL
TM1-FluA-MF forward primer (final conc. 400 nM)	2 μL
FluA-MR reverse primer (final conc. 400 nM)	2 μL
BHQ-TM1′ (final conc. 200 nM)	1 μL
FAM-TM1 (final conc. 100 nM)	0.5 μL
Water	17.5 μL

3. Add 2 μL of 0.2 ng/μL (starting concentration) template to PCR reaction mix, and pipette or vortex to mix. Centrifuge briefly in microfuge to push any droplets into the bulk reaction mix.
4. Follow the same procedure to prepare PCR mixes with other concentrations of templates.
5. Run the PCR reaction under the following condition: 95 °C for 10 min, followed by 45 cycles of 95 °C for 30 s, 45 °C for 30 s, and 72 °C for 30 s, and finish with a final extension of 95 °C for 1 min and 60 °C for 10 min. The fluorescence signal collecting point for each cycle is at the end of the step 45 °C for 30 s (*see* **Notes 18** and **20**).
6. After the amplification process, the standard curve for the reaction will be generated by the Rotor Gene-6000 software (*see* Fig. 2, Tables 3 and 4).

3.4.4 N. meningitidis ctrA and porA Genes in a Duplex Reaction in the PrimRGlo Real-Time System

1. Make serial dilutions of pGEM-T/NM-ctrA and pGEM-T/NM-porA plasmids from 2 to 2×10^{-6} ng/μL with DNase-free water as described above.
2. Prepare real-time PCR matrix in 0.2-mL tubes as follows:

ABI Taqman Universal PCR Master Mix (2×) (*see* **Note 19**)	25 μL
TM2-NM-PorAF forward primer (final conc. 400 nM)	2 μL
NM-PorAR reverse primer (final conc. 400 nM)	2 μL
BHQ-TM2′ (final conc. 200 nM)	1 μL
FAM-TM 2 (final conc. 100 nM)	0.5 μL
TM3-NM-CtrAF forward primer (final conc. 400 nM)	2 μL
NM-CtrAR reverse primer (final conc. 400 nM)	2 μL
BHQ-TM3′ (final conc. 200 nM)	1 μL
HEX-TM 3 (final conc. 100 nM)	0.5 μL
Water	12 μL

3. Add 2 μL of 0.2 ng/μL (starting concentration) porA and ctrA templates separately to the PCR reaction mix, and pipette or vortex to mix. Centrifuge briefly in a microfuge to push any droplets into the bulk reaction mix.

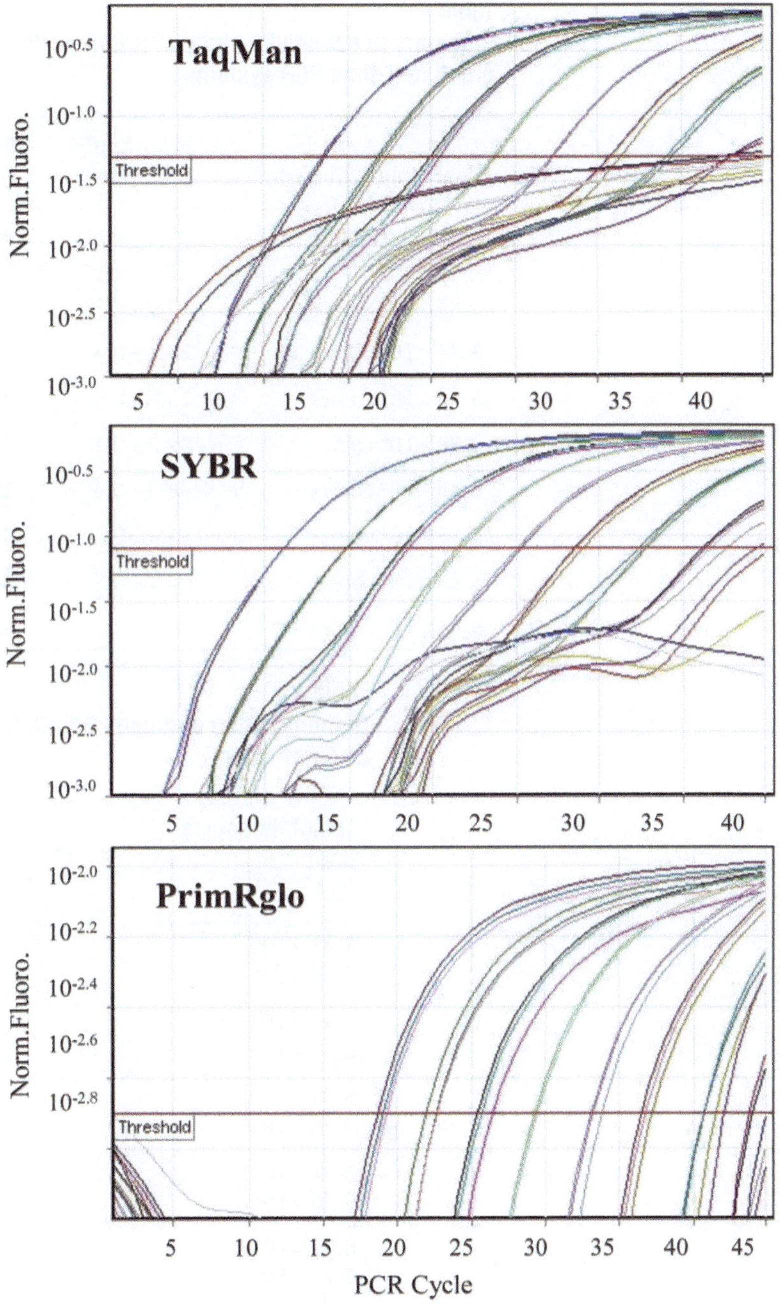

Fig. 2 PCR amplification curves for singleplex PCR reactions using the SYBRgreen, Taqman, or PrimRglo methods. These curves correspond to the Ct values in Table 3

4. Follow the same procedure to prepare PCR reaction mixes with the other concentrations of templates.
5. Run the PCR reaction under the following thermocycling conditions: 95 °C for 10 min, followed by 40 cycles of 95 °C for 30 s, 45 °C for 30 s, and 72 °C for 30 s, and finish with a final extension of 95 °C for 1 min and 60 °C for 10 min.

Table 3
Summary of sensitivity study of PrimRglo™, Taqman® probe, and SYBR® green real-time PCR systems

Real-time PCR system	PrimRglo™ (45 cycles)	Taqman® (40 cycles)	SYBR® green (40 cycles)
[pGEM-T/H1N1/ M plasmid]	Av C_t ± s.d. n = 3	Av C_t ± s.d. n = 3	Av C_t ± s.d. n = 3
5.85×10^7 copies	18.9 ± 0.3	13.6 ± 0.1	11.2 ± 0.1
5.85×10^6 copies	22.4 ± 0.4	17.1 ± 0.3	14.8 ± 0.0
5.85×10^5 copies	25.8 ± 0.5	20.3 ± 0.1	18.3 ± 0.2
5.85×10^4 copies	29.3 ± 0.1	23.8 ± 0.1	21.6 ± 0.1
5.85×10^3 copies	33.3 ± 0.4	27.3 ± 0.1	25.2 ± 0.1
5.85×10^2 copies	36.9 ± 0.4	31.0 ± 0.3	28.7 ± 0.3
5.85×10^1 copies	40.9 ± 0.5	34.2 ± 0.2	32.6 ± 0.3
5.85×10^0 copies	–	38.0 ± 0.2	36.2 ± 0.1

Table 4
Demonstration of duplex PrimRglo using porA and ctrA detection

Template copies *porA* *ctrA*	*porA/ctrA* **Av C_t ± s.d. (n = 3)**
5.91×10^8 5.75×10^8	12.7 ± 0.4 14.9 ± 1.0
5.91×10^7 5.75×10^7	15.7 ± 0.6 15.9 ± 0.7
5.91×10^6 5.75×10^6	20.5 ± 0.8 20.4 ± 1.5
5.91×10^5 5.75×10^5	24.5 ± 0.6 24.7 ± 0.9
5.91×10^4 5.75×10^4	28.0 ± 0.3 28.4 ± 0.3
5.91×10^3 5.75×10^3	31.3 ± 0.6 31.4 ± 0.4
5.91×10^2 5.75×10^2	34.6 ± 0.4 33.4 ± 0.3

The fluorescence signal measurement point for each cycle is at the end of the 45 °C for 30 s step.

6. After the amplification process, the standard curve for the reaction will be generated by the Rotor Gene-6000 software (*see* Fig. 3).

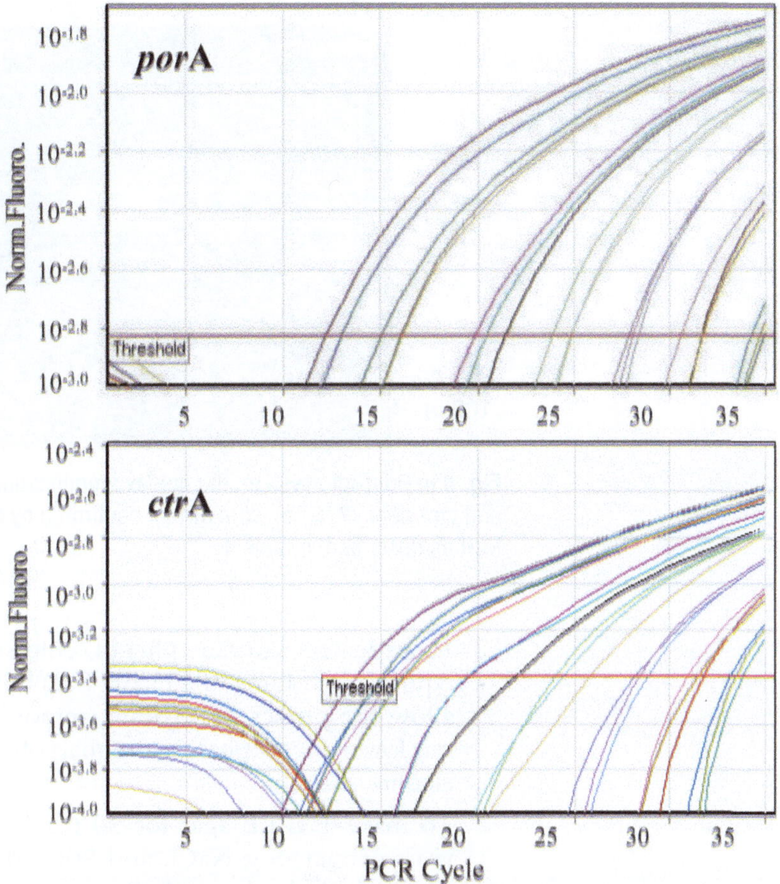

Fig. 3 PCR amplification curves for PrimRglo duplex reactions. When both ctrA and porA plasmids were added, amplification was achieved for both. These curves correspond to the Ct values in Table 4

7. The duplex amplification of products of the correct size for both porA (131 bp) and ctrA (210 bp) is confirmed by electrophoresis of 20 μL of the reaction mix in an agarose gel containing ethidium bromide (*see* Fig. 4). The molecular weight standard "ladder" in Fig. 4 is HyperLadder™ IV (Bioline, Alexandria, NSW 1435) ranging from 100 to 1,013 bp (5 μl loaded).

4 Notes

1. DNase-free water is prepared by autoclaving deionized water that has been certified to be free of endonuclease, exonuclease, and RNase activity (a commercial product). DEPC water is prepared by incubating 0.1 % v/v DEPC in the water for 1 h at 37 °C, and then autoclaving for at least 15 min. DEPC water is essential for the RNA extraction steps. DNase-free water is used for all other steps of this protocol.

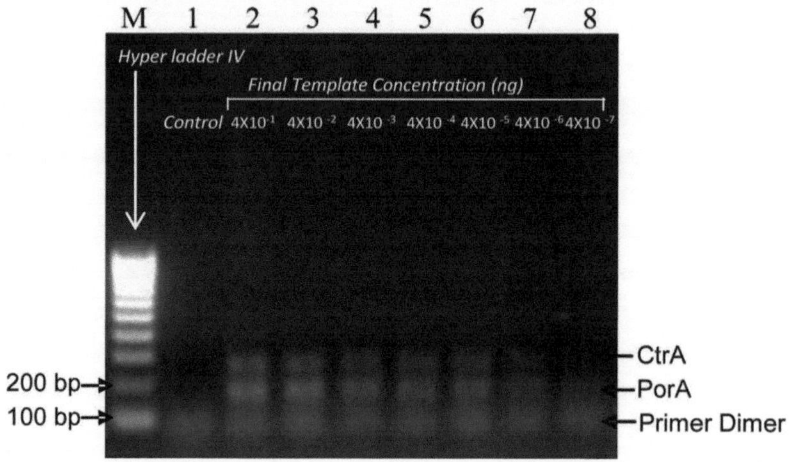

Fig. 4 In PrimRglo system, the duplex amplifications of both porA gene (131 bp) and ctrA gene (210 bp) were further confirmed by electrophoresis after real-time PCR as described in Table 4

2. In order to saturate phenol/chloroform with TE, add an equivalent volume of TE (pH 8.0) buffer into phenol, mix and allow the phases to separate, remove upper phase (TE), and mix lower phenol phase with equal volume of chloroform:isoamyl alcohol (24:1).
3. To make 1 L LB agar for 30 plates, add 10 g tryptone, 5 g yeast extract, 5 g NaCl, and 800 mL distilled water to a 2 L flask, stir and mix until all components are totally dissolved, adjust pH with 400 μL 5 N NaOH, continue adding distilled water up to 1,000 mL, add 15 g agar, and stir until dissolved. Then autoclave for 20 min at liquid setting, with a loose top or aluminum foil. Let agar cool to 40–50 °C, add 1 mL ampicillin (100 mg/mL) for a final concentration of 100 μg/mL, sterilize the flask mouth, pour agar onto plates until half-way full, let agar cool to its solid, and then equally spread 40 μL IPTG (100 mM) and 40 μL X-gal (40 mg/mL) on each plate. Label the plates and store at 4 °C upside down inside a bag.
4. RNA extracted can be stored for 1 month in RNA storage buffer at −80 °C. DNA isolated can be stored for half a year in water or TE buffer at −20 °C. DNA plasmids can be stored at 4 °C for short-term use, and −20 °C for 4 years.
5. The length of the universal "tag" strand of the PCR primer is designed to be similar in length to the other (3′) part of the PCR primer that is complementary to the target template. This ensures that the annealing temperature will be similar for (a) the interaction of the primer with the target template, and for (b) the interaction between the fluorescence–quencher and the universal "tag" sequence.

6. C3 spacer (IDT, VIC, Australia) was added to these two fluorescence/quencher sequences in order to inhibit possible extension during the extension step.
7. Available at http://www.idtdna.com/analyzer/Applications/OligoAnalyzer/.
8. High Pure Viral Nucleic Acid kit, cat. no. 11 858 874 01. The manufacturer's protocol is described in detail at https://e-labdoc.roche.com/LFR_PublicDocs/ras/11858874001_en_17.pdf.
9. The sample drop should fully cover the detection pore with no air bubbles in the drop.
10. QIAGEN OneStep RT-PCR Kit, cat. no. 210210. The manufacturer's protocol is described in detail at http://www.qiagen.com/Products/Catalog/Assay-Technologies/End-Point-PCR-and-RT-PCR-Reagents/QIAGEN-OneStep-RT-PCR-Kit#resources.
11. QIAquick PCR Purification Kit, cat. no. 28104. The manufacturer's protocol is described in detail at http://www.qiagen.com/Products/Catalog/Sample-Technologies/DNA-Sample-Technologies/DNA-Cleanup/QIAquick-PCR-Purification-Kit#resources.
12. QIAprep Spin Miniprep Kit, cat. no. 27104. The manufacturer's protocol is described in detail at http://www.qiagen.com/Products/Catalog/Sample-Technologies/DNA-Sample-Technologies/Plasmid-DNA/QIAprep-Spin-Miniprep-Kit#resources.
13. To be accurate, no less than 3 μL of a higher concentration solution should be applied to dilute into a lower concentration. Use no dilution more than tenfold between adjacent gradients.
14. This should be done physically isolated from the PCR mastermix preparation area to avoid contamination. In this case, two different rooms and sets of pipettes were used; always change gloves before template addition step.
15. The threshold value for the amplification plots was determined automatically by Rotor Gene-6000 software according to coefficient value.
16. A key element of the PrimRglo system is the relative concentration of the three oligonucleotides: (a) the universal "tag" sequence-modified PCR forward primer must not be in large excess; it should be readily consumed even when the starting template concentration is low; (b) based on the amount of forward primer, the quencher-modified complementary oligonucleotide should not be at such a high concentration in the reaction mix, as to completely block fluorescence signal through the whole reaction, but needs to be sufficiently high to quench fluorescence signals when the forward primer is consumed during the reaction; (c) the concentration of fluorescent dye-modified universal detection oligonucleotide (FAM-TM) should be less than that of its quencher sequence (BHQ-TM′). The optimal

ratio for FAM-TM1:BHQ-TM1′:FluA M forward primer in the reaction mixture is 1:2:4. The same ratio suits for all three genes tested in this system.

17. The temperature used during the annealing phase of the reaction is critical, because the hybridization of the fluorescent and quencher primer complex will reflect the consumption of PCR forward primers, and eventually the yield of target DNA product.
18. 45 cycles are used here to gain comparison data, over a wide range. Usually fewer cycles would be used in a diagnostic setting. The specificity of the reactions was verified by gel electrophoresis analysis.
19. The commercial PCR Master Mix should account for half the final volume of PCR reaction mix as per the manufacturer's instructions.
20. The amplification curve is inverted relative to traditional amplification curves (i.e., the raw fluorescence signal decreases, rather than increases, as the reaction progresses). This is inverted into a more customary pattern by the Rotor gene-6000 software.

References

1. Lai R et al (2012) PrimRglo: a multiplexable quantitative real-time polymerase chain reaction system for nucleic acid detection. Anal Biochem 422:89–95
2. Yao Y et al (2001) Sequences in Influenza A virus PB2 protein that determine productive infection for an avian influenza virus in mouse and human cell lines. J Virol 75:5410–5415
3. Alvarez AC et al (2008) A broad spectrum, one-step reverse-transcription PCR amplification of the neuraminidase gene from multiple subtypes of influenza A virus. Virol J 5:77–77
4. Bustin SA et al (2009) The MIQE guidelines: minimum information for publication of quantitative real-time PCR experiments. Clin Chem 55:611–622
5. Abd-Elsalam KA (2003) Bioinformatic tools and guideline for PCR primer design. Afr J Biotechnol 2:4
6. Rickert AM, Lehrach H, Sperling S (2004) Multiplexed real-time PCR using universal reporters. Clin Chem 50:1680–1683
7. Thornton B, Basu C (2011) Real-time PCR (qPCR) primer design using free online software. Biochem Mol Biol Educ 39:145–154
8. Raeymaekers L (1993) Quantitative PCR: theoretical considerations with practical implications. Anal Biochem 214:582–585
9. Wang X, Seed B (2003) A PCR primer bank for quantitative gene expression analysis. Nucleic Acids Res 31:e154
10. Whiley DM, Sloots TP (2005) A 5′-nuclease real-time reverse transcriptase–polymerase chain reaction assay for the detection of a broad range of influenza A subtypes, including H5N1. Diagn Microbiol Infect Dis 53:335–337
11. Whiley DM et al (2003) Detection of Neisseria Meningitidis in clinical samples by a duplex real-time PCR targeting the porA and ctrA genes. Mol Diagn 7:141–145

Chapter 5

Detection of SNP-Containing Human DNA Sequences Using a Split Sensor with a Universal Molecular Beacon Reporter

Yulia V. Gerasimova, Jack Ballantyne, and Dmitry M. Kolpashchikov

Abstract

Hybridization-based techniques have been extensively employed for the analysis of specific DNA/RNA sequences. Herein, we describe highly specific inexpensive smart hybridization-based sensor that takes advantage of a universal molecular beacon probe as a fluorescent reporter. The sensor has a straightforward design, and demonstrates improved selectivity and specificity of nucleic acid recognition. It is cost-efficient since it utilizes the same molecular beacon probe for the analysis of many nucleic acid sequences.

Key words Molecular beacon probe, High specificity, DNA junctions, Genotyping, Forensic SNPs

1 Introduction

Molecular beacon (MB) probe is a stem–loop folded oligonucleotide with a fluorophore and a dark quencher dye on the opposite termini of the hairpin [1]. The stem–loop structure of MB probe enables low fluorescent signal of the probe in the absence of a target nucleic acid due to the contact quenching of the fluorophore's fluorescence with a dark quencher. In the presence of the target, high fluorescent signal of MB probe is generated due to the removal of the fluorophore from the quencher in the probe–target duplex. A conformational constraint in the form of stem–loop improves the selectivity of MB probe in comparison with linear probes [2], which is an attractive property for single-nucleotide polymorphism (SNP) analysis.

SNPs are one of the types of genetic variation. In humans, SNPs can predetermine the risk of a disease development, the severity of the disease, and the person's response to treatment [3, 4]. In addition, SNP combination is unique for a group of people, which makes it serve as a "genetic fingerprint" for identity testing in forensic applications [5]. Therefore, assays for reliable SNP

Dmitry M. Kolpashchikov and Yulia V. Gerasimova (eds.), *Nucleic Acid Detection: Methods and Protocols*, Methods in Molecular Biology, vol. 1039, DOI 10.1007/978-1-62703-535-4_5, © Springer Science+Business Media New York 2013

genotyping benefit healthcare (a concept of personalized medicine) and law enforcement. Nowadays, more than three million SNPs are validated [6]. In most cases, multiple SNPs need to be analyzed simultaneously [7]. Application of MB probe for multiplex SNP genotyping assays, however, requires hundreds of expensive hairpin oligonucleotides with two dyes attached to their opposite termini. The structure of MB probe for each of the analyzed nucleic acid sequence should be thoroughly designed to avoid undesirable intramolecular hybridization between the stem and loop fragments of the probe, which can increase the background of the assay and impair the overall performance of MB probe. Sometimes, the problem of cross-hybridization between the stem and the loop is impossible to avoid, since the loop portion of MB probe is complementary to the analyzed DNA. In practice some nucleic acid sequences cannot be analyzed by MB probe at all. Moreover, precise temperature control is still required for accurate SNP identification.

In this chapter, we describe the design and performance of an MB-based hybridization sensor for SNP genotyping, which keeps the advantages of MB probe while improving the selectivity and lowering the cost of the probe in case of the analysis of multiple targets [8–12]. The sensor uses MB as a fluorescent signal reporter, which binds the analyzed nucleic acid indirectly, via two adaptor strands. Using the adaptor strands enables utilization of one "universal" MB reporter for the detection of practically any DNA or RNA. Such reporter can be easily optimized, since its structure is independent of the analyte sequences. The sensor offers great selectivity: differentiation between two analytes differing in one nucleotide can be achieved even at room temperature. This can be achieved by having analyte-binding arm of the adaptor strands complementary to the SNP site short (8–10 nt) and sensitive to even single-nucleotide mispairing.

Here, three SNP-containing fragments of human genomic DNA were analyzed in one tube. The polymorphic sites rs1490413, rs876724, and rs717302 were reported to be among SNP markers suitable for human identification [13]. The split probes targeting both alleles for each of the fragments are designed according to the same principle. One and the same MB probe is used as a reporter for all six sensors. The chapter describes in details the principles of the sensor design, provides the conditions for PCR amplification of the human DNA fragments that contain SNPs, gives the procedure for obtaining the single-stranded DNA analyte after amplification, and, finally, reports the protocol for the fluorescent assay for SNP genotyping of human DNA using the designed probes.

2 Materials

2.1 DNA Isolation

1. QIAamp DNA Investigator kit (Qiagen, USA).
2. 96 % ethanol.

2.2 DNA Amplification

1. PCR thermocycler.
2. Filtered pipette tips.
3. 1 μM Forward primers.
4. 1 μM Biotin-containing reverse primers.
5. PCR buffer II (10×) (Applied Biosystems).
6. 25 mM $MgCl_2$.
7. AmpliTaq Gold DNA polymerase (Applied Biosystems).
8. Nuclease-free H_2O.

2.3 Purification of Single-Stranded DNA

1. Microcon centrifugal filter YM-30 membranes.
2. Disposable empty polypropylene columns.
3. Streptavidin agarose resin.
4. Binding buffer: 50 mM Tris-HCl, pH 7.4, 140 mM NaCl, 5 mM KCl.
5. Eluting solution: 0.2 M NaOH.
6. Neutralizing solution: 2 M HCl.
7. Sodium acetate: 3 M sodium acetate, pH 5.5.

2.4 Probe Design

OligoAnalyzer (IDT).

2.5 SNP Detection

1. Detection buffer ×2: 100 mM Tris–HCl, pH 7.4, 100 mM $MgCl_2$.
2. Universal molecular beacon (UMB): 10 μM solution in water.
3. Oligonucleotide adaptor strands: 50 μM solutions in detection buffer (1×).
4. Synthetic oligonucleotide analytes: 250 nM solutions in detection buffer (1×).
5. Amplified fragments of human DNA: 250 nM solutions in detection buffer (1×).
6. UV spectrophotometer.

3 Methods

3.1 DNA Isolation

To isolate human DNA from blood or other sources, use Qiagen QIAamp DNA Investigator kit and follow the manufacturer's suggested guideline for isolation of total (genomic and mitochondrial) DNA from 1 to 100 μL of whole blood treated with EDTA, citrate, or heparin-based anticoagulants. At the final step, elute the isolated DNA with 100 μL of buffer.

3.2 DNA Amplification

Three fragments of human genomic DNA, which contain polymorphic sites rs1490413, rs876724, and rs717302, are amplified using three pairs of primers (Table 1). The design for the primers and the PCR conditions were previously optimized by Sanchez et al. [14]. The reverse primers are biotin labeled to enable separation of the two strands using streptavidin agarose resin.

1. Prepare the following mixture for PCR (total volume 500 μL) (*see* **Note 1**) (Table 2).

Table 1
Primers used for amplification of three fragments of human DNA [14]

Name[a]	Sequence	Amplicon size (bp)
rs14-F	5′-GTGTGGACTGGGCTGATGT-3′	68
Bio-rs14-R	5′-Bio-TTCTCACTAGTGTCCCTGCTCTG-3′	
rs87-F	5′-GCAGGCTCCATTTTTATACCACT-3′	83
Bio-rs87-R	5′-Bio-GAATATCTATGAGCAGGCAGTTAGC-3′	
rs71-F	5′-CTTTAGAAAGGCATATCGTATTAACTGTG-3′	86
Bio-rs71-R	5′-Bio-AACACAGAAAGAGGTTCATATGTTGG-3′	

[a]F and R denote forward and reverse primer, respectively; Bio—biotin label

Table 2
The composition of PCR mix

Reagent	Stock concentration	Volume (μL)	Final concentration
Human DNA	7 ng/μL	7	0.1 ng/μL
rs14-F	1 μM	30	0.06 μM
Bio-rs14-R	1 μM	30	0.06 μM
rs87-F	1 μM	10	0.02 μM
Bio-rs87-R	1 μM	10	0.02 μM
rs71-F	1 μM	20	0.04 μM
Bio-rs71-R	1 μM	20	0.04 μM
PCR buffer II ×10		50	×1
$MgCl_2$	25 mM	160	8 mM
dNTPs	10 mM	35	0.7 mM
AmpliTaq DNA polymerase	5 U/μL	8	0.08 U/μL
Nuclease-free H_2O		120	

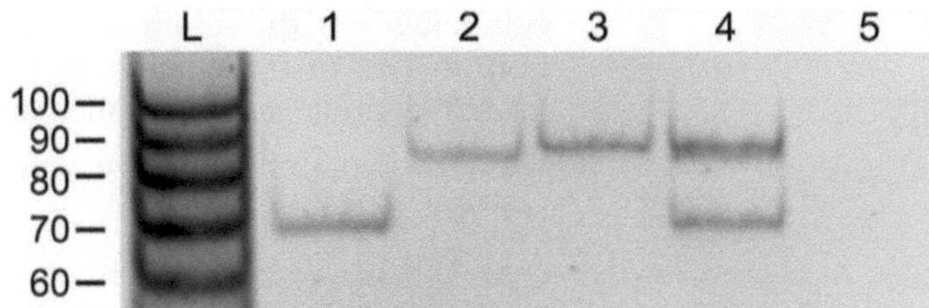

Fig. 1 10% native PAGE analysis of the products of PCR amplification of human genomic DNA. Three fragments containing polymorphic sites rs1490413, rs876724, and rs717302 were amplified either in separate tubes (*lanes 1–3*, respectively) or in one tube (*lane 4*). *Lane 5*—The control in the absence of the template DNA. *L*—Double-stranded DNA ladder; the sizes of the fragments are given in base pairs on the *left*

2. Split the prepared reaction mixture into five PCR tubes, 100 μL each.
3. Prepare a control mix (25 μL) without template DNA. Adjust the volume by adding H_2O accordingly (*see* **Note 2**).
4. Place the tubes into the PCR thermocycler. Run the following program.

   Initial denaturation at 94 °C for 5 min.

   95 °C for 30 s, 60 °C for 30 s, 65 °C for 30 s (35 cycles).

   65 °C for 7 min.
5. Combine the content of PCR tubes into one Eppendorf tube. Analyze the reaction products and the control using 10 % native PAGE (Fig. 1).

3.3 Purification of Single-Stranded DNA

1. After PCR-amplification of three regions of human DNA, purify DNA from the excess of the primers and dNTPs using Microcon centrifugal filter YM-30 membranes (*see* **Note 3**). For this purpose, load ~500 μL of the PCR-reaction mixture per a filter in the filter's sample reservoir and centrifuge for 12 min at 14,000 × *g* (*see* **Note 4**). Discard the filtrate. Wash the membrane with 500 μL of the binding buffer, centrifuge for 12 min at 14,000 × *g*, and discard the filtrate (*see* **Note 5**). To recover amplicon-containing retentate, place the sample reservoir in a new vial upside down and centrifuge for 3 min at 1,000 × *g*. Add binding buffer up to 200 μL.
2. Pack ~200 μL of slurried streptavidin agarose resin into an empty polypropylene column. Remove the excess of the storage solution by allowing it to flow through the column. Equilibrate the column with 600 μL of the binding buffer (*see* **Note 6**).
3. Load 200 μL of the amplicon-containing retentate in the binding buffer (*see* Subheading 3.3, **step 1**) onto the packed

column. Allow the solution to enter the resin bed. Close the bottom end of the column and incubate the resin with the loaded retentate at room temperature for 10 min.

4. Wash the column with five volumes of the binding buffer (*see* **Note 7**).
5. Add 150 μL of the eluting solution to the column. Slowly allow the solution to flow through the column while collecting the eluate (*see* **Note 7**).
6. Repeat Subheading 3.3, **step 5**, with a fresh portion of the elution solution.
7. Neutralize the collected eluate with 30 μL of the neutralizing solution.
8. Precipitate the eluted single-stranded DNA with ethanol. For this purpose, add 1/10 volume of 3 M sodium acetate to the neutralized eluate from Subheading 3.3, **step 7**. Add three volumes of ice-cold ethanol to the obtained aqueous solution. Mix and store on ice for 15 min (*see* **Note 8**). Collect the DNA pellet by centrifugation at the maximum speed for 15–20 min at room temperature or at 4 °C. Remove supernatant by pipetting off. Wash the pellet twice with 1 mL of 70 % ethanol. Centrifuge. Air-dry the pellet.
9. Dissolve the pellet in 50 μL of detection buffer (1×). Measure the concentration of the obtained DNA solution using a UV spectrophotometer (*see* **Note 9**).

3.4 Probe Design

Three sensors are designed to recognize fragments of human genomic DNA containing polymorphic sites rs1490413, rs717302, and rs876724. Each sensor contains a pair of adaptor strands and a UMB as a reporter (Fig. 2).

1. The design of the UMB reporter should meet the following requirements: short but stable stem for fast hybridization; A/T-rich loop to minimize nonspecific interactions with targets;

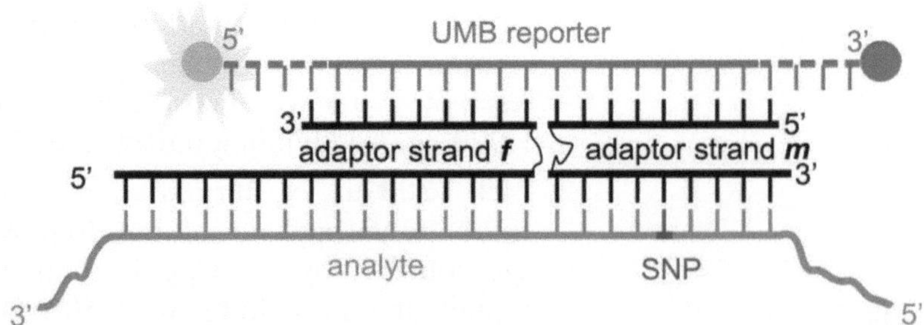

Fig. 2 Schematic representation of the tetrapartite complex between the adaptor strands **m** and **f**, UMB, and an analyte

absence of significant (three bases) complementarity between the stem and the loop sequences; and high absolute fluorescence for a stable and well-reproducible signal. The stem sequence of UMB consists of four alternating G–C base pairs. T_m of UMB under experimental conditions is ~66 °C, as predicted by Oligo Analyzer software. The loop sequence (16 nt long) contains mostly A and T to ensure the absence of complementarity with the G/C-containing stem (*see* **Note 10**). Fluorescein (FAM) and Black Hole 1 (BH1) are used as a fluorophore/quencher pair. FAM and BH1 are attached to the 5′- and 3′-end of the hairpin oligonucleotide, respectively.

2. Each of the two adaptor strands should have a sequence complementary to UMB (reporter-binding arm) and a sequence complementary to the analyzed target (analyte-binding arm). The arms are connected to each other via a triethylene glycol (TEG) linker (*see* **Note 11**).
3. To design the reporter-binding arms of the adaptor strands, make 3′-terminal fragment of strand **f** and 5′-terminal fragment of strand **m** complementary to nucleotides 4–12 and 13–21 of UMB, respectively (*see* **Note 12**) (Table 3). The length of the reporter-binding arms of the adaptor stands is nine nucleotides (*see* **Note 13**).
4. To design the analyte-binding arm of the adaptor strand **m**, make 9-nt 3′-terminal fragment of the strand complementary to the SNP-containing fragment of the target. Make sure that the SNP site is in the middle of the analyte-strand **m** duplex.
5. For strand **f**, design the 5′-terminal fragment of the strand to be complementary to the fragment of the analyte, which is adjacent to the 3′-end of the SNP-containing fragment of the analyte (Fig. 2). The length of the analyte-binding arm of strand **f** is 16–17 nucleotides (T_m for the hybrids between the analyte and the analyte-binding arm of strand **f** is in the range of 60–67 °C).
6. To be able to control the probe's performance, design synthetic analytes that mimic both alleles for each of the three sites.

   The sequences for UMB, adaptor strands, and synthetic analytes for SNPs rs1490413, rs717302, and rs876724 are given in Table 3.

3.5 SNP Detection

1. Prepare 800 μL of 200 nM solution of UMB in detection buffer ×1. For this purpose, mix 16 μL of 10 μM UMB solution with 400 μL of detection buffer (2×) and 384 μL H_2O.
2. Prepare 225 μL of a master mix for each pair of the adaptor strands (Table 4).

Table 3
The sequences for UMB, adaptor strands, and synthetic analytes

Name	Sequence[a]
UMB	FAM-5′-*CGC G*TT AAC ATA CAA TAG AT*C GCG*-BHQ1
AS rs14-*f*	5′-CCT GCT CTG AGG CCA GC/TEG/TAT GTT AAC
AS rs14*m*-A	5′-GAT CTA TTG/TEG/CAG TT**T** TGC
AS rs14*m*-G	5′-GAT CTA TTG/TEG/CAG TT**C** TGC
AS rs71-*f*	5′-ACC TAA TGA CAG ACG T/TEG/TAT GTT AAC
AS rs71*m*-A	5′-GAT CTA TTG/TEG/TCA C**T**A CAC
AS rs71*m*-G	5′-GAT CTA TTG/TEG/TCA C**C**A CAC
AS rs87-*f*	5′-TA GCA GAG TGT GAC AAA/TEG/TAT GTT AAC
AS rs87*m*-C	5′-GAT CTA TTG/TEG/AAA T**G**T ACT
AS rs87*m*-T	5′-GAT CTA TTG/TEG/AAA T**A**T ACT
rs1490413-G	5′-ACT GGG CTG ATG TGG GTT CTT TGC AGA ACT G GC TGG CCT CAG AG C AGG GA
rs1490413-A	5′-ACT GGG CTG ATG TGG GTT CTT TGC AAA ACT G GC TGG CCT CAG AG C AGG GA
rs717302-A	5′-GAA AGG CAT ATC GTA TTA ACT GTG TAG TGA ACG TCT GTC ATT AGG TTT AGC
rs717302-G	5′-GAA AGG CAT ATC GTA TTA ACT GTG TGG TGA ACG TCT GTC ATT AGG TTT AGC
rs876724-C	5′-ATA CCA CTG CAC TGA AGT ATA AGT ACA TTT TTT GTC ACA CTC TGC TA A CT
rs876724-T	5′-ATA CCA CTG CAC TGA AGT ATA AGT ATA TTT TTT GTC ACA CTC TGC TAA CT

[a]Stem-forming nucleotides of MB probe are shown in italics, the SNP sites and the positions complementary to SNPs are in bold; FAM and BHQ1 represent fluorescein and Black Hole quencher 1, respectively; "/TEG/" indicates triethylene glycol linkers

3. Split each master mix into four tubes (54 μL each). Add 6 μL of detection buffer (1×) into the first tube (control), 6 μL of synthetic analytes mimicking each allele into the second and third tube, respectively, and 6 μL of amplified human DNA into the fourth tube.
4. Incubate for 15 min at room temperature (20–25 °C).
5. Measure the fluorescence of all samples (tubes 1–4 for each of the six master mixes) at 517 nm while exciting at 485 nm.
6. Analyze the pattern of fluorescent signals produced by samples containing amplified human DNA (Fig. 3).

Table 4
The composition of master mix for SNP detection

Reagent	rs14-A mix	rs14-G mix	rs71-A mix	rs71-G mix	rs87-C mix	rs87-T mix
200 nM UMB in detection buffer	125 μL	125 μL	125 μL	125 μL	125 μL	125 μL
AS rs14-*f*	1.5 μL	1.5 μL	–	–	–	–
AS rs14*m*-A	1.5 μL	–	–	–	–	–
AS rs14*m*-G	–	1.5 μL	–	–	–	–
AS rs71-*f*	–	–	0.6 μL	0.6 μL	–	–
AS rs71*m*-A	–	–	0.6 μL	–	–	–
AS rs71*m*-G	–	–	–	0.6 μL	–	–
AS rs87-*f*	–	–	–	–	3 μL	5 μL
AS rs87*m*-C	–	–	–	–	3 μL	–
AS rs87*m*-T	–	–	–	–	–	5 μL
detection buffer (1×)	97 μL	97 μL	98.8 μL	98.8 μL	94 μL	90 μL

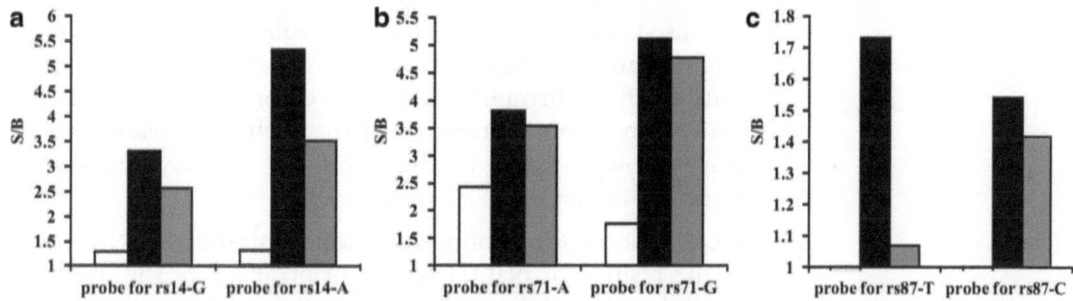

Fig. 3 Signal-to-background ratios for the probes specific for analytes containing SNPs rs1490413 (**a**), rs717302 (**b**), and rs876724 (**c**) in the presence of nonspecific synthetic analytes (*white bars*), specific synthetic analytes (*black bars*), or fragments from human DNA (*grey bars*). The concentration of synthetic analytes and human DNA fragments is 25 nM; the UMB concentration is 100 nM; the concentrations of the adaptor strands are 200 nM (for rs71-probes) or 300 nM (for rs14-probes and for rs87-probes). Based on the observed pattern, it can be suggested that human DNA contains both alleles of rs1490413, both alleles of rs717302 and the allele rs876724-C

4 Notes

1. Prepare the reaction mixtures on ice. To minimize contamination, use filtered pipette tips and add the template solution in the end of mixing.
2. It is convenient to prepare one master mix for both the reaction and the control and then to add appropriate amount of

DNA template into the reaction mixture or water into the control mixture.

3. Microcon centrifugal filter device YM-30 has a nucleotide cut-off (NCO) of 60 and 50 nucleotides for single- and double-stranded DNA, respectively. Since the primers are 19–29 nucleotides long, and the lengths of the amplified DNA fragments are 68–86 bp, this membrane size is optimal for the reported protocol. In case of different range of primers/DNA fragment lengths, different membrane size can provide better results.
4. According to the manufacturer's instructions, it is recommended not to touch the membrane while pipetting the sample into the sample reservoir. In addition, the spin speeds should not exceed the recommended value; otherwise, the membranes can be damaged.
5. Washing of the membrane helps to obtain purer amplicons. At the same time, additional spinning decreases the yield of the recovered double-stranded DNA. Overall, the step of removal of excess primers helps to recover more single-stranded DNA fragment at the next step (purification of the single-stranded target DNA fragments using streptavidin agarose resin, Subheading 3.3, **steps 2–8**).
6. We used gravity-flow chromatography method to obtain single-stranded DNA. However, it is possible to slightly accelerate the flow through the column by pressing the top end of the column with a finger or palm. Such modification of the gravity-flow method is acceptable for all the steps except elution (Subheading 3.3, **step 5**).
7. Incubation of the biotinylated double-stranded DNA bound to the resin with NaOH results in breakage of the hydrogen bonds between the strands. As a result, biotin-containing DNA strand is bound to the resin while the complementary strand is eluted.
8. Placing the tube at −20 °C for 1 h to overnight or at −80 °C for 20 min to 1 h may increase the amount of precipitated DNA.
9. To calculate the concentration of the amplified DNA, it is convenient to use the "averaged" molar extinction coefficient. Based on the data of Oligo Analyzer software, the molar extinction coefficients for the amplicons containing the SNP sites rs1490413, rs717302, and rs876724 at 260 nm are 657,500 M^{-1} cm^{-1}, 815,700 M^{-1} cm^{-1}, and 778,000 M^{-1} cm^{-1}, respectively. Correspondingly, the averaged molar extinction coefficient is 750,400 M^{-1} cm^{-1} (the differences in the coefficients for different alleles can be neglected).

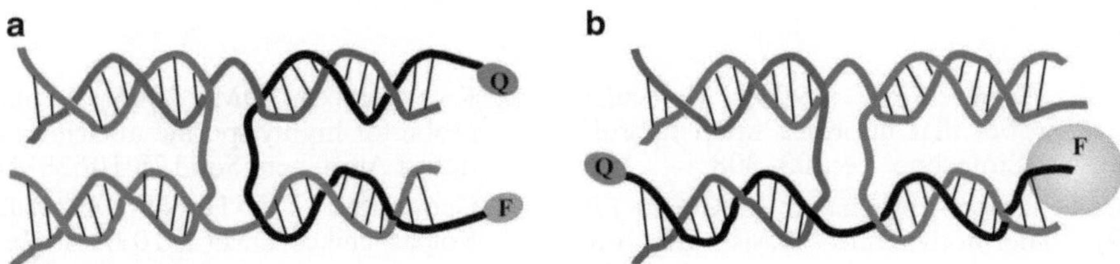

Fig. 4 The schematic representation of two possible conformations of the four-way junction-like structure formed in the presence of Mg^{2+} between MB reporter, two adaptor strands without TEG linkers and the analyzed nucleic acid

10. When designing a structure of UMB probe, it is important to avoid self-complementarity in the loop portion, as well as complementarity between the stem and the loop fragments of the hairpin (stem interference). In case of loop self-complementarity, formation of additional base pairs stabilizes the closed conformation of MB, thus making it more difficult for the adaptor strands to unwind the reporter in the presence of the target. In case of stem interference, the formation of alternative structure for the reporter is possible. In this structure, a fluorophore may not be efficiently quenched by the quencher dye. Moreover, the loop sequence should not be complementary to the analyzed DNA. In our design, UMB has A/T-rich sequence of the loop, which provides the reduced affinity to nonspecific nucleic acids. At the same time, the short and stable G/C-rich stem ensures high hybridization rate and low background of the assay.
11. The presence of the TEG linkers in the adaptor strands helps to bend the strands to ensure that in the final quadripartite complex UMB–**m**–**f**–analyte UMB acquires a conformation, in which the fluorophore is removed from the quencher (Fig. 4, structure b).
12. Notice that one nucleotide of the reporter-binding arms of the adaptor strands is complementary to the stem portion of UMB. This facilitates the opening of UMB upon hybridization with the adaptor strands and the target.
13. Under the experimental conditions, 9-nt hybrids between UMB and the reporter-binding arms of each of the adaptor strands are not stable enough to open up the stem–loop structure of the molecular beacon probe in the absence of the analyte. Indeed, based on the theoretical predictions using Oligo Analyzer software, the melting temperatures for the hybrids lied in the range of 24–25 °C. This ensures low background of the assay. In addition, the amount of the complex formed between UMB and the adaptor strands can be adjusted by varying the concentration of each oligonucleotide.

References

1. Tyagi S, Kramer FR (1996) Molecular beacons: probes that fluoresce upon hybridization. Nat Biotechnol 14:303–308
2. Bonnet G, Tyagi S, Libchaber A, Kramer FR (1999) Thermodynamic basis of the enhanced specificity of structured DNA probes. Proc Natl Acad Sci USA 96: 6171–6176
3. Johnson GC, Todd JA (2000) Strategies in complex disease mapping. Curr Opin Genet Dev 10:330–334
4. Güzey C, Spigset O (2004) Genotyping as a tool to predict adverse drug reactions. Curr Top Med Chem 4:1411–1421
5. Risch NJ (2000) Searching for genetic determinants in the new millennium. Nature 405: 847–856
6. Altshuler D, Brooks LD, Chakravarti A, Collins FS, Daly MJ, Donnelly P (2001) A haplotype map of the human genome. Nature 409:928–933
7. Ragoussis J (2009) Genotyping technologies for genetic research. Annu Rev Genomics Hum Genet 10:117–133
8. Kolpashchikov DM (2006) A binary DNA probe for highly specific nucleic acid recognition. J Am Chem Soc 128:10625–10628
9. Gerasimova YV, Hayson A, Ballantyne J, Kolpashchikov DM (2010) A single molecular beacon probe is sufficient for the analysis of multiple nucleic acid sequences. Chembiochem 11:1762–1768
10. Grimes J, Gerasimova YV, Kolpashchikov DM (2010) Real-time SNP analysis in secondary structure-folded nucleic acids. Angew Chem Int Ed Engl 49:8950–8953
11. Nguyen C, Grimes J, Gerasimova YV, Kolpashchikov DM (2011) Molecular beacon-based tricomponent probe for SNP analysis in folded nucleic acids. Chemistry 17: 13052–13058
12. Gerasimova YV, Kolpashchikovn DM (2012) Detection of bacterial 16S rRNA using a molecular beacon-based X sensor. Biosens Bioelectron 41:386–390
13. Sanchez J, Phillips C, Børsting C, Balogh K, Bogus M, Fondevila M et al (2006) A multiplex assay with 52 single nucleotide polymorphisms for human identification. Electrophoresis 27:1713–1724

Chapter 6

SNP Analysis Using a Molecular Beacon-Based Operating Cooperatively (OC) Sensor

Evan M. Cornett and Dmitry M. Kolpashchikov

Abstract

Analysis of single-nucleotide polymorphisms (SNPs) is important for diagnosis of infectious and genetic diseases, for environment and population studies, as well as in forensic applications. Herein is a detailed description to design an "operating cooperatively" (OC) sensor for highly specific SNP analysis. OC sensors use two unmodified DNA adaptor strands and a molecular beacon probe to detect a nucleic acid targets with exceptional specificity towards SNPs. Genotyping can be accomplished at room temperature in a homogenous assay. The approach is easily adaptable for any nucleic acid target, and has been successfully used for analysis of targets with complex secondary structures. Additionally, OC sensors are an easy-to-design and cost-effective method for SNP analysis and nucleic acid detection.

Key words Single-nucleotide polymorphism, Nucleic acid detection, Molecular beacon, Fluorescent probe, DNA four-way junction

1 Introduction

Single-nucleotide polymorphisms (SNPs) have been used to analyze nucleic acids to determine drug resistance in parasites [1] and bacteria [2], estimate cancer prognosis [3], predict mental illness prevalence [4], and for many other disease- and diagnostic related purposes. Additionally SNPs can be used for human identification in forensics applications [5]. Indeed, in the post-genomic era, as additional SNPs and their physiological outcomes are rapidly identified, SNP genotyping will continue to be an important task in biomedical science. Several enzymatic methods to identify SNPs are well established, including primer extension, ligation, and invasive cleavage assays. However, these techniques require expensive equipment and reagents. Alternatively, hybridization probes are more cost effective and easier to design. The most common hybridization probe is the molecular beacon (MB) probe, first introduced by Tyagi and Kramer [6]. The MB probe is a stem–loop folded oligonucleotide that contains a fluorophore and quencher

Dmitry M. Kolpashchikov and Yulia V. Gerasimova (eds.), *Nucleic Acid Detection: Methods and Protocols*, Methods in Molecular Biology, vol. 1039, DOI 10.1007/978-1-62703-535-4_6, © Springer Science+Business Media New York 2013

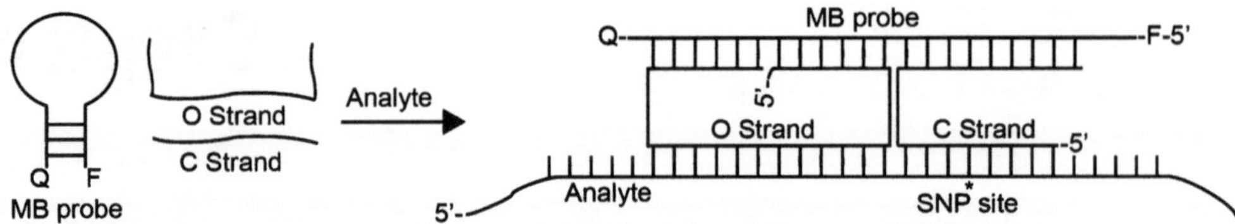

Fig. 1 Schematic of OC sensor for SNP-specific nucleic acid detection. An MB probe, O, and C adaptor strands hybridize with a target nucleic acid to form a tetra partite fluorescent complex

dyes, conjugated on opposite ends. Upon hybridization to a target nucleic acid, the fluorophore is separated from the quencher and emits a detectable fluorescent signal. While successful, SNP genotyping using an MB probe requires analysis of MB-analyte melting temperatures to correctly identify matched and mismatched analytes bound to the MB probe [7]. Such analysis requires expensive instruments, and it is difficult to interpret the results. Alternatively, the MB probe has been adapted for use in several alternative hybridization probe strategies, which allow for SNP genotyping at room temperature [8–14]. The method detailed within this chapter is one of these MB probe-based binary DNA probes, called "operating cooperatively" OC sensor.

The OC sensor uses two adaptor strands, each of which hybridizes to the target analyte and to an MB probe, thus opening the MB probe and producing a tetrapartite fluorescent complex, which is easily detected using a fluorescent spectrophotometer in a homogenous assay at room temperature (Fig. 1).

OC sensors contain several advantages over an MB probe alone, including ease of design, multiplex analysis of complex SNP samples, and room-temperature analysis. The OC sensor design also has a unique cost advantage over other indirect MB probe-based designs: the O and C adapter strands do not require any modifications. The OC sensor design was successfully used to detect five different SNPs indicated to cause rifampin resistance in *Mycobacterium tuberculosis* [15]. Herein, the design and analysis of an OC sensor for one of these SNPs are described in detail. More recently we demonstrated how OC sensor can be used to analyze a secondary structure forming analytes [16].

2 Materials

2.1 Reagents

All solutions are prepared in DNA-grade sterile water.

1. Buffer C: 50 mM $MgCl_2$, 50 mM Tris–HCl, pH 7.4.
2. MB probe.
3. OC adaptor strands.

2.2 Instrument and Equipment

1. Fluorescent spectrophotometer.
2. Quartz cuvette.

3 Methods

3.1 Design of OC Sensor Adaptor Strands

1. O and C adaptor strands are each designed to recognize the specific SNP-containing target sequence and the MB probe (Fig. 1) (*see* **Note 1**).
2. To design an O strand, select 12 nucleotides (nts) near the 3′ end of the MB probe, which will serve as the MB-binding region. The 5′ end of the O strand begins at the midpoint of the selected 12-nt region and has 6 nts complementary to the MB. The next 12 nts of the O strand are complementary to the target sequence, starting 5 nts upstream of the SNP and proceeding away from the SNP. The last 6 nts of the O strand are complementary to the remaining portion of the 12-nt MB region (Fig. 2) (*see* **Note 2**).
3. Design the C strand to have a 9-nt MB probe-binding region and an 8-nt analyte-binding arm. The 5′ portion of the C strand is complementary to the analyte. The C strand is responsible for correct SNP discrimination, and the SNP should be situated in the middle of the analyte-binding arm. The C strand binds the analyte and MB probe at the abutting positions to the O strand (Fig. 3).

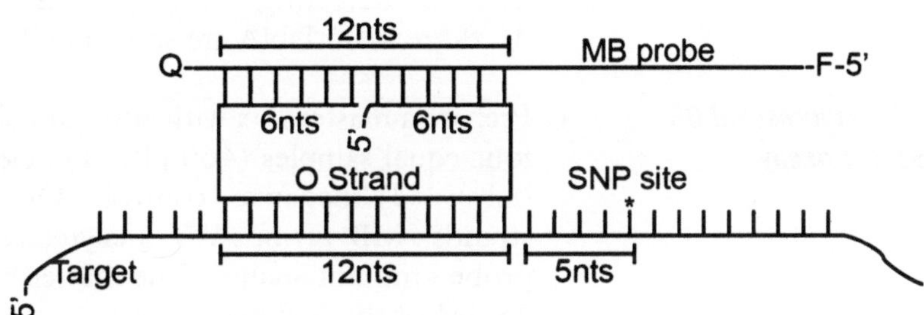

Fig. 2 Design of O strand for an OC sensor

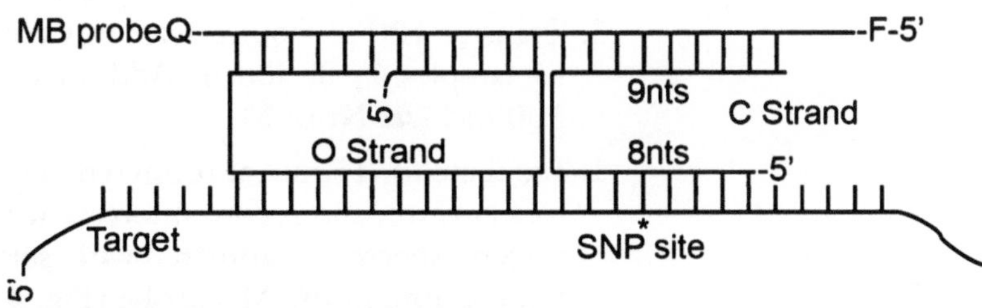

Fig. 3 Design of C strand for an OC sensor

Table 1
Sequences of an OC sensor and *M. tuberculosis* rpoB gene fragment

	Name	Sequences
Analytes	WT (synthetic)	5′-GC ACC AGC CAG CTG AGC CAA TTC ATG GAC CAG AAC AAC CCG CTG TCG GGG TTG ACC CAC AAG CGC CGA CTG TCG GCG CTG
	M1 (synthetic)	5′-GC ACC AGC CAG CTG AGC CAA TTC ATG GAC CAG AAC AAC CCG CTG TCG GGG TTG ACC CAC AAG CGC CGA CTG TTG GCG CTG
O and	C-M1	5′-GCC AAC A ATG TTA ACG
C strands	O-M1	5′-TAT TGT GT CGG CGCT T GT GCG ATC
MB probe		5′-/FAM/GCG TTA ACA TAC AAT AGA TCG C/ BHQ/-3′

SNP is underlined in analyte sequences

4. Confirm that O and C strands are not self-complementary and form no hairpin structures (*see* **Note 3**).
5. Example O and C strand, MB probe, and target sequences used for detecting a rifampin resistance conferring SNP in *M. tuberculosis* DNA are shown in Table 1.

3.2 Fluorescent OC Sensor Assay

1. Prepare a master mix with enough volume to be separated into four equal samples (460 μl): (1) negative control (MB probe only); (2) negative control (OC strand + MB); (3) OC strands + MB probe + WT analyte; and (4) OC strands + MB probe + mutant analyte. The master mix contains buffer C and 25 nM of the MB probe.
2. Transfer 120 μl from the master mix into sample 1. Add O and C strands to the remaining master mix at final concentrations of 200 and 100 nM, respectively (*see* **Note 4**).
3. Transfer 120 μl from master mix containing O and C strands to samples 2, 3, and 4. Add analyte to samples 3 and 4 at 500 nM (*see* **Note 5**).
4. Incubate for 15 min at room temperature, transfer each sample to a cuvette, and record fluorescence of all samples using fluorescent spectrophotometer with settings appropriate for the fluorophore in the MB probe (Fig. 4).

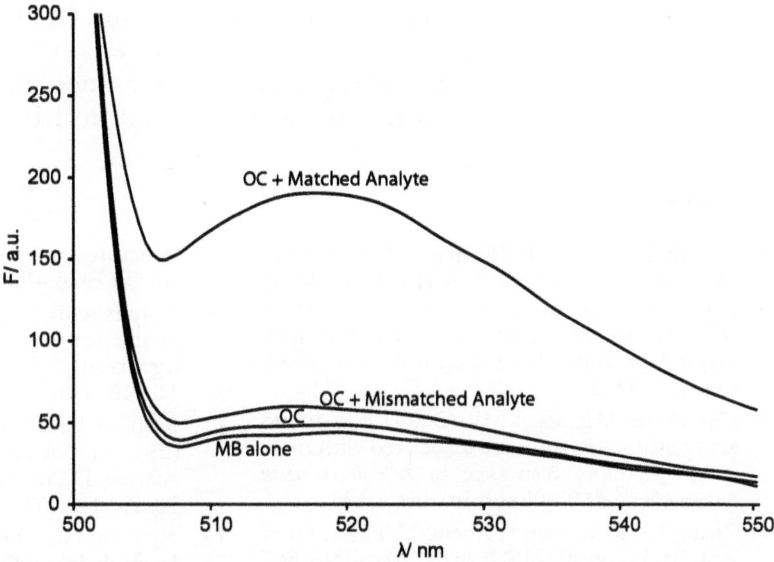

Fig. 4 Fluorescent intensity scan of MB alone, OC sensor, and OC sensor with matched and mismatched analytes after 15-min room-temperature incubation. MB, O, C, and analytes are included at 25, 200, 100, and 500 nM, respectively, in buffer C

4 Notes

1. The OC sensor design allows one universal MB probe to be used for analysis of several different target sequences. Once a successful design of the MB-binding portions of O and C strands has been found, only the analyte-binding portion needs to be changed when designing an OC sensor for a new analyte that will use the same MB probe. The design of MB probes was thoroughly reviewed previously [17].
2. We found that 6 nts for the short portions of the O strand was sufficient for complex formation; however the O strands can be designed to be larger for specific applications (analytes with secondary structures) as long as the overall MB- and analyte-binding regions are of the same size.
3. Avoid designing O or C strands which contain complementary nucleotides across the junction. For example, the first nucleotide of the C strand MB-binding arm should not be complementary to one nucleotide upstream of the last nucleotide in the C strand analyte-binding arm.
4. The ratio of O and C strand concentration can be optimized to obtain optimal SNP discrimination. Lowering the concentration of the C strand typically improves SNP discrimination.

5. The limit of detection of the OC sensor is similar to an MB probe, around 2 nM. However, OC sensor will perform better at higher analyte concentrations, and is still able to correctly discriminate an SNP at high analyte concentrations (Fig. 4).

References

1. Veiga MI, Ferreira PE, Jörnhagen L, Malmberg M, Kone A, Schmidt BA et al (2011) Novel polymorphisms in *Plasmodium falciparum* ABC transporter genes are associated with major ACT antimalarial drug resistance. PLoS One 6:e20212
2. Cummings MP, Segal MR (2004) Few amino acid positions in rpoB are associated with most of the rifampin resistance in *Mycobacterium tuberculosis*. BMC Bioinformatics 5:137
3. Zhang L, Liu Y, Song F, Zheng H, Hu L, Lu H et al (2011) Functional SNP in the microRNA-367 binding site in the 3′UTR of the calcium channel ryanodine receptor gene 3 (RYR3) affects breast cancer risk and calcification. Proc Natl Acad Sci USA 108:13653–13658
4. Lee SH, DeCandia TR, Ripke S, Yang J (2012) Estimating the proportion of variation in susceptibility to schizophrenia captured by common SNPs. Nat Genet 44:247–250
5. Sanchez JJ, Phillips C, Borsting C, Balogh K, Bogus M, Fondevila M et al (2006) A multiplex assay with 52 single nucleotide polymorphisms for human identification. Electrophoresis 27:1713–1724
6. Tyagi S, Kramer FR (1996) Molecular beacons: probes that fluoresce upon hybridization. Nat Biotechnol 14:303–308
7. Reed GH, Kent JO, Wittwer CT (2007) High-resolution DNA melting analysis for simple and efficient molecular diagnostics. Pharmacogenomics 8:597–608
8. Gerasimova YV, Hayson A, Ballantyne J, Kolpashchikov DM (2010) A single molecular beacon probe is sufficient for the analysis of multiple nucleic acid sequences. Chembiochem 11:1762–1768
9. Grimes J, Gerasimova YV, Kolpashchikov DM (2010) Real-time SNP analysis in secondary-structure-folded nucleic acids. Angew Chem Int Ed Engl 49:8950–8953
10. Kolpashchikov DM (2006) A binary DNA probe for highly specific nucleic acid recognition. J Am Chem Soc 128: 10625–10628
11. Kolpashchikov DM, Gerasimova YV, Khan MS (2011) DNA nanotechnology for nucleic acid analysis: DX motif-based sensor. Chembiochem 12:2564–2567
12. Nguyen C, Grimes J, Gerasimova YV, Kolpashchikov DM (2011) Molecular-beacon-based tricomponent probe for SNP analysis in folded nucleic acids. Chemistry 17: 13052–13058
13. Li J, Qi XJ, Du YY, Fu HE, Chen GN, Yang HH (2012) Efficient detection of secondary structure folded nucleic acids related to Alzheimer's disease based on junction probes. Biosens Bioelectron 36:142–146
14. Nakayama S, Yan L, Sintim HO (2008) Junction probes—sequence specific detection of nucleic acids via template enhanced hybridization processes. J Am Chem Soc 130: 12560–12561
15. Cornett EM, Campbell EA, Gulenay G, Peterson E, Bhaskar N, Kolpashchikov DM (2012) Molecular logic gates for DNA analysis: detection of rifampin resistance in *M. tuberculosis* DNA. Angew Chem Int Ed Engl 51:9075–9077
16. Cornett EM, O'steen MR, Kolpashchikov DM (2013) Operating cooperatively (OC) sensor for highly specific recognition of nucleic acids. PLOS one 8, e55919
17. Wang K, Tang Z, Yang CJ, Kim Y, Fang X, Li W et al (2009) Molecular engineering of DNA: molecular beacons. Angew Chem Int Ed Engl 48:856–870

Part III

Isothermal Amplification Methods for Nucleic Acid Detection

Chapter 7

Detection of *rpoB* Gene Mutations Using Helicase-Dependent Amplification

Wanyuan Ao and Robert Jenison

Abstract

For patients infected with tuberculosis, detection of *rpoB* gene mutations is critical for diagnosing drug-resistant strains of the causative pathogen, *Mycobacterium tuberculosis* (MTB). Traditional approaches to drug resistance include culture, which is very slow. Recently described real-time polymerase chain reaction approaches have addressed turnaround time but at relatively high cost. Here, we describe a novel amplification method, termed blocked-primer helicase-dependent amplification, for amplifying *rpoB* gene sequences in MTB. Resultant amplicon is hybridized to a probe set arrayed on a modified silicon-based chip to determine if there is any mutation in that region. Using this method, we could detect the majority of clinically relevant mutations in *rpoB* gene.

Key words *rpoB*, *Mycobacterium tuberculosis*, Helicase-dependent amplification, Gene mutation, Mutation detection, Drug resistance

1 Introduction

TB infection, particularly with increasing drug-resistant TB cases, is still a major healthcare concern in the world [1]. Mutations in the *rpoB* gene have been shown to be clinically relevant to the diagnosis of *Mycobacterium tuberculosis* (MTB) strains resistant to the front-line drug, rifampin [2–8]. The effective treatment and preventing the spread of this disease require a highly sensitive and specific detection of MTB in infected patients and mutations in the *rpoB* gene [9–12]. Traditional microscopy lacks the sensitivity and real-time polymerase chain reaction suffers from high cost or poor ease of use [13–17].

Our novel approach described below using the bpHDA method could improve both sensitivity and specificity in the *rpoB* gene mutation detection [18]. This approach utilizes the isothermal amplification method, helicase-dependent amplification (HDA), coupled with blocked primers and RNase H2-mediated

Dmitry M. Kolpashchikov and Yulia V. Gerasimova (eds.), *Nucleic Acid Detection: Methods and Protocols*, Methods in Molecular Biology, vol. 1039, DOI 10.1007/978-1-62703-535-4_7, © Springer Science+Business Media New York 2013

target-specific "hot-start" to exponentially amplify the target *rpoB* gene sequence [19]. The blocked primers are constructed with a single ribonucleotide inserted 4 or 5 bases upstream of a 3′-end block. Blocked primers can hybridize to the complementary target sequence, but cannot be extended until thermal-stable RNase H2 is activated to cleave the ribonucleotide linkage in the primer present in the duplex DNA. RNase H2, isolated from *Pyrococcus abyssi*, has no or little activity at 40 °C or lower temperatures, but highly active at around 65 °C, which is the optimal temperature for the helicase enzyme to function for HDA amplification. As the temperature is raised to 65 °C, RNase H2 cleaves the ribonucleotide present in the primer and the short fragment of the primer 3′ of the ribonucleotide dissociates from the duplex DNA, liberating the block to create a free 3′-hydroxyl available for primer extension. Since the primer/primer hybrids are not stable at elevated temperatures, little or no primer artifact will be amplified in blocked-primer HDA (bpHDA) and thus this dramatically improves the sensitivity for the detection [18, 19]. After amplification, the resultant amplicons are hybridized to a probe set that is spotted on a silicon-based chip and designed to detect mutations in the amplified region of the *rpoB* gene using a sensitive colorimetric visual detection system [20].

In summary, we have DNA probes spotted on chemically coated wafers to produce chips, MTB samples amplified by bpHDA to generate amplicons, and resultant amplicons hybridized to chips to determine if there is a mutation in the MTB *rpoB* gene region. Using this method, we unambiguously detected 28/29 (96.5 %) clinically relevant single mutations in the *rpoB* gene using full-length synthetic templates covering the amplicon region [18]. In addition, we have also verified these results using genomic DNA from clinical specimens. In that study, we correctly classified 11/11 rifampin-sensitive and 25/26 rifampin-resistant clinical isolates [18].

2 Materials

2.1 Primers and Probes

1. DNA suspension buffer: 10 mM Tris–HCl, pH 8.0, 0.1 mM EDTA. Store at 4 °C.
2. Blocked primers.

   Forward: 5′-cgatcaaggagttcttcggcraccag/iSpC3-3′.

   Reverse: 5′/5BioTEG/ggcacgctcacgtgacagarccgcc/iSpC3-3′.
3. Capture probes (also *see* ref. [18]).

   Probe 1: 5′-/5Ilink12//iSp18/caccagccagctgagc-3′.

   Probe 2: 5′-/5Ilink12//iSp18/gagccaattcatggacc-3′.

   Probe 3: 5′-/5Ilink12//iSp18/catggaccagaacaaccc-3′.

Probe 4: 5′-/5Ilink12//iSp18/agaacaacccgctgtcgg-3′.

Probe 5: 5′-/5Ilink12//iSp18/cggggttgacccacaagcg-3′.

Probe 6: 5′-/5Ilink12//iSp18/gactgtcggcgctgggg-3′.

These probes are designed to perfectly match the wild-type *rpoB* sequence of MTB (GenBank accession no. L27989, ref. [21]).

4. All primers and probes are chemically synthesized and dissolved in DNA suspension buffer at the final concentration of 100 μM. Store at −20 °C (*see* **Note 1**).

2.2 DNA Templates

1. Genomic DNA templates: Purified from wild-type (ATCC strain: H37Ra or H37Rv) or clinical isolate mutant TB strains.
2. Synthetic DNA templates: Chemically synthesized and PAGE purified. The wild-type synthetic template WT128 is a 128 bp single-stranded DNA fragment (5′-cg atc aag gag ttc ttc ggc acc agc cag ctg agc caa ttc atg gac cag aac aac ccg ctg tcg ggg ttg acc cac aag cgc cga ctg tcg gcg ctg ggg ccc ggc ggt ctg tca cgt gag cgt gcc-3′), which covers the amino acid codes from 502atc to 544gcc with two additional bases at the 5′ end (5′-cg). The mutant synthetic templates have the exact same sequence as WT128 except with one single-nucleotide mutation (capital as listed below) in one of the amino acid codes or two-nucleotide mutations (only for 522CAg). Those mutant templates (total 29) are assigned as 508Gcc, 509aCc, 510caT, 511cCg, 511cGg, 513cCa, 513Gaa, 513Aaa, 513cTa, 515Gtg, 516gGc, 516Tac, 516gTc, 518aTc, 522tTg, 522CAg, 522tGg, 526cCc, 526Tac, 526Gac, 526caG, 526Aac, 526cGc, 526cTc, 531tTg, 531tGg, 531Ccg, 533cCg, and 533Gtg. They all are clinically relevant mutations (*see* ref. [18]).
3. All DNA templates are dissolved in DNA suspension buffer at the final concentration of 1.5×10^5 copies/μl. Store at −70 °C (*see* **Note 1**).

2.3 Wafer and Chip Preparation Components

1. TSPS coated silicon wafer (Great Basin).
2. Wafer scriber (Dyna Tek).
3. PBS buffer, pH 6.
4. PPL solution: 50 mg/mL polypeptide poly(phenylalanine-lysine), 2 M NaCl in PBS buffer ×1.
5. SFB solution: 10 μM succinimidyl-4-formyl benzoate, 0.1 M sodium borate buffer, pH 8.2.
6. Stripping buffer: 0.1 % sodium dodecyl sulfate (SDS).
7. Spotting buffer: 0.1 M phosphate buffer, pH 7.8, 10 % glycerol.
8. Nitrogen gas.

9. Detection control: 5′-ILink12/AAAAAAAAAAAAAAAAA/3BioTEG3′, 100 μM.
10. Hybridization control (intmut1): 5′-ILink12//iSp18/GAGCATCTTAGGAGGTC-3′, 100 μM.

2.4 HDA Amplification Components

1. 10× ABII buffer (BioHelix).
2. 100 mM $MgSO_4$ and 500 mM NaCl.
3. IsoAmp dNTPs (BioHelix).
4. IsoAmp enzyme mixture (BioHelix).
5. 160× RNase H2 (Great Basin).
6. 1 % Tween-20 and Triton-100.
7. 20× EvaGreen dye (for real-time amplification).
8. 2.7× Stabilizer mix (Great Basin).
9. 5 μM forward primer.
10. 5 μM reverse primer.
11. Template DNA (stock stored at 1.5×10^5 copies/μl in DNA suspension buffer at −70 °C).
12. LightCycler 480 (Roche) plates and instrument.

2.5 Chip Hybridization Assay Components

1. 96-well plates with flat, square-bottom wells (Whatman).
2. Hybridization buffer: SSC ×5, 5 mg/ml alkaline-treated casein, 0.05 % Tween-20, 0.03 % ProClin300 preservative.
3. Two hybridization ovens (set at 95 and 53 °C).
4. Wash buffer A: SSC 0.1×, 0.1 % SDS.
5. Wash buffer B: SSC 0.1×, 0.01 % Tween-20.
6. Conjugate solution: 75 mM sodium citrate, 50 mM NaCl, 10 % v/v fetal bovine serum, 5 mg/μl alkaline-treated casein, 0.03 % v/v ProClin300, 1 μg/μl peroxidase-conjugated monoclonal mouse antibody against biotin.
7. TMB (3,3′,5,5′-tetramethylbenzidine) substrate (One Component HRP Member Substrate, SurModics, Inc., product number: TMBM-0100-01).
8. CCD camera.
9. Complementary sequence to hybridization control (intmut1): 5′/BioTEG/GACCTCCTAAGATGCTC-3′, 250 pM.

3 Methods

3.1 Wafer Preparation

1. Soak the wafers in PPL solution overnight at room temperature (*see* **Note 2**).
2. Transfer the wafers into SFB solution and soak for 2 h at room temperature.

3. Wash the wafer extensively with water and PBS.
4. Dry the wafers with a stream of nitrogen gas.
5. Store the wafers dry in a sealed bag purged with nitrogen gas.

3.2 Chip Spotting

1. Spot the probes at the following concentrations in the spotting buffer on the wafer using BioDot Dispenser (model AD5000):
   Probe 1: 350 nM.
   Probe 2: 250 nM.
   Probe 3: 250 nM.
   Probe 4: 500 nM.
   Probe 5: 700 nM.
   Probe 6: 450 nM.
2. Spot the biotin-labeled detection control (DC) probe at 50 nM, which serves as a control for the activity of the anti-biotin antibody/HRP conjugate and TMB performance.
3. Spot the hybridization control (HC) at 50 nM, which serves as a control for the stringency of the hybridization step by reacting with a biotin-labeled complementary probe present in the hybridization buffer.
4. Incubate the spotted wafers overnight in a humidified chamber at room temperature.
5. Incubate the wafers in the stripping buffer at 37 °C for 60 min followed with extensive rinsing with water.
6. Dry the wafers using a stream of nitrogen gas.
7. Scribe the wafers into 6.5 mm^2 chips and store in nitrogen-purged bags.

3.2.1 Manual Spotting

1. Scribe the wafers into 6.5 mm^2 chips before spotting.
2. Manually apply 0.5–1.0 μl of the probes (total 6) in the same pattern on each chip using micropipette in a humidified hood at room temperature (*see* **Note 3**).
3. Incubate the spotted wafers overnight in a humidified chamber at room temperature.
4. Rinse the wafer with sterile water (*see* **Note 4**).
5. Fill a petri dish with the stripping buffer and make sure that the wafer is completely covered by the buffer. Incubate the wafer in a 37 °C oven for 60 min.
6. Rinse the wafer with sterile water (*see* **Note 4**).
7. Dry the wafer using a stream of nitrogen gas.
8. Store the chip in a dry and nitrogen-purged bag.

3.3 HDA Amplification

1. Thaw all components and store them on ice.
2. Mix the following components in order as HDA master mixture (for 10× reaction, total 240 μl) in a 1.5-ml Eppendorf tube on ice:

   35.95 μl of water (molecular grade).

   26 μl of 10× ABII buffer.

   11.55 μl of 100 mM $MgSO_4$.

   24 μl of 500 mM NaCl.

   21 μl of IsoAmp dNTPs.

   3 μl of 1 % Tween-20.

   3 μl of 1 % Triton-100.

   3 μl of 20× EvaGreen dye.

   90 μl of 2.7× Stabilizer mix.

   21 μl of IsoAmp enzyme mixture.

   1.5 μl of 160× RNase H2.

   Keep cold until use (*see* **Note 5**).
3. Make 40 μl of the primer mixture in total as following in a 0.5-ml PCR tube for 10× reactions: Mix 12 μl of forward primer, 24 μl of reverse primer, and 4 μl of 10× ABII buffer.
4. For each reaction in a total volume of 30 μl, add 24 μl of HDA master mixture into one well of the LightCycler plate followed by 4 μl of primer mixture, and then add 2 μl of DNA template (*see* **Note 6**).
5. Mix briefly in the well and spin at 4,000 RCF for 30 s in a plate centrifuge.
6. Then, immediately place the plate into the pre-warmed LightCycler 480 and start to run the reaction at 65 °C for 60–90 min.
7. After the amplification process, a "melting curve" program could be run to determine the melting temperature of the amplicon. The melting temperature of our *rpoB* gene amplicon is 90.05 °C.

Figure 1 shows a typical real-time amplification curve of the HDA reaction (a) and the melting temperature of the *rpoB* gene amplicon (b).

3.4 Chip Hybridization Assay

1. Tape the chips in the wells of the 96-well assay plate with flat and square bottom using double-sided tape.
2. Add 80 μl of hybridization buffer into each chip.
3. Add 20 μl of amplicon into the hybridization buffer and briefly mix by pipetting up and down a few times. Cover the plate with a plate sealer.

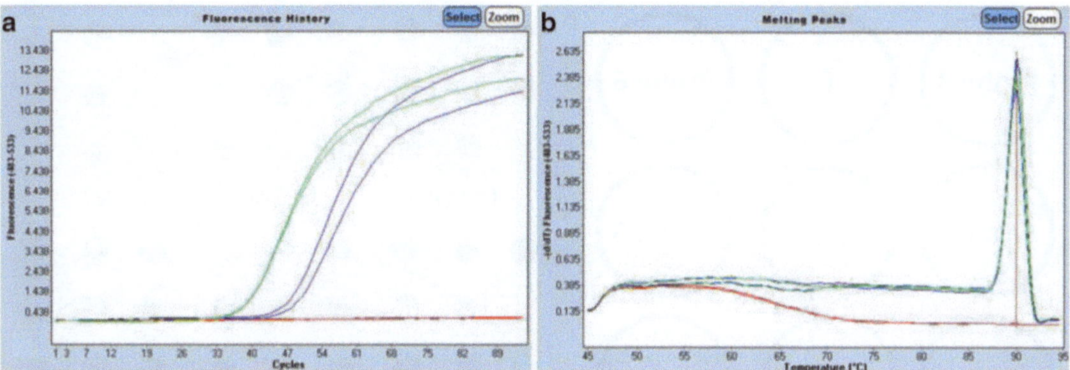

Fig. 1 TB *rpoB* gene bpHDA amplification. (**a**) Amplification curves (duplicates): *green*—3,000 copies of wild-type TB template input; *blue*—300 copies of wild-type TB template input; *red*—no template input control. (**b**) Melting curves of (**a**): the melting temperature is 90.05 °C

4. Place the plate in a hybridization oven at 95 °C (*see* **Note 7**).
5. Incubate the samples in the plate for 6 min.
6. Then, immediately transfer the plate into another hybridization oven at 53 °C (*see* **Note 7**).
7. Incubate the samples on chips for 10 min.
8. When the hybridization is done, take out the plate from the oven and immediately remove the plate sealer and pipette out the hybridization mixture.
9. Then, add 200 μl of wash buffer A into each well to wash the chip briefly and then remove the wash buffer (*see* **Note 8**).
10. Repeat **step 9** with wash buffer A two more times (*see* **Note 9**).
11. Then, add 200 μl of wash buffer B into each well to wash the chip briefly and then remove the wash buffer.
12. Repeat **step 11** with wash buffer B two more times.
13. After the wash steps, add 100 μl of conjugate solution directly onto the chip and incubate at room temperature for 10 min.
14. Remove the conjugate solution and wash the chips with 200 μl of wash buffer B.
15. Repeat this step with wash buffer B two more times.
16. After wash, add 100 μl of TMB and incubate at room temperature for 5 min.
17. Remove TMB.
18. Wash the chips twice briefly with distilled water (*see* **Note 10**).
19. Wash the chips additionally twice with ethanol to dry the chips (*see* **Note 10**).
20. Further dry the chips gently using a stream of compressed air/nitrogen gas.

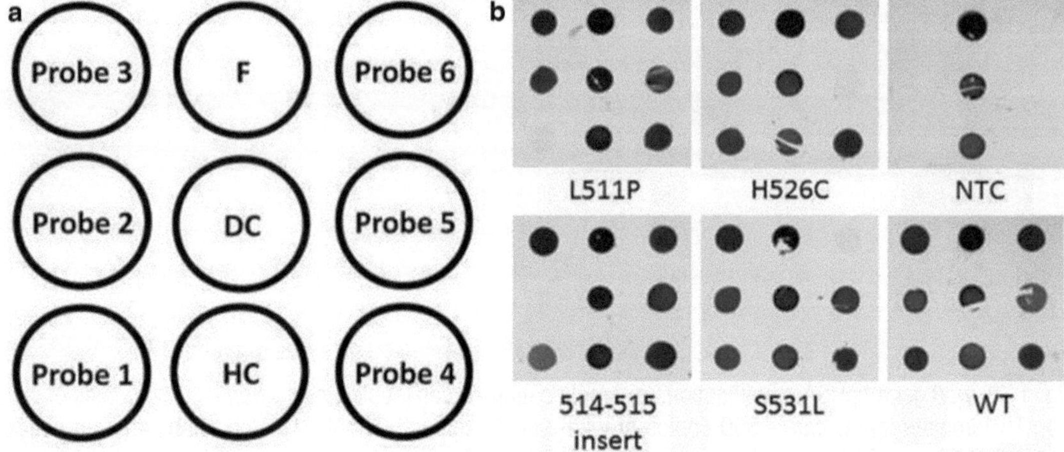

Fig. 2 TB *rpoB* gene mutation detection. Genomic templates from clinical isolates were amplified using bpHDA, hybridized to the chip, and a CCD camera image was taken. (**a**) Array map. *HC* hybridization control, *DC* detect control, *F* fiducial marker. (**b**) Representative images of wild-type and some mutant clinical samples (*WT* wild-type, *NTC* no template control)

21. Visually make a call based on the signal intensity of the spots.
22. Image the chip using a CDD camera.

Figure 2 shows a few typical patterns of the chips: Nine reacted spots indicate a wild-type *rpoB* gene in the region; three reacted control spots only show that no *rpoB* gene is detected in the sample. If the three reacted control spots are present but the signal is absent or very weak compared with other spots for one or more of the six *rpoB*-specific capture probes, that indicates one or more mutations in that specific probe region in the *rpoB* gene.

4 Notes

1. To avoid multiple freezing/thaw, the 100 μM stock can be aliquoted into 5–10 μl each in multiple tubes and use one each time.
2. Make sure that the wafer is completely covered by the solution.
3. This could be done with a petri dish in a small hood with a piece of wet paper towel around inside the edge of the dish and additional humidity provided by a beak of water in the hood.
4. The uniform wash or rinse of the wafer surface is important.
5. Make sure to mix gently but thoroughly.

6. The DNA template is added in the last step. Once it is added, proceed with the next step immediately.
7. Place the plate in the middle of the metal rack of a conventional forced-air oven. The temperature may vary a bit in different locations of the oven.
8. Try to keep the chip covered with fluid or minimize the dry time for each wash or fluid exchange step in this whole process. If the chip dries out, some background issue may arise at the end.
9. For these wash steps, the wash buffer is added to cover the whole chip and then immediately (within 30–60 s) the buffer is removed, and proceed to the next step.
10. A squirt bottle could be used for this step as a quick process. Water or ethanol is squirted into the wells and then dumped out into the sink quickly.

Acknowledgments

We thank Adrianne Clifford and all other members of the Research Group at Great Basin Corporation for technical assistance and Steve Aldous for financial support for this project.

References

1. World Health Organization: Anti-*Tuberculosis* Drug Resistance in the World: Fourth Global Report (2008) The WHO/IUATLD global project on anti-*tuberculosis* drug resistance surveillance. World Health Organization, Geneva, Switzerland, Publication no. WHO/HTM/TB/2008.394
2. Caws M, Duy PM, Tho DQ, Lan NT, Hoa DV, Farrar J (2006) Mutations prevalent among rifampin- and isoniazid-resistant *Mycobacterium tuberculosis* isolates from a hospital in Vietnam. J Clin Microbiol 44: 2333–2337
3. Chan RC, Hui M, Chan EW, Au TK, Chin ML, Yip CK et al (2007) Genetic and phenotypic characterization of drug-resistant *Mycobacterium tuberculosis* isolates in Hong Kong. J Antimicrob Chemother 59:866–873
4. Heep M, Brandstätter B, Rieger U, Lehn N, Richter E, Rüsch-Gerdes S et al (2001) Frequency of *rpoB* mutations inside and outside the cluster I region in rifampin-resistant clinical *Mycobacterium tuberculosis* isolates. J Clin Microbiol 39:107–110
5. Herrera L, Jiménez S, Valverde A, García-Aranda MA, Sáez-Nieto JA (2003) Molecular analysis of rifampin-resistant *Mycobacterium tuberculosis* isolated in Spain (1996-2001). Description of new mutations in the *rpoB* gene and review of the literature. Int J Antimicrob Agents 21:403–408
6. Mani C, Selvakumar N, Kumar V, Narayanan S, Narayanan PR (2003) Comparison of DNA sequencing, PCR-SSCP and PhaB assays with indirect sensitivity testing for detection of rifampin resistance in *Mycobacterium tuberculosis*. Int J Tuberc Lung Dis 7:652–659
7. Rossau R, Traore H, De Beenhouwer H, Mijs W, Jannes G, De Rijk P et al (1997) Evaluation of the INNO-LiPA Rif. TB assay, a reverse hybridization assay for the simultaneous detection of *Mycobacterium tuberculosis* complex and its resistance to rifampin. Antimicrob Agents Chemother 41:2093–2098
8. Yang B, Koga H, Ohno H, Ogawa K, Fukuda M, Hirakata Y et al (1998) Relationship between antimycobacterial activities of rifampin, rifabutin and KRM-1648 and *rpoB* mutations of *Mycobacterium tuberculosis*. J Antimicrob Chemother 42:621–628
9. Millen SJ, Uys PW, Hargrove J, van Helden PD, Williams BG (2008) The effect of diagnostic delays on the drop-out rate and the total delay to diagnosis of *tuberculosis*. PLoS One 3:e1933

10. Mitnick CD, Appeton SC, Shin SS (2008) Epidemiology and treatment of multidrug resistant *tuberculosis*. Semin Respir Crit Care Med 29:499–524
11. Mitnick CD, Shin SS, Seung KJ, Rich ML, Atwood SS, Furin JJ et al (2008) Comprehensive treatment of extensively drug-resistant *tuberculosis*. N Engl J Med 359: 563–574
12. Perkins MD, Cunningham J (2007) Facing the crisis: improving the diagnosis of *tuberculosis* in the HIV era. J Infect Dis 196:S15–S27
13. Boehme CC, Nabeta P, Hillemann D, Nicol MP, Shenai S, Krapp F et al (2010) Rapid molecular detection of *tuberculosis* and rifampin resistance. N Engl J Med 363:1005–1015
14. Nordhoek GT, van Embden JD, Kolk AH (1996) Reliability of nucleic acid amplification for detection of *Mycobacterium tuberculosis*: an international collaborative quality control study among 30 laboratories. J Clin Microbiol 34:2522–2525
15. Pai M, Kalantri S, Dheda K (2006) New tools and emerging technologies for the diagnosis of *tuberculosis*: Part II. Active tuberculosis and drug resistance. Expert Rev Mol Diagn 6:423–432
16. Suffys P, Palomino JC, Cardaso Leao S, Espitia C, Cataldi A, Alito A et al (2000) Evaluation of the polymerase chain reaction for the detection of *Mycobaterium tuberculosis*. Int J Tuberc Lung Dis 4:179–183
17. Telenti A, Imboden P, Marchesi F, Lowrie D, Cole S, Colston MJ et al (1993) Detection of rifampin-resistance mutations in *Mycobacterium tuberculosis*. Lancet 341:647–650
18. Ao W, Aldous S, Woodruff E, Hicke B, Larry R, Kreiswirth B et al (2012) Rapid detection of *rpoB* gene mutations conferring rifampin resistance in *Mycobacterium tuberculosis*. J Clin Microbiol 50:2433–2440
19. Dobosy JR, Rose SD, Beltz KR, Rupp SM, Powers KM, Behlke MA et al (2011) RNase H-dependent PCR (rhPCR): improved specificity and single nucleotide polymorphism detection using blocked cleavable primers. BMC Biotechnol 11:80
20. Jenison R, La H, Haeberli A, Ostroff R, Polisky B (2001) Silicon-based biosensors for rapid detection of protein or nucleic acid targets. Clin Chem 47:1894–1900
21. Miller LP, Crawford JT, Shinnick TM (1994) The *rpoB* gene of *Mycobacterium tuberculosis*. Antimicrob Agents Chemother 38:805–811

Chapter 8

Rapid Detection of *Brucella* spp. Using Loop-Mediated Isothermal Amplification (LAMP)

Shouyi Chen, Xunde Li, Juntao Li, and Edward R. Atwill

Abstract

Brucella spp. are facultative intracellular bacteria that cause zoonotic disease of brucellosis worldwide. Livestock that are most vulnerable to brucellosis include cattle, goats, and pigs. *Brucella* spp. cause serious health problems to humans and animals and economic losses to the livestock industry. Traditional methods for detection of *Brucella* spp. take 48–72 h (Kumar et al., J Commun Dis 29:131–137, 1997; Barrouin-Melo et al., Res Vet Sci 83:340–346, 2007) that do not meet the food industry's need of rapid detection. Therefore, there is an urgent need of fast, specific, sensitive, and inexpensive method for diagnosing of *Brucella* spp. Loop-mediated isothermal amplification (LAMP) is a method to amplify nucleic acid at constant temperatures. Amplification can be detected by visual detection, fluorescent stain, turbidity, and electrophoresis. We targeted at the *Brucella*-specific gene *omp*25 and designed LAMP primers for detection of *Brucella* spp. Amplification of DNA with Bst DNA polymerase can be completed at 65 °C in 60 min. Amplified products can be detected by SYBR Green I stain and 2.0 % agarose gel electrophoresis. The LAMP method is feasible for detection of *Brucella* spp. from blood and milk samples.

Key words Loop-mediated isothermal amplification, LAMP, Detection, *Brucella*, Milk, Blood

1 Introduction

Loop-mediated isothermal amplification (LAMP) is a novel approach to nucleic acid amplification developed by Notomi et al. in 2000 [3]. LAMP is characterized by the use of four different primers specifically designed to recognize six distinct regions on the target gene. The LAMP reaction uses a single constant temperature (about 65 °C) incubation, thereby obviating the need for expensive thermal cyclers. It provides high amplification efficiency, with DNA being amplified 10^9–10^{10} times in 15–60 min. Detection of amplification products can be achieved by visual detection of white sediments, photometry of turbidity, fluorescent stain, and gel electrophoresis. LAMP is a fast, specific, sensitive, and simple method which is widely used for detection of bacteria, parasites, and viruses [5–8].

Dmitry M. Kolpashchikov and Yulia V. Gerasimova (eds.), *Nucleic Acid Detection: Methods and Protocols*, Methods in Molecular Biology, vol. 1039, DOI 10.1007/978-1-62703-535-4_8,

1.1 Design of LAMP Primers

The primers for LAMP include a set of four different primers comprising two outer and two internal primers specifically designed to recognize six distinct regions of the target gene: the F3c, F2c, and F1c regions at the 3′ side and the B1, B2, and B3 regions at the 5′ side. The four primers used are as follows (Fig. 1a) [9].

1. Forward inner primer (FIP): The FIP consists of an F2 region at the 3′end and an F1c region at the 5′end. The F2 region is complementary to the F2c region of the template sequence. The F1c region is identical to the F1c region of the template sequence.
2. F3 (forward outer primer (FOP)): The F3 (FOP) consists of an F3 region, which is complementary to the F3c region of the template sequence.
3. Backward inner primer (BIP): The BIP consists of a B2 region at the 3′ end and a B1c region at the 5′ end. The B2 region is complementary to the B2c region of the template sequence. The B1c region is identical to the B1c region of the template sequence.
4. B3 (backward outer primer (BOP)): The B3 (BOP) consists of a B3 region which is complementary to the B3c region of the template sequence.

1.2 LAMP Reactions

Double-stranded DNA is in the dynamic equilibrium condition at the temperature around 65 °C. LAMP amplification is initiated by an inner primer containing both sense strands of target DNA. Strand displacement for DNA synthesis is followed by an outer primer that releases a single-stranded DNA. Then, DNA synthesis takes place by the second inner and outer primers using the single-strand DNA template. The second inner and outer primers hybridize to the other end of the target DNA and synthesize a stem–loop DNA structure (Fig. 1b) [3, 9].

Cyclic amplification is initiated by one inner primer, which hybridizes to the loop of the stem–loop DNA structure. The amplification produces the same stem–loop DNA as original as well as a new one with a stem but elongated structure. The cycling amplification repeats and results in accumulated copies of DNA (approximately 10^9) in less than 60 min. The final products are stem–loop DNAs with several repeats of the target DNA (Fig. 1c) [3, 9].

2 Materials

2.1 Equipment

1. Water bath.
2. Thermostat.
3. Biosafety cabinet.
4. Electrophoresis apparatus.

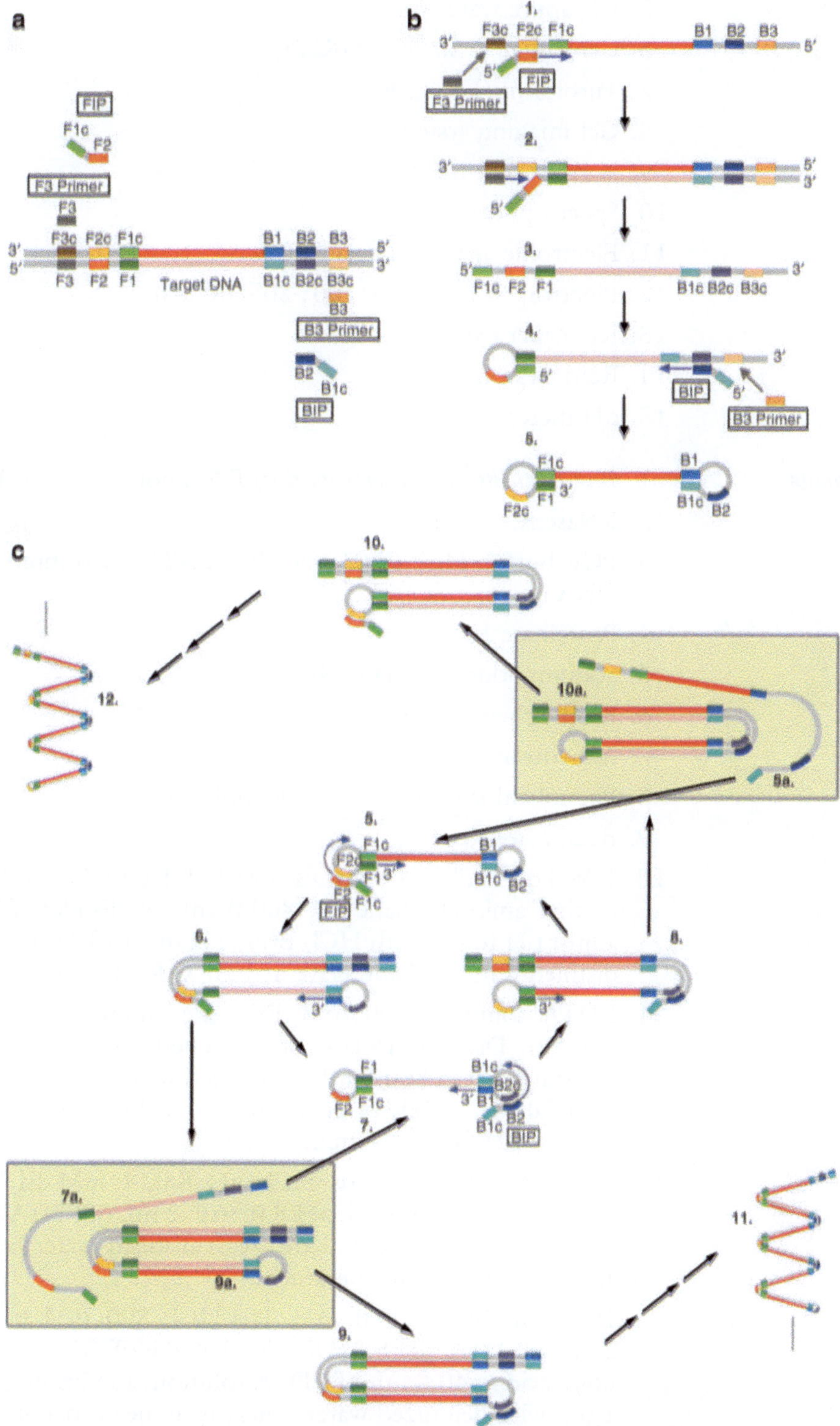

Fig. 1 Principle of LAMP. Adapted from Eiken Chemical CO., LTD with permission; Tomita et al., *Nature Protocols* 2008;3(5):877–882 [9]

5. Ultrapure water system.
6. Centrifuge (−20 °C) (MR23i).
7. Turbiditor (LA-320C).
8. Gel-imaging system.
9. Vortex (WH-861).
10. Spectrophotometer.
11. Electronic analytical balance.
12. Pipettors (0.5–10; 10–100; 20–200 μL).
13. Ice maker (SIM-F140).
14. Refrigerator.
15. pH meter.

2.2 Reagents

1. *Bst* (*Bacillus stearothermophilus*) DNA polymerase (8 U/μL).
2. RNase A.
3. PCR buffer: Mg^{2+} (8.0 mmol/L), dNTPs (1.6 mmol/L), *Bst* DNA polymerase (8 U).
4. Proteinase K.
5. Sodium dodecyl sulfate (SDS).
6. Bromophenol blue.
7. Anhydrous ethanol.
8. Phenol:chloroform:isoamyl alcohol (25:24:1).
9. Betain (*see* **Note 1**)
10. 1 M Tris–HCl, pH 8.0. Dissolve 121.1 g of Tris (trihydroxymethyl aminomethane) in 800.0 mL of distilled water and adjust pH to 8.0 with HCl. Bring volume to 1,000.0 mL with distilled water and autoclave at 121 °C for 15 min.
11. EDTA solution: 500 mM Ethylenediaminetetraacetic acid, pH 8.0. Dissolve 18.6 g of ethylenediaminetetraacetic acid disodium salt in 800.0 mL of distilled water. Adjust pH to 8.0 with 10 M NaOH, bring volume to 1,000.0 mL, and autoclave at 121 °C for 15 min.
12. TE buffer: 10 mM Tris–HCl, pH 8.0, 1 mM EDTA. Mix 10 mL of 1 M Tris–HCl, pH 8.0, with 2 mL of EDTA solution. Bring volume to 1,000.0 mL with distilled water, autoclave, and store at room temperature.
13. 10× TBE buffer: 890 mM Tris-HCl, 890 mM boric acid, 20 mM EDTA, pH 8.3. Dissolve 108.0 g of Tris and 55.0 g of boric acid in 40.0 mL of EDTA solution, and bring volume to 1.0 L with deionized water. There is no need to adjust pH.
14. Electrophoresis running buffer: 0.5× TBE. Dilute 10× TBE 20-fold with deionized water.

15. Loading buffer (6×): 30 mM EDTA, 36 % (v/v) glycerol, 0.05 % (w/v) xylene cyanol FF, 0.05 % (w/v) bromophenol blue.
16. EB solution: 10 g/L Ethidium bromide. Dissolve 1.0 g of EB in 100.0 mL of double-distilled water using a magnetic stir for several hours to ensure complete dissolving. Store in a brown glass bottle at 4 °C.
17. Lysis buffer: 10 % SDS, pH 7.2. Mix 10.0 g of SDS and 80.0 mL of double-distilled water in a flask and heat to 68 °C to dissolve. Adjust pH to 7.2 with HCl, bring volume to 100.0 mL, and store at 4 °C (*see* **Note 2**).
18. NET buffer: 50 mM Tris–HCl, pH 8.0, 125 mM EDTA, 50 mM NaCl.
19. Fresh thawed milk.
20. Blood serum.

3 Methods

3.1 Primers

1. Primers are designed using online software "Primer Explorer version 3" (https://Primerexplorer.jpamp3.0.o/index.htmL). The omp25 gene of *Brucella abortus* (GenBank access no. X79284.1) was used for designing LAMP primers. Among the hundreds of sets of primers generated by the software, inner and outer primers were selected based on stability of 5′-, 3′-ends, GC contents, and Tm value and verified for second structure by DNAstar software. The two inner primers are FIP and BIP. FIP should be composed of F1c sequence, TT connector, and F2 sequence. BIP should be composed of B1c sequence, TT connector, and B2 sequence. The two outer primers are F3 and B3 primers (*see* **Note 3**). The inner and outer primers selected are shown in Table 1.
2. Prepare F3/B3 primers at a concentration of 10 μM and FIP/BIP primers at a concentration of 40 μmol/L. Centrifuge

Table 1
Sequences of primers

Primer	Sequence
F3	CAAGACCAGCACCGTTGG
B3	GGTTCAGGTCGTAGCCGA
FIP	GGTCCTGCTGGAAGTTCCAGCAGCATC AAGCCTGACGATTG
BIP	CGGTGTTGAAGGTGATGCAGGTTTCAAA GCCCTGCTTGACTT

the stock solutions of the primers at 5,000 × *g* for 1 min. Use ddH_2O for dilutions. Mix the solutions thoroughly, centrifuge again at 5,000 × *g* for 30 s, and store at −20 °C.

3.2 Extraction of DNA from Milk and Blood Samples

1. Place 500 μL of fresh milk or thawed milk, or 200 μL of blood serum, into a centrifuge tube and add 100 μL of NET buffer (*see* **Note 4**).
2. Add 100 μL of lysis buffer (final concentration of SDS is 3.4 %) and mix thoroughly.
3. Incubate at 95–100 °C for 10 min and immediately place on ice for cooling for 10–15 min.
4. Add RNase A to a final concentration of 75 μg/μL and incubate at 50 °C for 2 h.
5. Add Proteinase K to a final concentration of 325 μg/μL and incubate at 50 °C for 2 h.
6. Add one volume of phenol:chloroform:isoamyl alcohol (25:24:1) to the tube of the extraction solution. Reverse the tube 2–3 times and mix thoroughly.
7. Centrifuge at 3,000 × *g* for 10 min at 4 °C and transfer the aqueous phase to a new tube.
8. Repeat **step 6**.
9. Combine aqueous phases, add 2.5× volume of cold anhydrous ethanol, and incubate at −20 °C for 30 min for DNA precipitation.
10. Centrifuge at 5,000 × *g* for 10 min, and discard all supernatants.
11. Rinse the pellets 2–3 times with 1.0 mL of 70 % ethanol, centrifuge at 5,000 × *g* for 2 min, and discard supernatants.
12. Dry the pellets in vacuum at room temperature.
13. Dissolve the pellets in 25 μL of sterilized double-distilled water and store at −20 °C.

3.3 LAMP Reaction

1. Prepare 50 μL of LAMP reaction mixture.

   0.2 μM outer primers.

   1.6 μM inner primers (*see* **Note 5**).

   1.6 mM dNTPs (*see* **Note 6**).

   8 U *Bst* DNA polymerase (*see* **Note 7**).

   8 mM Mg^{2+} (*see* **Note 8**).

   1× PCR buffer.
2. Incubate the reaction mixture at 65 °C for 60 min in a water bath.
3. Incubate the reaction mixture at 80 °C for 10 min to terminate the reaction (*see* **Notes 9** and **10**).

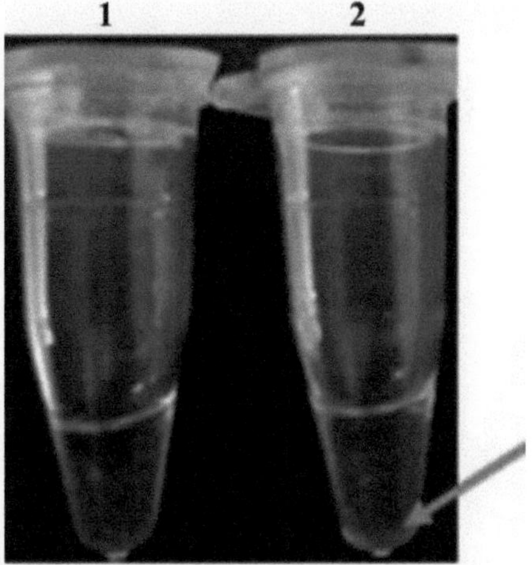

Fig. 2 Visual detection of LAMP. Tube 1 represents negative reaction; tube 2 represents positive reaction (*arrowhead* points at white sediments)

3.4 Detection of LAMP Products

1. The appearance of LAMP products can be visually detected by observing a white pellet of the reaction by-product magnesium pyrophosphate (Fig. 2). When large quantities of a nucleic acid are synthesized during the LAMP reaction, the pyrophosphate ions released from dNTPs conjunct with Mg^{2+} in the reaction system and form magnesium pyrophosphate as white sediments [10]. Therefore, the presence of white sediments indicates that the LAMP reaction occurred (*see* **Note 11**).
2. Alternatively, LAMP reactions can be detected by the changes of fluorescence intensity upon the reaction of the amplified products with fluorescent stains, such as EB and SYBR Green I [10] (*see* **Note 12**). To fluorescently detect the LAMP products, add approximately 2 μL of 1,000× SYBR Green I to a 50 μL reaction system. The color of SYBR Green I will change from orange to green if the LAMP reaction occurred (Fig. 3). In the meantime, the white sediments could still be observed.
3. The final products of LAMP reactions are DNAs with stem–loop structure. These products can be analyzed by electrophoresis in a 2 % agarose gel. Mix 5 μL of the amplified products with 1 μL of loading buffer and run electrophoresis on 2 % agarose gel. Stain the gel with EB solution. LAMP reactions can be verified by the bands of electrophoresis (Fig. 4).
4. The amount of magnesium pyrophosphate in the reaction system is proportionate to the amount of amplified products [4, 11]. Therefore, a real-time detection method based on the turbidity of the reaction mixture can be used. At consistent reaction temperatures, real-time and quantitative detection of amplified products can be done by using the LA-320c turbidimeter. Hence, the whole process of the LAMP reaction can be monitored, and amplification and detection can be completed within one tube (Fig. 5).

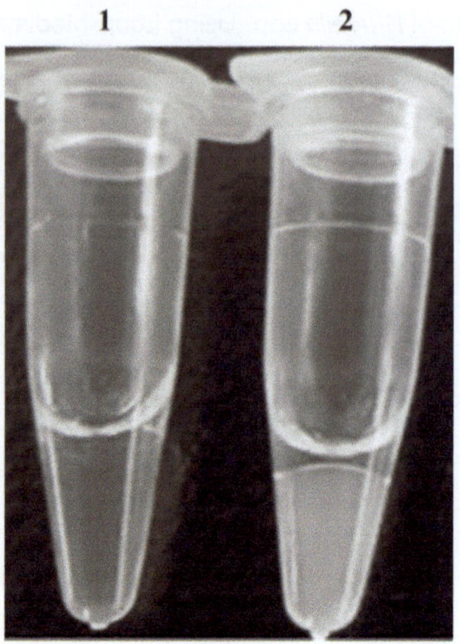

Fig. 3 SYBR Green I staining of LAMP products. Tube 1 represents negative reaction (*orange*) while tube 2 represents positive reaction (*green*)

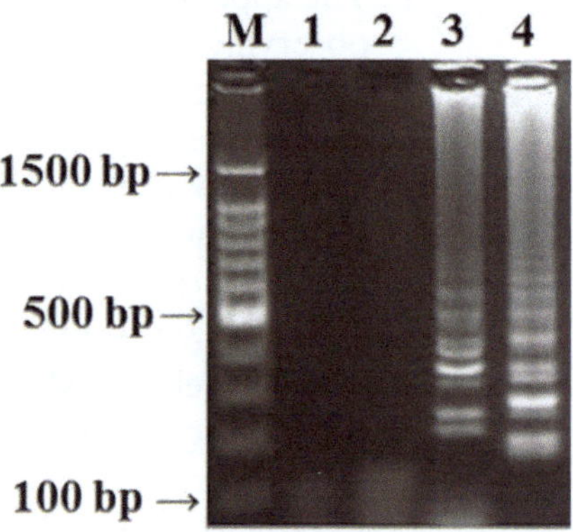

Fig. 4 Electrophoresis: *Lane M* represents DNA molecular weight markers; *lanes 1* and *2* represent negative reaction and *lanes 3* and *4* represent positive reactions

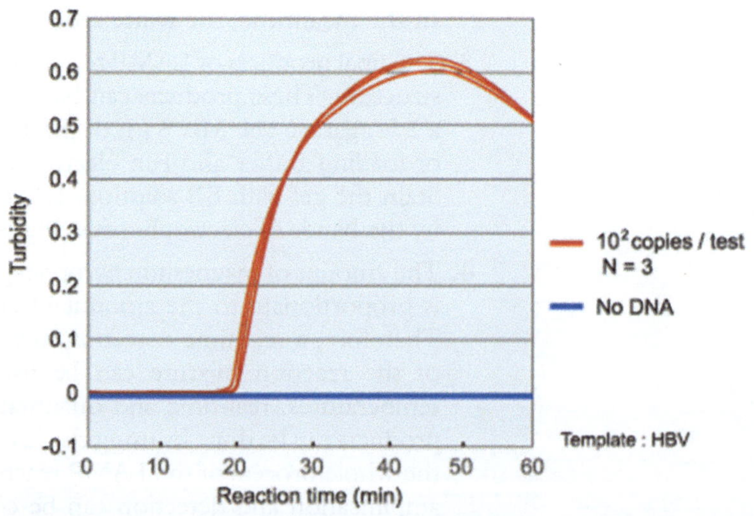

Fig. 5 Real-time detection of turbidity of LAMP reaction (http://loopamp.eiken.co.jp/e/tech/detect_real.html)

4 Notes

1. Effects of betain on LAMP reaction. Amplification products can be detected with or without betain in the reaction system. However, more clear bands were observed in electrophoresis when betain presented in the system at concentrations of 1.4 mmol/L.
2. SDS precipitates at 4 °C. Therefore, the lysis buffer needs to be warmed prior to use.
3. We targeted the *Brucella*-specific gene *omp*25 (*B. abortus omp*25 gene X79284.1) and designed primers using Primer Explorer Version V3 (Https://Primerexplorer.jpamp3.0/index). Among the hundreds of pairs of primers generated by the software, several pairs of primers were selected according to stability of 5′- and 3′-ends, contents of GC, and Tm value. Secondary structures of the selected primers were verified with DNAstar software, and finally three pairs of primers were selected (Table 1).
4. Fresh milk can be stored at either 4 °C for analysis in the same day or −20 °C for future analysis.
5. Ratio of inner and outer primers. A ratio of 1:8 of inner and outer primers (for example, 0.2 μmol/L inner primers and 1.6 μmol/L outer primers) results in best amplification.
6. Concentrations of dNTPs. Amplifications occur at the dNTP concentrations ranged 0.4–2.0 mmol/L. Maximum amplification occurs at dNTP concentration of 1.6 mmol/L.
7. Concentrations of *Bst* DNA polymerase. Amplification can be observed starting at Bst DNA polymerase concentration of 2.4 U. Maximum amplification was observed when Bst DNA polymerase concentration was 4.0 U and no difference was observed at higher concentrations. Therefore, the best Bst DNA polymerase concentration was 4.0 U in the LAMP for detection of *Brucella*.
8. Concentrations of Mg^{2+}. The best concentration of Mg^{2+} was determined to be 8.0 mmol/L by both electrophoresis and fluorescent stain.
9. Temperature for reaction. Clear bands of amplification were observed on amplifications at 63 and 65 °C. However, brightest fluorescent stains were observed on amplification at 65 °C. Therefore, 65 °C was determined as the best temperature for detection of *Brucella* using LAMP.
10. Reaction times. Bands can be observed after 30-min reaction but not very clearly. Clear bands were observed after 60-min reaction in electrophoresis. Therefore, 60 min was the best time for LAMP reaction for detection of *Brucella*.
11. Variations on visual detection exist among individuals.

12. For example, SYBR Green I only conjuncts to the minor groove of double-stranded DNA. After the conjunction, the SYBR Green I generates fluorescence that is 800~1,000 stronger than that before conjunction. When large amount of amplified DNA is present, the SYBR Green I will automatically combine to double-stranded DNA.

Acknowledgment

This work was partially supported by Guangzhou Medicine and Hygiene Foundation No. 2006-zdi-11.

References

1. Kumar P, Singh DK, Barbuddhe SB (1997) Sero-prevalence of Brucellosis abattoir personnel of Delhi-prahadl Kuma. J Commun Dis 29:131–137
2. Barrouin-Melo SM, Poester FP, Ribeiro MB, de Alcântara AC, Aguiar PH, Nascimento IL et al (2007) Diagnosis of canine brucellosis by ELISA using an antigen obtained from wild *Brucella canis*. Res Vet Sci 83:340–346
3. Notomi T, Okayama H, Mabuchi H, Yonekawa T, Watanabe K, Amino N et al (2000) Loop-mediated isothermal amplification of DNA. Nucleic Acids Res 28:e63
4. Mori Y, Nagamine K, Tomita N, Notomi T (2001) Detection of loop-mediated isothermal amplification reaction by turbidity derived from magnesium pyrophosphate formation. Biochem Biophys Res Commun 289:150–154
5. Techathuvanan C, Draughon FA, D'Souza DH (2010) Loop-mediated isothermal amplification (LAMP) for the rapid and sensitive detection of *Salmonella Typhimurium* from pork. J Food Sci 75:165–172
6. Kono T, Savan R, Sakai M, Itami T (2004) Detection of white spot syndrome virus in shrimp by loop-mediated isothermal amplification. J Virol Methods 115:59–65
7. Mekata T, Kono T, Savan R, Sakai M, Kasornchandra J, Yoshida T et al (2006) Detection of yellow head virus in shrimp by loop-mediated isothermal amplification (LAMP). J Virol Methods 135:151–156
8. Ikadai H, Tanaka H, Shibahara N, Matsuu A, Uechi M, Itoh N et al (2004) Molecular evidence of infections with *Babesia gibsoni* parasites in Japan and evaluation of the diagnostic potential of a loop-mediated isothermal amplification method. J Clin Microbiol 42:2465–2469
9. Tomita N, Mori Y, Kanda H, Notomi T (2008) Loop-mediated isothermal amplification (LAMP) of gene sequences and simple visual detection of products. Nat Protoc 3:877–882
10. Iwamoto T, Sonobe T, Hayashi K (2003) Loop-mediated isothermal amplification for direct detection of Mycobacterium tuberculosis complex, *M. avium*, and *M. intracellulare* in sputum samples. J Clin Microbiol 41:2616–2622
11. Mori N, Motegi Y, Shimamura Y, Ezaki T, Natsumeda T, Yonekawa T et al (2006) Development of a new method for diagnosis of rubella virus infection by reverse transcription-loop-mediated isothermal amplification. J Clin Microbiol 44:3268–3273

Chapter 9

Loop-Mediated Isothermal Amplification Method for a Differential Identification of Human *Taenia* Tapeworms

Yasuhito Sako, Agathe Nkouawa, Tetsuya Yanagida, and Akira Ito

Abstract

Loop-mediated isothermal amplification (LAMP), which employs a *Bst* DNA polymerase with strand-displacement activity and four primers (two inner primers and two outer primers) recognizing six distinct regions on the target DNA, is a highly sensitive, specific, simple, and rapid nucleotide amplification method. Moreover, because the *Bst* DNA polymerase resists much DNA polymerase inhibitors present in biological specimens, the LAMP method is suitable for the detection of infectious agents from clinical material such as fecal samples. Here, we describe the LAMP method which can differentially detect and identify human *Taenia* tapeworms, *Taenia solium*, *T. saginata*, and *T. asiatica*, using DNA specimens prepared from parasite tissue and human fecal sample.

Key words Loop-mediated isothermal amplification, *Taenia solium*, *T. saginata*, *T. asiatica*, Differential detection, Proglottid, Fecal sample

1 Introduction

Human taeniasis is an infection caused by the adult tapeworm of *Taenia solium*, *T. saginata*, *or T. asiatica* [1]. The adult worm residing in the small intestine produces embryonated eggs infective to the suitable intermediate mammalian hosts, and accidental ingestion of eggs causes cysticercosis (=metacestodiasis) in parenteral tissues of the intermediate mammalian hosts. The intermediate hosts for *T. solium*, *T. saginata*, and *T. asiatica* are human and swine, bovine, and swine, respectively. Although taeniasis is relatively innocuous, human cysticercosis caused by *T. solium* is one of the most serious zoonotic parasitic diseases worldwide especially in developing countries [2]. In addition, swine and bovine cysticercosis lead to economic losses due to condemnation of livestock [3]. Therefore, differential detection of these *Taenia* species is a key point for control and prevention of taeniasis/cysticercosis in endemic areas. Especially, the sensitive detection of

Dmitry M. Kolpashchikov and Yulia V. Gerasimova (eds.), *Nucleic Acid Detection: Methods and Protocols*, Methods in Molecular Biology, vol. 1039, DOI 10.1007/978-1-62703-535-4_9, © Springer Science+Business Media New York 2013

the carriers of *T. solium* tapeworms is important for the prevention of lethal cysticercosis in human.

Diagnosis of taeniasis by fecal examination to detect taeniid eggs lacks both sensitivity and specificity because the eggs of *Taenia* species are morphologically indistinguishable. Moreover, the identification of *Taenia* species relying on comparative morphology of proglottids also lacks specificity. Therefore, various molecular biological techniques for detection of parasites based on PCR have been developed for sensitive differential examination of taeniid cestodes [4–7]. However, it is not easy to exploit PCR techniques in the laboratories of developing countries because they require sophisticated equipment such as a thermal cycler. Furthermore, *Taq* DNA polymerase is often inactivated by inhibitors present in biological samples such as fecal samples, which sometimes cause problems in its sensitivity and reproducibility [8].

Recently developed loop-mediated isothermal amplification (LAMP) method amplifies DNA with high specificity, sensitivity, and rapidity under isothermal conditions [9]. This method requires a *Bst* DNA polymerase with strand-displacement activity, four primers recognizing six regions on the target DNA and simple incubators such as water bath, block heater, or thermos bottle. Unlike the *Taq* DNA polymerase used in PCR, the *Bst* DNA polymerase resists much enzyme inhibitors in biological specimens. It means that it is suitable to use against clinical specimens. In addition, a large amount of white precipitate of magnesium pyrophosphate is produced as a by-product, which enables visual judgment of amplification by the naked eyes without the need of electrophoresis [10]. These features strongly make us focus on the LAMP method for the detection of infectious agents as a clinical test [11–14].

For a differential detection and identification of *Taenia* species, we have developed the LAMP method targeting cytochrome *c* oxidase subunit 1 (cox1) and cathepsin L-like cysteine peptidase (clp) genes [15]. This method has proved to be highly sensitive and specific for differential detection of *Taenia* species by using DNA specimens prepared from proglottids, cysticerci, and fecal samples with or without eggs of taeniasis patients. The evaluation using fecal samples from taeniasis patients has revealed higher sensitivity of the LAMP method than that of PCR [16].

In this chapter, at first we introduce several efficient DNA extraction methods which should be selected properly depending on the size of parasite tissues, the source of samples (e.g., human fecal samples), or working situation. And then the procedure of specific DNA amplification by LAMP method to differentially detect and identify human *Taenia* species is described.

2 Materials

2.1 Specimen

1. Proglottid from a carrier.
2. Fecal sample from a patient and a suspected person.

2.2 Extraction of DNA from Parasite Tissue and Fecal Sample

2.2.1 Extraction of DNA from Parasite Tissue

1. QIAamp DNA mini kit (Qiagen).
2. Ethanol (96–100 %) (*see* **Note 1**).
3. Autoclaved ultrapure water.
4. Phosphate-buffered saline (PBS).
5. 0.05 N NaOH.
6. 1.5, 2.0 mL microcentrifuge tubes.
7. Pipette tips with aerosol barrier.
8. Vortex mixer.
9. Water bath or block heater.

2.2.2 Extraction of DNA from a Fecal Sample

1. QIAamp DNA Stool kit (Qiagen).
2. Ethanol (96–100 %).
3. 1.5, 2.0 mL microcentrifuge tubes.
4. Pipette tips with aerosol barrier.
5. Glass beads (φ3 mm), sterilized by dry-heating.
6. Parafilm.
7. Vortex mixer.
8. Water bath or block heater.

2.3 LAMP Reaction

1. LAMP buffer (10×): 200 mM Tris–HCl, pH 8.8, 100 mM KCl, 100 mM $(NH4)_2SO_4$, 1.0 % Tween 20.
2. $MgSO_4$ (10×): 80 mM $MgSO_4$.
3. LAMP primers (*see* **Note 2**): 5 μM working solution of outer primers (F3 and B3) and 40 μM working solution of inner primers (FIP and BIP). Locations and sequences of primer recognition sites are shown in Figs. 1 and 2, and Tables 1 and 2.
4. 100 mM dNTPs.
5. 5 M Betaine.
6. LAMP buffer (2×): The components are shown in Table 3.
7. *Bst* DNA polymerase, large fragment (New England Biolabs).
8. NEBuffer 2 (10×) (New England Biolabs).
9. Restriction enzyme, *Hin*f I (10,000 U/mL) (New England Biolabs).

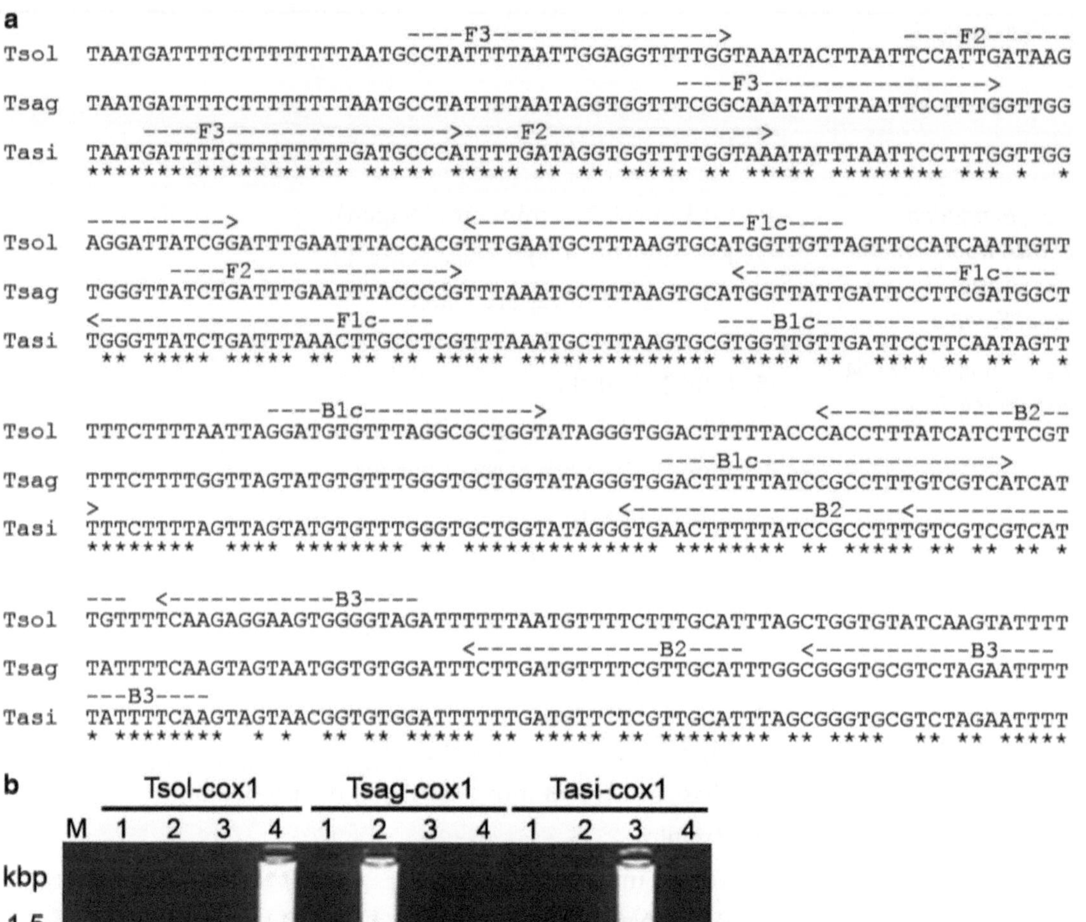

Fig. 1 Nucleotide sequence alignment of the target region of cox1 genes (**a**) and LAMP results (**b**). (**a**) The locations of the primer recognition sites are indicated by *arrows*. (**b**) The LAMP products were run on a 2 % agarose gel. *Lane M*, 100-bp DNA ladder marker (Promega); *lane 1*, negative control; *lane 2*, *T. saginata* genomic DNA; *lane 3*, T. asiatica genomic DNA; *lane 4*, T. solium genomic DNA. *Tsol-cox1*, results of LAMP with primer set Tsol-cox1; *Tsag-cox1*, results of LAMP with primer set Tsag-cox1; *Tasi-cox1*, results of LAMP with primer set Tasi-cox1

10. 1.5, 2.0 mL microcentrifuge tubes.
11. 0.5, 0.2 mL PCR tubes.
12. Pipette tips with aerosol barrier.
13. Vortex mixer.
14. Water bath or block heater.

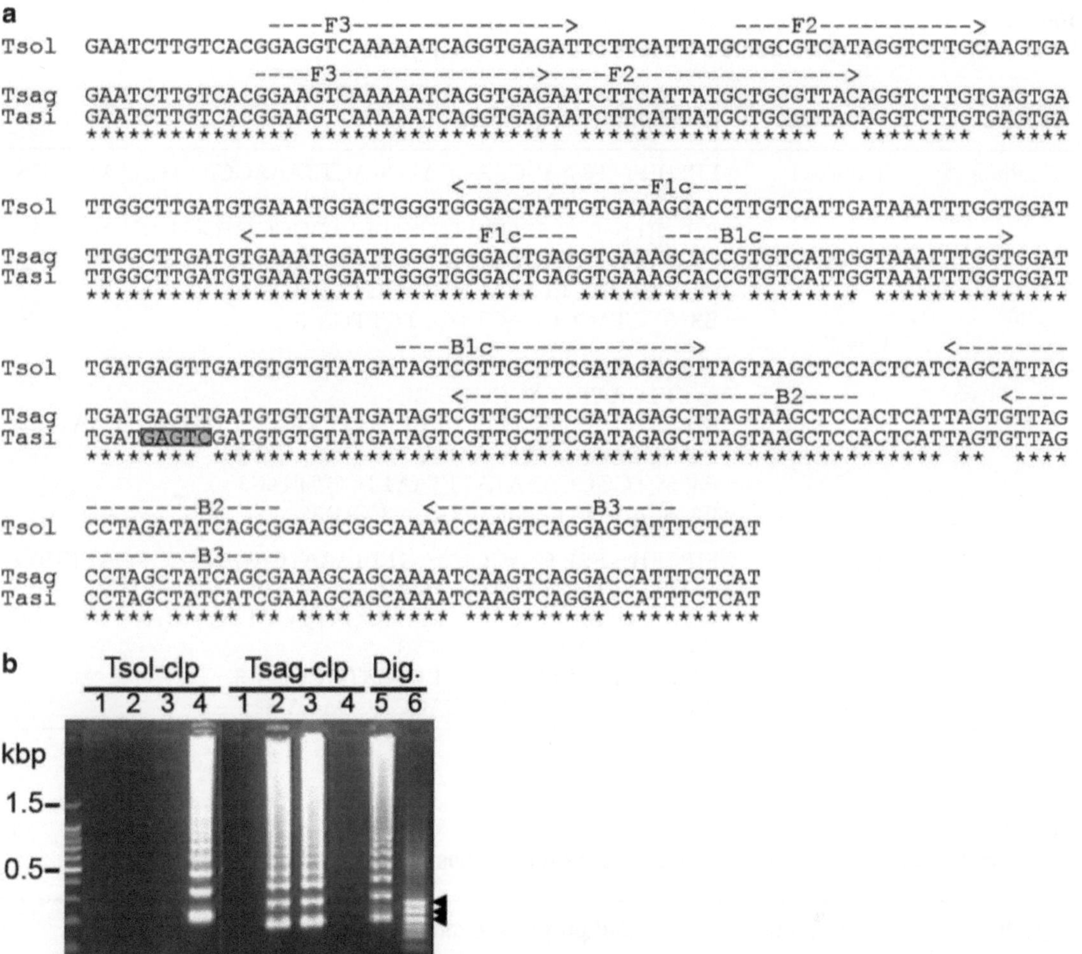

Fig. 2 Nucleotide sequence alignment of the target region of clp genes (**a**) and LAMP results (**b**). (**a**) The locations of the primer recognition sites are indicated by *arrows* and the restriction enzyme *Hinf* I recognition site in the LAMP products of *T. asiatica* clp gene is *gray boxed*. (**b**) The LAMP products and *Hinf* I digestion products were run on a 2 % agarose gel. *Lane M*, 100-bp DNA ladder marker (Promega*); lane 1*, negative control; *lane 2*, *T. saginata* genomic DNA; *lane 3*, *T. asiatica* genomic DNA; *lane 4*, *T. solium* genomic DNA; *lane 5*, *Hinf* I digestion of LAMP products from *T. saginata* genomic DNA with primer set Tsag-clp; *lane 6*, *Hinf* I digestion of LAMP products from *T. asiatica* genomic DNA with primer set Tsag-clp; *Tsol-clp*, results of LAMP with primer set Tsol-clp; *Tsag-clp*, results of LAMP with primer set Tsag-clp; *Dig.*, results of *Hinf* I-digestion. The DNA fragments generated after digestion with *Hinf* I were indicated by *arrowheads*

3 Methods

Three different DNA extraction methods from parasite tissues and human fecal samples are described (*see* **Note 3**). We have to use them properly depending on the materials and the situation of test.

Table 1
LAMP primer sets for cytochrome *c* oxidase subunit 1 (cox1) genes

Species	Primer set	LAMP primer sequences
T. solium	Tsol-cox1	FIP (F1c+F2): 5′-CAACCATGCACTTAAAGCATTCAAATTCCA TTGATAAGAGGATTATCGG-3′ BIP (B1c+B2): 5′-GGATGTGTTTAGGCGCTGGTACAACGAA GATGATAAAGGTG-3′ F3: 5′CCTATTTTAATTGGAGGTTTTGG-3′ B3: 5′-CTACCCCACTTCCTCTTGA-3′
T. saginata	Tsag-cox1	FIP (F1c+F2): 5′-GCCATCGAAGGAATCAATAACCAATCTGA TTTGAATTTACCCCG-3′ BIP (B1c+B2): 5′-GACTTTTTATCCGCCTTTGTCGTCAATGC AACGAAAACATCAAGA-3′ F3: 5′-TCGGCAAATATTTAATTCCTTTG-3′ B3: 5′-AAATTCTAGACGCACCCG-3′
T. asiatica	Tasi-cox1	FIP (F1c+F2): 5′-AGGCAAGTTTAAATCAGATAACCCATTTTG ATAGGTGGTTTTGGTAA-3′ BIP (B1c+B2): 5′-GTGGTTGTTGATTCCTTCAATAGTTTAAGG CGGATAAAAAGTTCAC-3′ F3: 5′-GATTTTCTTTTTTTTGATGCCCA-3′ B3: 5′-TTGAAAATAATGACGACGACA-3′

Table 2
LAMP primer sets for cathepsin L-like peptidase (clp) genes

Species	Primer set	LAMP primer sequences
T. solium	Tsol-clp	FIP (F1c+F2): 5′-AGGTGCTTTCACAATAGTCCCCTGCG TCATAGGTCTTGC-3′ BIP (B1c+B2): 5′-TAGTCGTTGCTTCGATAGAGCTCGCT GATATCTAGGCTAATGCTG-3′ F3: 5′-GAGGTCAAAAATCAGGTGAGAT-3′ B3: 5′-AATGCTCCTGACTTGGTT-3′
T. saginata *T. asiatica*	Tsag-clp	FIP (F1c+F2): 5′-CTCAGTCCCACCCAATCCATTTCAAAT CTTCATTATGCTGCGTTAC-3′ BIP (B1c+B2): 5′-GCACCGTGTCATTGGTAAATTTG GTGGAGCTTACTAAGCTCTATCG-3′ F3: 5′-GGAAGTCAAAAATCAGGTGAG-3′ B3: 5′-CGCTGATAGCTAGGCTAAC-3′

3.1 DNA Extraction from Parasite Tissue with QIAamp DNA Mini Kit

In case that parasite tissue with enough size like *Taenia* proglottids is obtained, standard method (e.g., phenol/chloroform extraction method) or commercially available DNA extraction kit can be used. We usually use QIAamp DNA Mini kit for this purpose because quality and quantity of DNA extracted by this kit are steady.

Table 3
Components of 2× LAMP buffer

Reagents	Volume (μL)	Final concentration
10× LAMP buffer	200	2×
10× $MgSO_4$	200	2×
5 M Betaine	320	1.6 M
100 mM dATP	28	2.8 mM
100 mM dCTP	28	2.8 mM
100 mM dGTP	28	2.8 mM
100 mM dTTP	28	2.8 mM
Ultrapure water	168	–
Total	1,000	–

1. Rinse a proglottid expelled from the carrier with ultrapure water or PBS.
2. Cut proglottid (<25 mg) into small pieces and place them into a 1.5-mL microcentrifuge tube. Or homogenize proglottid with disposable plastic pestle in a 1.5-mL microcentrifuge tube.
3. Add 180 μL of Buffer ATL from the kit into the tube containing parasite tissue.
4. Add 20 μL of proteinase K, mix by vortexing, and incubate at 56 °C until the tissue is completely lysed. The overnight incubation with shaking is recommended for an efficient digestion.
5. Perform the further purification steps according to the manufacture's protocol.
6. Elute DNA in 100 μL of Buffer AE from the kit.
7. Keep DNA sample at −20 °C until use.

3.2 DNA Extraction from Parasite Tissues with NaOH

In case that the size of parasite specimen is small or DNA extraction must be carried out in the field, easy but efficient DNA extraction method with NaOH can be used.

1. Cut 5–7 mg of proglottid and put into 0.5 mL or PCR tubes.
2. Add 20 μL of 0.05 N NaOH into the tube containing parasite tissue.
3. Incubate at 95 °C for 10–30 min in the heat block or water bath or thermal cycler (*see* **Note 4**).

4. Spin down the tube at full speed in the microcentrifuge at room temperature for 1 min to remove debris.
5. Transfer supernatant to a new tube.
6. Keep DNA sample at −20 °C for a short time.

3.3 DNA Extraction from Fecal Samples with QIAamp DNA Stool Mini Kit

In case of the fecal sample, we usually use QIAamp DNA Stool Mini Kit. It is sometimes difficult to extract DNA of parasites because *Taenia* eggs have thick shell that prevents lysing parasite cells completely. Therefore disruption of eggs shell to release parasite cells is an imperative step to obtain DNA as much as possible [17]. We described the procedure of DNA extraction following the instruction manuals with some modifications (*see* **Note 5**).

1. Put about 10–12 glass beads (φ3 mm) into a 2 mL microcentrifuge tube.
2. Weigh 230 mg of stool and place it into a 2 mL microcentrifuge tube containing the glass beads. The starting amount of fecal is increased because a small amount of fecal will remain in the tube during the transfer.
3. Add 1 mL of Buffer ASL from the kit to each stool sample and seal the tube with Parafilm to avoid leakage of solution during shaking. Vortex continuously for 1 min or until the stool sample is thoroughly homogenized.
4. Attach the tube together with the rubber tray of the vortex mixer using the adhesive tape and shake at maximum speed on the vortex mixer for 20–30 min at room temperature.
5. Cut the end of the pipet tip to make pipetting easier and transfer all the fecal samples into a new 2 mL tube. Rinse the tubes containing the glass beads with 0.4 mL of Buffer ASL to make a total volume of 1.4 mL Buffer ASL and pipet it into the same new 2 mL tube.
6. Vortex the suspension and heat for 5 min at 95 °C, not 70 °C. The heating temperature is increased in order to increase the DNA yield and to allow the lysis of cells that are difficult to lyse.
7. Perform the further purification steps according to the manufacture's protocol.
8. Elute DNA in 100 μL of Buffer AE from the kit.
9. Keep DNA sample at −20 °C until use.

3.4 LAMP Reaction

Multiplex-LAMP with several primer sets in a single tube is far from practical use such as multiplex-PCR [5] at present. Therefore, one primer set needs one reaction mixture. To differentially detect *Taenia* species, three reaction mixtures containing *T. solium*-, *T. saginata*-, and *T. asiatica*-primer set separately must be set up if

Table 4
LAMP reaction mixture without DNA template

Reagents	Volume per reaction (μL)
2× LAMP buffer	12.5
40 μM FIP primer	1.0
40 μM BIP primer	1.0
5 μM F3 primer	1.0
5 μM B3 primer	1.0
Bst DNA polymerase (8 U/μL)	1.0
Ultrapure water	6.5
Total	24.0

the target gene is cox1 gene, on the other hand, two reaction mixtures containing *T. solium*- and *T. saginata*-primer set separately must be prepared in case of clp gene. When the fecal samples are examined, the LAMP with cox1 primer sets shows higher sensitivity than that of the LAMP with clp primer sets [16]. The number of copies of each target gene within samples may be responsible for the differences in sensitivity since a large number of mitochondria DNA exist in a cell. Thus, the LAMP with cox1 primer sets is recommended against fecal samples as the first-choice test.

1. Prepare the master mixture (*see* **Note 6**) in a 1.5 or 0.5 mL microcentrifuge tube. The components of one reaction mixture are shown in Table 4. First calculate the volume of each reagent and primer for the number of reaction mixtures including a positive and negative control, then add the required volume of each reagent into a 1.5 or 0.5 mL microcentrifuge tube and mix thoroughly (*see* **Note 7**).
2. Centrifuge the tube briefly.
3. Dispense 24 μL of the master mixture into 0.2 mL PCR tubes (*see* **Note 8**).
4. Add 1 μL of DNA template to the master mixture.
5. Mix the mixture by pipetting or tapping.
6. Incubate the reaction mixture for 60 min at 60 °C for cox1-primer sets or 63 °C for clp primer sets, and then incubate for 5 min at 80 °C to inactivate the *Bst* DNA polymerase. For DNA samples from fecal samples, 90 min to 2 h incubation is required (*see* **Note 9**).
7. When a white precipitate, magnesium pyrophosphate, is visible, the reaction is considered to be positive as shown in Fig. 3.

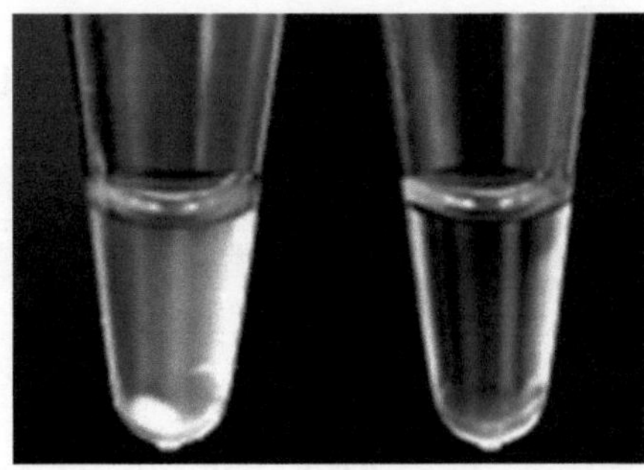

Fig. 3 Results of LAMP reaction can be assessed by naked eye without electrophoresis based on the turbidity of the reaction mixture. The white precipitate of magnesium pyrophosphate produced as by-products during the reaction is visible in the *left* tube (positive reaction), whereas the *right* tube remain clean (negative reaction)

Table 5
Components of *Hinf* I-digestion mixture

Reagents	Volume (μL)
10× NEBuffer 2	2.0
LAMP products	1.5
Ultrapure water	16.2
Hinf I (10 U/μL)	0.3
Total	20.0

If difficult to judge, the centrifugation of the tube at full speed for 5 min helps the white precipitate to be easily visible at the bottom of tube.

8. If needed (*see* **Note 10**), the LAMP products (1.5 μL) are electrophoresed in a 2.0 % agarose gel and detected by ethidium bromide staining. As shown Figs. 1 and 2, the LAMP products are detected as a ladder of multiple bands on the gel due to the formation of a mixture of stem-loop DNAs with various stem lengths and cauliflower-like structures, with multiple loops formed by annealing between alternately inverted repeats of the target sequence in the same strand.

3.5 Restriction Enzyme *Hinf* I-Digestion of LAMP Products with Tsag-clp Primer Set

1. When a positive result is obtained with Tsag-clp primer set, treat the LAMP products with *Hinf* I to differentiate *T. saginata* and *T. asiatica*.
2. Prepare the digestion mixture as shown in Table 5.
3. Incubate the digestion mixture for 2 h at 37 °C.

4. Analyze the LAMP products (1.5 μL) and *Hinf* I-digested products (10 μL) by electrophoresis using a 2.0 % agarose gel and subsequent ethidium bromide staining. Because there is a recognition site for *Hinf* I in the amplified region of *T. asiatica* clp gene, restriction enzyme digestion of the LAMP products from *T. asiatica* genomic DNA produces three bands with the predicted size of 179, 217, and 255 bp as shown in Fig. 2. In contrast, the electrophoresed-bands pattern of the LAMP products from *T. saginata* treated with *Hinf* I shows no changes.

4 Notes

1. Do not use denatured alcohol, which contains other substances such as methanol or methylethylketone.
2. LAMP primers were designed using PrimerExplorer V4 software (http://primerexplorer.jp/elamp4.0.0/index.html). All primers were designed from sequence data deposited in GenBank (*T. solium* mitochondrial DNA, AB086256; *T. saginata* mitochondrial DNA, AY684274; *T. asiatica* mitochondrial DNA, AF445798; *T. solium* clp gene, AB441815; *T. saginata* clp gene, AB441816; *T. asiatica* clp gene, AB441817).
3. Parasite tissues and fecal samples in sealed containers can be stored in the refrigerator for a few days. If the extraction of DNA cannot be performed immediately, they must be kept in ethanol (>70 %). Storing in formalin is not recommended because formalin makes them difficult to extract DNA efficiently.
4. We have succeeded in DNA extraction by this method using hot water and thermos bottle in the field [18].
5. **Steps 2–5** must be carried out in the fume hood. Wear gloves throughout the entire procedure, in case of contact between gloves and sample, change gloves immediately.
6. Because the LAMP method is extremely sensitive, any contamination of *Taenia* genomic DNA or amplified products from other examinations causes incorrect interpretation. To avoid possible contamination, preparations of reagents and samples in a specifically designed room or in a clean bench are recommended. If possible, use dedicated micropipettes only for preparation of reaction mixture.
7. Mix the master mixture gently by pipetting or tapping, because the *Bst* DNA polymerase loses activity by vortex mixing.
8. Keep all reagents on ice until the staring the reaction.

9. Do not put the tube on incubator until the temperature of incubator reaches the required value.
10. To avoid the release of the LAMP products into the air as a source of the contamination, this step should not be performed if at all possible.

References

1. Ito A, Nakao M, Wandra T (2003) Human taeniasis and cysticercosis in Asia. Lancet 362:1918–1920
2. Garcia HH, Gonzalez AE, Evans CA, Gilman RH (2003) *Taenia solium* cysticercosis. Lancet 362:547–556
3. Schantz PM, Cruz M, Sarti E, Pawlowski Z (1993) Potential eradicability of Taeniasis and cysticercosis. Bull Pan Am Health Organ 27:397–403
4. Gonzalez LM, Montero E, Harrison LJ, Parkhouse RM, Garate T (2000) Differential diagnosis of *Taenia saginata* and *Taenia solium* infection by PCR. J Clin Microbiol 38:737–744
5. Yamasaki H, Allan JC, Sato MO, Nakao M, Sako Y, Nakaya K et al (2004) DNA differential diagnosis of taeniasis and cysticercosis by multiplex PCR. J Clin Microbiol 42:548–553
6. Nunes CM, Dias AK, Dias FE, Aoki SM, de Paula HB, Lima LG et al (2005) *Taenia saginata*: differential diagnosis of human taeniasis by polymerase chain reaction-restriction fragment length polymorphism assay. Exp Parasitol 110:412–415
7. Mayta H, Gilman RH, Prendergast E, Castillo JP, Tinoco YO, Garcia HH et al (2008) Nested PCR for specific diagnosis of *Taenia solium* taeniasis. J Clin Microbiol 46:286–289
8. Abu AlSoud W, Radstrom P (2000) Effects of amplification facilitators on diagnostic PCR in the presence of blood, feces, and meat. J Clin Microbiol 38:4463–4470
9. Notomi T, Okayama H, Masubuchi H, Yonekawa T, Watanabe K, Amino N et al (2000) Loop-mediated isothermal amplification of DNA. Nucleic Acids Res 28:E63
10. Mori Y, Nagamine K, Tomita N, Notomi T (2001) Detection of loop-mediated isothermal amplification reaction by turbidity derived from magnesium pyrophosphate formation. Biochem Biophys Res Commun 289: 150–154
11. Bakheit MA, Torra D, Palomino LA, Thekisoe OM, Mbati PA, Ongerth J et al (2008) Sensitive and specific detection of *Cryptosporidium* species in PCR-negative samples by loop-mediated isothermal DNA amplification and confirmation of generated LAMP products by sequencing. Vet Parasitol 158:11–22
12. Liang SY, Chan YH, Hsia KT, Lee JL, Kuo MC, Hwa KY et al (2009) Development of loop-mediated isothermal amplification assay for detection of *Entamoeba histolytica*. J Clin Microbiol 47:1892–1895
13. Macuhova K, Kumagai T, Akao N, Ohta N (2010) Loop-mediated isothermal amplification assay for detection and discrimination of *Toxocara canis* and *Toxocara cati* eggs directly from sand samples. J Parasitol 96:1224–1227
14. Mori Y, Notomi T (2009) Loop-mediated isothermal amplification (LAMP): a rapid, accurate, and cost-effective diagnostic method for infectious diseases. J Infect Chemother 15:62–69
15. Nkouawa A, Sako Y, Nakao M, Nakaya K, Ito A (2009) Loop-mediated isothermal amplification method for differentiation and rapid detection of *Taenia* species. J Clin Microbiol 47:168–174
16. Nkouawa A, Sako Y, Li T, Chen X, Wandra T, Swastika IK et al (2010) Evaluation of a loop-mediated isothermal amplification method using fecal specimens for differential detection of *Taenia* species from humans. J Clin Microbiol 48:3350–3352
17. Nunes CM, Lima LG, Manoel CS, Pereira RN, Nakano MM, Garcia JF (2006) Fecal specimens preparation methods for PCR diagnosis of human taeniosis. Rev Inst Med Trop Sao Paulo 48:45–47
18. Nkouawa A, Sako Y, Li T, Chen X, Nakao M, Yanagida T et al (2012) A loop-mediated isothermal amplification method for a differential identification of *Taenia* tapeworms from human: application to a field survey. Parasitol Int 61:723–725

Chapter 10

Detection of Mutation by Allele-Specific Loop-Mediated Isothermal Amplification (AS-LAMP)

Hiroka Aonuma, Athanase Badolo, Kiyoshi Okado, and Hirotaka Kanuka

Abstract

For effective control of pathogen-transmitting mosquitoes, precise surveillance data of mosquito distribution are essential. Recently, an increase of insecticide resistance due to the *kdr* mutation in *Anopheles gambiae*, a mosquito that transmits the malaria parasite, has been reported. With the aim of developing a simple and effective method for surveying resistant mosquitoes, LAMP was applied to the allele-specific detection of the *kdr* gene in *An. gambiae*. Allele-specific LAMP (AS-LAMP) method successfully distinguished the *kdr* homozygote from the heterozygote and the wild type. The robustness of AS-LAMP suggests its usefulness for routine identification of insects, not only mosquitoes but also other vectors and agricultural pests. Here we describe the method of AS-LAMP to detect mutation in *Anopheles* mosquitoes.

Key words LAMP, Mosquito, Allele-specific, *kdr*

1 Introduction

Mosquitoes are one of the most important arthropod vectors that transmit various infectious diseases. To prevent epidemics of mosquito-transmitted diseases, in particular malaria, indoor residual insecticide spray (IRS) and insecticide-treated nets (ITNs) have been widely used [1]. However, a recent study reported that the number of mosquitoes carrying pyrethroid knockdown resistance (*kdr*) mutation in the malaria vector, *Anopheles* mosquito, has been increasing [1]. These mosquitoes are resistant to pyrethroid and DDT, commonly used insecticides, and are an urgent issue for the Global Malaria Programme that promotes the use of long-lasting insecticidal nets (LLINs). Moreover, the importance of monitoring insecticide resistance was emphasized in The Global Plan for Insecticide Resistance Management for malaria vectors (GPIRM) [2].

Dmitry M. Kolpashchikov and Yulia V. Gerasimova (eds.), *Nucleic Acid Detection: Methods and Protocols*, Methods in Molecular Biology, vol. 1039, DOI 10.1007/978-1-62703-535-4_10, © Springer Science+Business Media New York 2013

To date, two types of *kdr* mutation have been reported. A single point mutation, from A to T, in a gene encoding a sodium channel has been revealed to be the cause of West African-type *kdr* mutation (*kdr-w*) [3]. This mutation involves the replacement of a leucine residue (TTA) with phenylalanine (TTT) at position 1014 (L1014F), which reduces sensitivity of the insect nervous system to pyrethroid. The other type of *kdr* mutation, East African-type *kdr* mutation (*kdr-e*), is due to another mutation in the same nucleotide (L1014S) [4].

With the aim of developing a quick and reliable method to survey *kdr* mutations in *An. gambiae*, LAMP (loop-mediated isothermal amplification) was applied. LAMP, a DNA amplification method, was developed in 2000 and has been applied for the detection of various species in the last decade. It has the distinguishing feature that only a single enzyme, *Bst* polymerase, is required and that the reaction proceeds under isothermal conditions [5]. *Bst* DNA polymerase can synthesize a new strand of DNA while simultaneously displacing the complementary strand, thereby enabling DNA amplification at a single temperature. LAMP reaction requires four specific primers, FIP (F1c-F2), BIP (B1c-B2), F3, and B3. F3 and B3 contribute to the formation of a stem-loop structure, while the other two primers, FIP and BIP, designed complementary to the inner sequence of the stem-loop structure, are used to amplify the target sequence, thus providing higher specificity and sensitivity to the reaction. LAMP has been effectively applied to the identification of pathogen-carrying mosquitoes [6–8]. Additionally, the detection of gene point mutation using LAMP has also been reported [9–11]. Owing to its specificity, allele-specific LAMP successfully detected West African-type *kdr* mutation (*kdr-w*) in field-collected mosquitoes [12].

Here we describe the method for allele-specific detection of mutation in mosquitoes.

2 Materials

Ultrapure water should be used for all solutions unless otherwise indicated.

2.1 Extraction of Genomic DNA

1. 1.5 mL microtubes (Treff AG, Degersheim, Switzerland).
2. Buffer A for homogenization: 0.1 M Tris–HCl (pH 9.0), 0.1 M EDTA, 1 % SDS, and 0.5 % DEPC (*see* **Note 1**).
3. Pellet mixer for microtubes (Treff AG, Degersheim, Switzerland).
4. 5 M KOAc.
5. Isopropanol.
6. 70 % ethanol: Dilute ethanol with water to 70 %.

7. TE (pH 7.4) for dissolving DNA: 10 mM Tris–HCl (pH 7.4), 1 mM EDTA (pH 8.0). Prepare and mix 1 M Tris–HCl (pH 7.4) and 0.5 M EDTA (pH 8.0).

2.2 Amplification

1. Primers for LAMP reaction: FIP (40 pmol/μL), BIP (40 pmol/μL), F3 (5 pmol/μL), and B3 (5 pmol/μL) (*see* **Note 2**).
2. Loopamp DNA Amplification Kit (Eiken Chemical, Co., Ltd., Tokyo, Japan): 2× Reaction Mix, *Bst* DNA polymerase, and distilled water.
3. Loopamp Realtime Turbidimeter (Eiken Chemical, Co., Ltd., Tokyo, Japan) (*see* **Note 3**).
4. Loopamp reaction tube (Eiken Chemical, Co., Ltd., Tokyo, Japan).
5. Sterilized tip with filter (Axygen, Inc., Union City, CA, USA).

2.3 Agarose Gel Electrophoresis and Detection

1. Agarose-RE (Nacalai Tesque, Inc., Kyoto, Japan).
2. 1× Tris-acetate–EDTA (TAE) buffer (prepared as described elsewhere).
3. Loading buffer containing bromophenol blue.
4. DNA molecular weight marker.
5. Ethidium bromide in H_2O.
6. UV Transilluminator (FAS-IV, Nippon Genetics Co., Ltd., Tokyo, Japan).

3 Methods

3.1 Design and Preparation of Primers

3.1.1 General Primers for LAMP

Appropriate primers for amplifying the target gene are essential for detection. Four specific primers (FIP, BIP, F3, and B3) recognizing six specific sequences are designed basically with the software Primer Explorer (Fujitsu Systems East Ltd., Tokyo, Japan). Primers, at least FIP and BIP, should be prepared at HPLC grade or better. Design several sets of primers and screen for the best set. The appropriate amplification temperature (60–65 °C) and appropriate time (30–90 min) with the primers must be examined and determined beforehand (*see* **Note 4**).

3.1.2 Primers for Allele-Specific LAMP Reaction

Design four specific primers using the software Primer Explorer, as mentioned above. Redesign B2 so that the mutation will be on the 3′ end of B2, which will be the 5′ end of the BIP primer. Change the second nucleotide from the 5′ end of BIP as an additional mismatch for specificity (*see* **Note 5**). Design two specific BIP primers for wild type or mutant (*see* **Note 6**) (Fig. 1).

3.2 Extraction of DNA

1. Collect individual mosquitoes in 1.5 mL microtubes one by one.
2. Add 100 μL of Buffer A and homogenize with a pellet mixer.

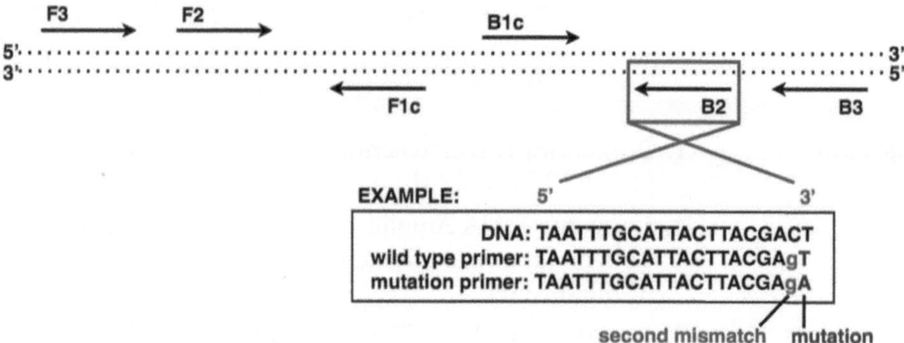

Fig. 1 Primer design for allele-specific detection. B2 is redesigned so that the mutation will be at the 3′ end (nucleotide with *blue upper-case letter*). The second nucleotide from the mutation needs to be changed as an additional mismatch for specificity (nucleotide with *red lower-case letter*) (color figure online)

3. Incubate homogenate at 70 °C for 30 min.
4. Add 22.4 μL of 5 M KOAc and mix by vortexing.
5. Chill for 30 min on ice.
6. Centrifuge at 20,000 × *g* for 15 min at 4 °C.
7. Take 90 μL of DNA-containing supernatant and transfer to a new microtube.
8. Add 45 μL of isopropanol and mix by vortexing.
9. Centrifuge at 20,000 × *g* for 20 min at 4 °C.
10. Discard the supernatant using a pipette.
11. Rinse the DNA pellet with 70 % ethanol.
12. Centrifuge at 20,000 × *g* for 5 min at 4 °C.
13. Collect the DNA pellet by carefully discarding the supernatant using a pipette.
14. Dry the DNA pellet at room temperature.
15. Dissolve each DNA pellet in 20–30 μL of TE buffer. 2–4 μg of DNA can be collected from each mosquito.

3.3 LAMP Amplification

Perform LAMP reaction in accordance with the manufacturer's instructions (Eiken Chemical Co., Ltd.). Use filter-tip-attached pipettes in all procedures (*see* **Note 7**). Keep on ice during preparation of the reaction mixture (*see* **Note 8**).

1. Amplifications are performed in a reaction mixture of 12.5 μL. Mix 1 μL of extracted DNA (approximately 50–200 ng), 2.5 pmol of each F3 and B3 primer, 20 pmol of each FIP and BIP primer, 6.25 μL of 2× Reaction Mix, and distilled water in Loopamp Reaction Tubes on ice. 2× Reaction Mix and distilled water are enclosed in the Loopamp DNA Amplification Kit. Add 0.5 μL of *Bst* DNA polymerase, enclosed in the kit, and mix gently using a pipette.

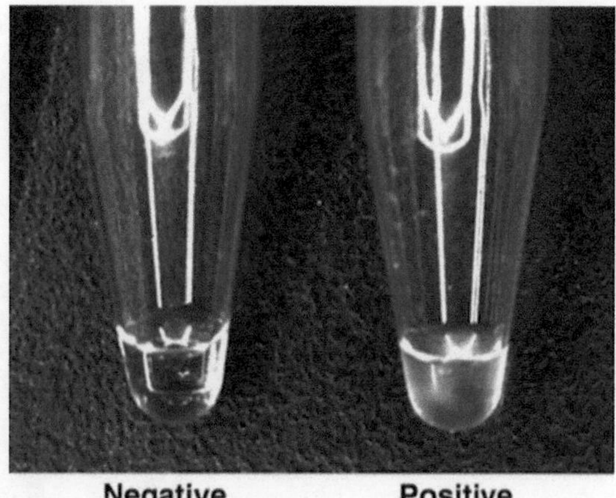

Fig. 2 White turbidity indicates the presence of amplified DNA and is observed in positive samples but not in negative ones

2. Immediately after mixing the polymerase, incubate the reaction mixture at 60–65 °C for 60–90 min using Loopamp Realtime Turbidimeter (*see* **Note 3**) and monitor the turbidity.
3. Terminate the reaction by heating at 95 °C for 2 min.

3.4 Detection

3.4.1 Monitoring Turbidity

Amplification of the DNA in the LAMP reaction causes turbidity due to the accumulation of magnesium pyrophosphate, a by-product of the amplification reaction. The turbidity can be monitored using Loopamp Realtime Turbidimeter (Eiken Chemical Co., Ltd., Tokyo, Japan).

3.4.2 Visual Detection

Amplification of the target sequence can be observed as white turbidity (magnesium pyrophosphate, a by-product of the amplification reaction) by the naked eye (Fig. 2).

3.4.3 Electrophoresis

1. Add an appropriate amount of loading buffer to the amplified product and analyze by electrophoresis on a 2 % agarose gel in 1× TAE buffer at 100 V.
2. Stain the gel with ethidium bromide and examine under UV light (Fig. 3).

4 Notes

1. Buffer A must be prepared freshly every time before use. Prepare and store a higher concentration of each solution. Mix and adjust to the final concentration with water when needed.
2. FIP and BIP primers should be purified to HPLC grade or better.

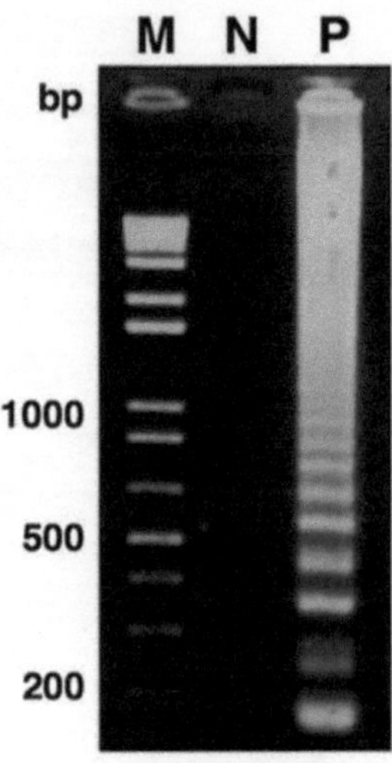

Fig. 3 Detection of reaction products by LAMP using electrophoresis. Amplification of the target sequence is detected as a ladder pattern

3. The incubator with a hot bonnet (temperature accuracy within ±0.5 °C) is a substitute.
4. Choose the primer set that maximizes the turbidity, as a result of the amplification, in a shorter time.
5. An additional mismatched nucleotide increases the specificity by stopping the reaction, which may be overcome with a single mismatch.
6. Screen appropriate primer sets by the LAMP reaction to detect mutation in the plasmids. Prepare plasmids inserted with the wild-type or the mutant gene of the target. Confirm the specificity by amplifying the target sequence with the designed primer sets.
7. A false result can be caused by a low level of DNA contamination because of extra-high sensitivity of LAMP. Take great care with work using a pipette.
8. A 96-well aluminum rack can be used to keep the reaction mixture cool.

Acknowledgments

This study was supported in part by Grants-in-Aid for Scientific Research from the Japanese Ministry of Education, Science, Sports, Culture and Technology to H.K., and the Funding Program for the Next Generation of World-Leading Researchers (NEXT Program) to H.K. H.A. is a research fellow of the Japan Society for the Promotion of Science.

References

1. World Health Organization (2011) Vector control. World malaria report, Chapter 4. World Health Organization, Geneva
2. World Health Organization (2012) Launch of the global plan for insecticide resistance management in malaria vectors. Event report. World Health Organization, Geneva
3. Martinez-Torres D, Chandre F, Williamson MS, Darriet F, Bergé JB, Devonshire AL, Guillet P, Pasteur N, Pauron D (1998) Molecular characterization of pyrethroid knockdown resistance (*kdr*) in the major malaria vector *Anopheles gambiae s.s.* Insect Mol Biol 7:179–184
4. Ranson H, Jensen B, Vulule JM, Wang X, Hemingway J, Collins FH (2000) Identification of a point mutation in the voltage-gated sodium channel gene of Kenyan *Anopheles gambiae* associated with resistance to DDT and pyrethroids. Insect Mol Biol 9:491–497
5. Notomi T, Okayama H, Masubuchi H, Yonekawa T, Watanabe K, Amino N, Hase T (2000) Loop-mediated isothermal amplification of DNA. Nucleic Acids Res 28:E63
6. Aonuma H, Suzuki M, Iseki H, Perera N, Nelson B, Igarashi I, Yagi T, Kanuka H, Fukumoto S (2008) Rapid identification of *Plasmodium*-carrying mosquitoes using loop-mediated isothermal amplification. Biochem Biophys Res Commun 376:671–676
7. Aonuma H, Yoshimura A, Perera N, Shinzawa N, Bando H, Oshiro S, Nelson B, Fukumoto S, Kanuka H (2009) Loop-mediated isothermal amplification applied to filarial parasites detection in the mosquito vectors: *Dirofilaria immitis* as a study model. Parasit Vectors 2:15
8. Perera N, Aonuma H, Yoshimura A, Teramoto T, Iseki H, Nelson B, Igarashi I, Yagi T, Fukumoto S, Kanuka H (2009) Rapid identification of virus-carrying mosquitoes using reverse transcription-loop-mediated isothermal amplification. J Virol Methods 156: 32–36
9. Iwasaki M, Yonekawa T, Otsuka K, Suzuki W, Nagamine K, Hase T, Tatsumi K, Horigome T, Notomi T, Kanda H (2003) Validation of the loop-mediated isothermal amplification method for single nucleotide polymorphism genotyping with whole blood. Genome Lett 2:119–126
10. Fukuta S, Mizukami Y, Ishida A, Kanbe M (2006) Development of loop-mediated isothermal amplification (LAMP)-based SNP markers for shelf-life in melon (*Cucumis melo* L.). J Appl Genet 47:303–308
11. Ikeda S, Takabe K, Inagaki M, Funakoshi N, Suzuki K (2007) Detection of gene point mutation in paraffin sections using *in situ* loop-mediated isothermal amplification. Pathol Int 57:594–599
12. Badolo A, Okado K, Guelbeogo WM, Aonuma H, Bando H, Fukumoto S, Sagnon N, Kanuka H (2012) Development of an allele-specific, loop-mediated, isothermal amplification method (AS-LAMP) to detect the L1014F *kdr-w* mutation in *Anopheles gambiae s. l.* Malar J 11:227

References

Part IV

Signal Amplification Approach

Chapter 11

DNA Detection by Cascade Enzymatic Signal Amplification

Bingjie Zou, Yinjiao Ma, and Guohua Zhou

Abstract

Although many approaches based on template replication were developed and applied in DNA detection, cross-contamination from amplicons is always a vexing problem. Thus, signal amplification is preferable for DNA detection due to its low risk of cross-contamination from amplicons. Here, we proposed a cascade enzymatic signal amplification (termed as CESA) by coupling *Afu* flap endonuclease with nicking endonuclease, including three steps: invasive signal amplification, flap ligation, and nicking endonuclease signal amplification. Because of the advantages of low risk of contamination, no sequence requirement of target DNA, and the universal reaction conditions for any target detection, CESA has a great potential in clinical diagnosis.

Key words Signal amplification, *Afu* flap endonuclease, Nicking endonuclease, Molecular beacon

1 Introduction

Detection of specific sequences in target DNA at an ultra low level is very critical to clinical diagnosis, many approaches based on various principles were thus developed [1–4]. However, those methods based on template replication will increase the risk of cross-contamination from amplicons, so false-positive results frequently occur. Amplification of target DNA-specific signals instead of target DNA itself is an effective way to solve this problem [5–7]. Here, we proposed a cascade enzymatic signal amplification (termed as CESA) [8] based on *Afu* flap endonuclease [9] coupled with nicking endonuclease [10] for ultrasensitive DNA detection.

The principle of CESA is illustrated in Fig. 1, including three steps: invasive signal amplification [11–13], flap ligation, and nicking endonuclease signal amplification (NESA) [14, 15]. Firstly, design a downstream probe (dp) and an upstream probe (up) specific to the target of interest (T1). If the upstream probe–target DNA duplex overlaps the downstream probe–target DNA duplex by 1 bp, 5′ flap of the downstream probe is cleaved by *Afu* flap

Dmitry M. Kolpashchikov and Yulia V. Gerasimova (eds.), *Nucleic Acid Detection: Methods and Protocols*,
Methods in Molecular Biology, vol. 1039, DOI 10.1007/978-1-62703-535-4_11, © Springer Science+Business Media New York 2013

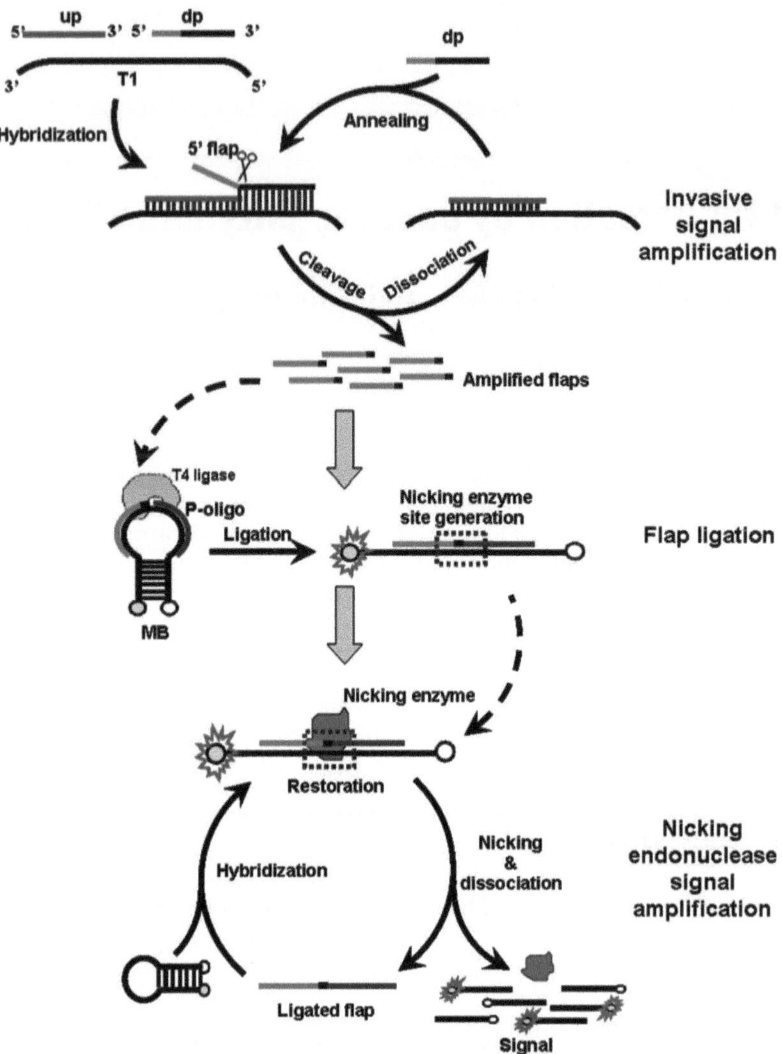

Fig. 1 The principle of cascade enzymatic signal amplification (CESA) for target DNA detection. There are three steps for CESA: invasive signal amplification to generate amplified flaps by *Afu* endonuclease, flap ligation to form a nicking site by T4 ligase, and nicking reaction to produce amplified signals by NEase [8]

endonuclease. Because cleavage reaction is carried out at a temperature around the T_m of the downstream probe, the cleaved downstream probe can rapidly dissociate from the target; as a much higher concentration than target DNAs, the intact downstream probe predominates over the cleaved probe in annealing to the target DNAs. Flap accumulation is thus initiated by the cycle of cleavage–dissociation–annealing. At this stage, several thousands of flaps (signals) can be produced from one target molecule [13]. As no requirement of endonuclease-specific recognition sequence in a target DNA, any target can be detected. Secondly, use a molecular beacon (MB) hybridizing with a 5′ phosphorylated oligonucleotide (p-oligo) in the loop of MB to capture the amplified flaps; ligation reaction occurs when the flaps hybridize to the MB at a position adjacent to the 5′ end of the p-oligo. The ligated flap opens the stem of MB [16], forming the full recognition site of the nicking endonuclease (termed as NEase, e.g., Nb.BsmI). Thirdly, add a nicking endonuclease to nick the MB strand in an MB-ligated flap duplex. After nicking reaction at 60 °C, the cleaved MB pieces spontaneously dissociate from the ligated flap strand, and the fluorescently labeled oligonucleotides produce signals. Subsequently, the ligated flap strand will be captured by another intact MB, generating the cycle of nicking–dissociation–hybridization. At this step, ligated flaps are templates of nicking reaction, and each ligated flap leads to around 1,000 fluorescently labeled oligonucleotides. Therefore, the cascade signal amplification by the two kinds of endonucleases results in a significant increase in the sensitivity of DNA detection; theoretically, one target DNA molecule can produce at least one million cleaved MB by CESA.

CESA does not need the target of interest having a NEase recognition sequence, so any target can be detected. On the other hand, CESA is cost-effective as reagents for flap ligation and nicking amplification are universal for any DNA sequence detection; in case of a new target DNA, only two dye-free target-specific oligonucleotides (downstream and upstream probes) are needed to be synthesized before detection. Moreover, the operation of CESA is quite simple because we only need to open the tube for adding reagents at two steps: flap ligation and nicking reaction. Although CESA is not a close tube detection system, there is no risk of amplicon cross-contamination because CESA is based on "real" signal amplification; thus the application of CESA to the in situ rapid diagnosis of infectious disease out of a regular lab is much suitable.

2 Materials

1. Invasive reaction buffer (10×): 100 mM MOPS, pH 7.5, 0.5 % Tween-20, 0.5 % Nonidet P40, 75 mM $MgCl_2$. Stored at −20 °C.
2. 71 ng/μL *Afu* flap endonuclease (prepared in our lab [8]).
3. Ligation buffer (2×): 132 mM Tris–HCl, pH 7.6, 20 mM $MgCl_2$, 2 mM DTT, 2 mM ATP, 15 % PEG 6000. Stored at −20 °C (*see* **Note 1**).
4. 350 U/μL T4 ligase (TaKaRa, China).
5. Nicking buffer (10×): 1 M NaCl, 100 mM Tris–HCl, pH 7.9, 100 mM $MgCl_2$, and 10 mM DTT.
6. Nicking endonuclease: 10 U/μL Nb.BsmI (New England Biolabs, USA).

7. Oligonucleotides: The upstream probes, the downstream probes, and the p-oligo (Invitrogen company, China) and the MB (TaKaRa, China). (*see* **Note 2**)
8. MJ Opticon 2 thermal cycler with a continuous fluorescence detector.

3 Methods

3.1 Design of Upstream and Downstream Probes, MB, and P-oligo for DNA Detection

1. The downstream probe contains a 5′ flap and a target-specific region. The T_m of the target-specific region should be around 63 °C (*see* **Note 3**). The sequences of the 5′ flap are independent from the target sequences but complementary to the loop region of MB at the position adjacent to the 5′ end of the p-oligo. The 3′ end of the cleaved flap is at the position of nicking endonuclease recognition site.
2. The upstream probe is complementary to the target, except for its 3′ terminal nucleotide, which overlaps with the downstream probe's target-specific region by one base and should not be complementary to the target at the overlapping site. The most suitable base type of the 3′ terminal nucleotide is T and C, followed by A and G. The length of the invasive probe is chosen so that the T_m of a probe–target duplex is 73–78 °C or is 10–15 °C higher than that of a downstream probe.
3. The MB has a universal sequence which contains the sequence of one strand of a nicking endonuclease (Nb.BsmI) recognizing site (GCATTC) in the middle of the loop region. The loop region of the MB is complementary to the cleaved flap and the p-oligo.
4. The p-oligo is complementary to the loop region of MB at the position adjacent to the 3′ end of the cleaved flap (*see* **Note 4**). The 5′ terminal base of p-oligo is at the position of nicking endonuclease recognition site and phosphorylated, so that it can be ligated with the 3′OH of the cleaved flap to form a full recognition site of the nicking endonuclease Nb.BsmI.

3.2 Invasive Reaction

1. Add 1 μL of a DNA sample (*see* **Note 5**) to 8 μL of invasive reaction mixture containing 1 μL of 10× invasive reaction buffer, 1 μL of 1 μM upstream probe, 1 μL of 10 μM downstream probe, and 5 μL of water.
2. Incubate the mixture at 95 °C for 5 min.
3. Add 1 μL of 71 ng/μL of *Afu* flap endonuclease to the above mixture (total volume 10 μL).
4. Incubate the mixture at 63 °C for 2 h.

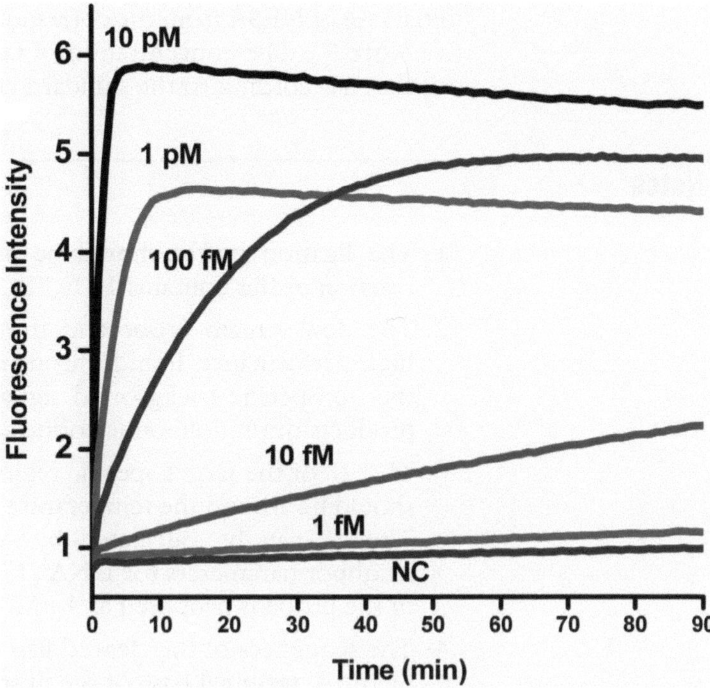

Fig. 2 Time-courses of CESA of target DNA at various concentrations (10 pM, 1 pM, 100 fM, 10 fM, 1 fM). NC curve is time-courses of CESA without target DNA [8]

3.3 Ligation Reaction

1. After invasive reaction, add 39.5 μL of ligation mixture containing 25 μL of 2× ligation buffer, 4.8 μL of 10 μM p-oligo, 4.0 μL of 10 μM MB, and 5.7 μL of water.
2. Incubate the obtained mixture at 45 °C for 5 min.
3. Add 0.5 μL of T4 ligase (17.5 U/μL) (*see* **Note 6**) at 26 °C.
4. Keep the mixture at 26 °C for 15 min.

3.4 Nicking Reaction

1. After ligation, add 50 μL of 2× nicking buffer containing 50 U of Nb.BsmI into ligated products in a volume of 100 μL.
2. Incubate at 60 °C for 90 min. Perform the readout every 1 or 2 min during incubation by using MJ Opticon 2 thermal cycler with a continuous fluorescence detector.

3.5 Data Analysis

The typical results of different concentrations of target DNA detected by CESA are shown in Fig. 2. When the signal from a sample is obviously higher than that from a negative control (without target DNA), the sample can be determined as a positive sample. Quantification of a target DNA by CESA can be achieved by performing CESA on targets with serially diluted concentrations, and a standard curve can be readily obtained by measuring the initial rate of NESA from the corresponding time-courses in CESA (*see* **Note 7**). The concentration of target DNA in sample can be calculated according to the standard curve.

4 Notes

1. The ligation buffer should be mixed carefully, because the reaction buffer contains 16 % PEG, which may not mix well.
2. The downstream probe and the MB should be purified by high-performance liquid chromatography (HPLC) to avoid the nonspecific background signal, which caused by the by-products during oligonucleotide synthesis.
3. The T_m of the target-specific region of the downstream probe should be around the temperature of invasive reaction (63 °C). The T_m can be calculated by any program using nearest-neighbor parameters for DNA [17, 18] and the concentration of the probe is proposed as 1 μM.
4. The sequences of the cleaved flap contain the base overlapped by the 3′ terminal base of the upstream probe.
5. DNA sample could be DNAs extracted from raw biological samples or the PCR products, and the amount of target DNA in a sample added in the invasive reaction should be more than 0.1 amol, which is the detection limit of CESA.
6. T4 ligase should be diluted by 1× ligation buffer to 17.5 U/μL before subjected to ligation reaction.
7. As the steady-state reaction rate of either invasive amplification or NESA is proportional to the target concentration [15, 19], reaction rates can be employed to quantify a target DNA. If the initial concentration of the target is $[C]_0$, and the cycling cleavage rate (the number of flaps produced from a target molecule per unit of time) in invasive reaction is defined as α, the initial cleavage rate of the reaction can be expressed as $d[C_{flap}]/dt = \alpha[C]_0$ [19]. In case of ultrasensitive detection, $[C]_0$ is much lower than the concentration of downstream probe, so no plateau of the invasive reaction will be reached at a given time period (usually <2 h); the final concentration of cleaved flaps can thus be expressed as $[C_{flap}] = \alpha[C]_0 T$ after the reaction time of T unit. Provided that the efficiency of ligation reaction is β which is constant at a fixed ligation condition, the concentration of ligated flaps subjected to nicking reaction should be $\beta[C_{flap}]$. According to the kinetics of NESA [15], the initial rate of NESA (V_0) can be expressed as $V_0 = \alpha'\beta[C_{flap}] = \alpha'\beta\alpha[C]_0 T$, where α' is a pseudo first-order rate constant of nicking reaction [15], V_0 is hence proportional to $[C]_0$ (the initial concentration of target). V_0 at various target concentrations ($[C]_0$)

can be measured from the corresponding time-courses in CESA, a standard curve can thus be readily obtained. The concentration of target DNA in sample can be calculated according to the standard curve.

Acknowledgment

This work was supported by the National Natural Science Foundation of China (Grant 20975113, 31200638, and 21275161).

References

1. Compton J (1991) Nucleic acid sequence-based amplification. Nature 350:91–92
2. Lizardi PM, Huang X, Zhu Z, Bray-Ward P, Thomas DC, Ward DC (1998) Mutation detection and single-molecule counting using isothermal rolling-circle amplification. Nat Genet 19:225–232
3. Mullis KB, Faloona FA (1987) Specific synthesis of DNA in vitro via a polymerase-catalyzed chain reaction. Methods Enzymol 155: 335–350
4. Notomi T, Okayama H, Masubuchi H, Yonekawa T, Watanabe K, Amino N et al (2000) Loop-mediated isothermal amplification of DNA. Nucleic Acids Res 28:E63
5. Zou B, Ma Y, Wu H, Zhou G (2012) Signal amplification by rolling circle amplification on universal flaps yielded from target-specific invasive reaction. Analyst 137:729–734
6. Schwartz JR, Sarvaiya PJ, Leiva LE, Velez MC, Singleton TC, Yu LC et al (2012) A facile, branched DNA assay to quantitatively measure glucocorticoid receptor auto-regulation in T-cell acute lymphoblastic leukemia. Chin J Cancer 31:381–391
7. Baumeister MA, Zhang N, Beas H, Brooks JR, Canchola JA, Cosenza C et al (2012) A sensitive branched DNA HIV-1 signal amplification viral load assay with single day turnaround. PLoS One 7:e33295
8. Zou B, Ma Y, Wu H, Zhou G (2011) Ultrasensitive DNA detection by cascade enzymatic signal amplification based on Afu flap endonuclease coupled with nicking endonuclease. Angew Chem Int Ed Engl 50:7395–7398
9. Kaiser MW, Lyamicheva N, Ma W, Miller C, Neri B, Fors L et al (1999) A comparison of eubacterial and archaeal structure-specific 5′-exonucleases. J Biol Chem 274: 21387–21394
10. Zhu Z, Samuelson JC, Zhou J, Dore A, Xu SY (2004) Engineering strand-specific DNA nicking enzymes from the type IIS restriction endonucleases BsaI, BsmBI, and BsmAI. J Mol Biol 337:573–583
11. Hall JG, Eis PS, Law SM, Reynaldo LP, Prudent JR, Marshall DJ et al (2000) Sensitive detection of DNA polymorphisms by the serial invasive signal amplification reaction. Proc Natl Acad Sci U S A 97:8272–8277
12. Lyamichev V, Brow MA, Dahlberg JE (1993) Structure-specific endonucleolytic cleavage of nucleic acids by eubacterial DNA polymerases. Science 260:778–783
13. Lyamichev V, Mast AL, Hall JG, Prudent JR, Kaiser MW, Takova T et al (1999) Polymorphism identification and quantitative detection of genomic DNA by invasive cleavage of oligonucleotide probes. Nat Biotechnol 17:292–296
14. Kiesling T, Cox K, Davidson EA, Dretchen K, Grater G, Hibbard S et al (2007) Sequence specific detection of DNA using nicking endonuclease signal amplification (NESA). Nucleic Acids Res 35:e117
15. Li JJ, Chu Y, Lee BY, Xie XS (2008) Enzymatic signal amplification of molecular beacons for sensitive DNA detection. Nucleic Acids Res 36:e36
16. Tang Z, Wang K, Tan W, Li J, Liu L, Guo Q et al (2003) Real-time monitoring of nucleic acid ligation in homogenous solutions using molecular beacons. Nucleic Acids Res 31:e148
17. Allawi HT, SantaLucia J Jr (1997) Thermodynamics and NMR of internal G.T mismatches in DNA. Biochemistry 36: 10581–10594
18. SantaLucia J Jr (1998) A unified view of polymer, dumbbell, and oligonucleotide DNA nearest-neighbor thermodynamics. Proc Natl Acad Sci U S A 95:1460–1465
19. Lyamichev VI, Kaiser MW, Lyamicheva NE, Vologodskii AV, Hall JG, Ma WP et al (2000) Experimental and theoretical analysis of the invasive signal amplification reaction. Biochemistry 39:9523–9532

Part V

Non-fluorescent Detection Formats

Chapter 12

Visual DNA Detection and SNP Genotyping Using Asymmetric PCR and Split DNA Enzymes

Jia Ling Neo and Mahesh Uttamchandani

Abstract

We describe a method to detect DNA sequences visually through a color change reaction using DNAzymes. We successfully applied the assay for the detection of Salmonella and Mycobacterium DNA, as well as for genotyping single base differences from within human genomic DNA samples. Our approach adopts a split probe targeting system, designed with G-rich sequences, which reassembles in the presence of target DNA, producing G-quadruplexes with catalytic activity. Asymmetric PCR is first performed to amplify the target region into single-stranded copies, with primer ratios tailored for optimum amplification. This is followed by direct addition of the visual probes, substrates, and reagents to produce a color change within 15 min should the desired target sequences be present. This approach hence offers a rapid readout, ease-of-use, and handling convenience, especially at the point-of-care.

Key words Visual DNA detection, Genotyping, Single-nucleotide polymorphism, DNAzymes, G-quadruplex, Pathogen detection

1 Introduction

In 1996, Sen and his colleagues discovered G-rich DNA sequences capable of binding hemin, through an in vitro selection procedure (SELEX) [1]. The DNA–hemin complex was found to catalyze oxidation reactions with a 250 magnitude greater than just hemin alone and has since found numerous applications [2]. X-ray diffraction data revealed that the G-rich sequences, held by π–π interactions, formed looped four-stranded structures. These structures contained two or more G-quartets stacked upon each other, interacting via Hoogsteen base pairing (Fig. 1). This G-rich complex is referred to as the G-quadruplex DNAzyme [3].

The G-quadruplex DNAzymes can be employed not only with hybridization probes for targeting nucleic acid sequences but also with aptamers for detecting various cations (Ag^{+}, Hg^{2+}, Pb^{2+}, K^{+}, Cu^{2+}, and Tb^{3+}) and ligands (thrombin, ATP, RNaseH, and cocaine) [4].

Dmitry M. Kolpashchikov and Yulia V. Gerasimova (eds.), *Nucleic Acid Detection: Methods and Protocols*,
Methods in Molecular Biology, vol. 1039, DOI 10.1007/978-1-62703-535-4_12,

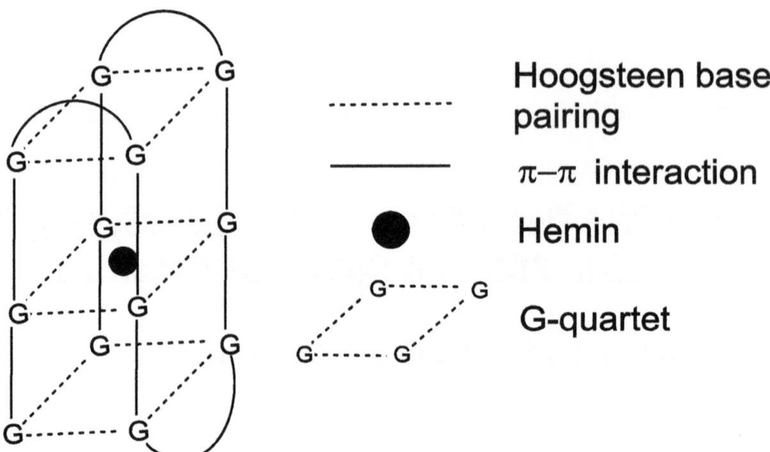

Fig. 1 G-rich sequences held by π–π interaction forming G-quadruplex structure consisting three G-quartets interacting via Hoogsteen base pairing

Various studies continue to improve the sensitivity and selectivity of target interaction, offering a promising visual assay for detection of these substances. The first application of G-quadruplexes towards visual single-nucleotide polymorphism (SNP) genotyping was demonstrated by Kolpashchikov et al. in 2008 [5]. This was achieved by designing a special set of split binary probes linked to G-rich sequences, which upon binding to adjacent regions on the target strand, brought the G-rich sequences together to form the G-quadruplex structure. This structure binds strongly to hemin, which catalyzes the reduction of hydrogen peroxide and subsequent oxidation of diaminobenzidine tetrahydrochloride (DAB) or 2,2′-azino-bis(3-ethylbenzothiazoline-6-sulphonic acid) (ABTS) producing a brown or green product, respectively. This color change is easily discernible by eye. The protocol presented here extends this concept by utilizing actual genomic DNA not synthetic single-stranded targets. This is first illustrated with the pathogenic bacteria, Salmonella and Mycobacterium [6] (Fig. 2). Salmonella is a major cause of foodborne illness throughout the world, with symptoms such as diarrhea, fever, and abdominal cramps, whereas Mycobacterium causes diseases such as tuberculosis and leprosy. In addition, this protocol was also successful in discerning single base differences of the human SNP rs2304682 from human genomic DNA with all three possible genotypes (GG, GC, and CC) [7] (Fig. 3). The selected SNP lies in the elastin microfibril interfacer 1 (Emilin 1) gene in Chromosome 2 and is reported to regulate blood pressure [8, 9]. Overall, this DNA detection protocol is rapid, cheap, requires minimal manipulation, and is suitable for on-site testing.

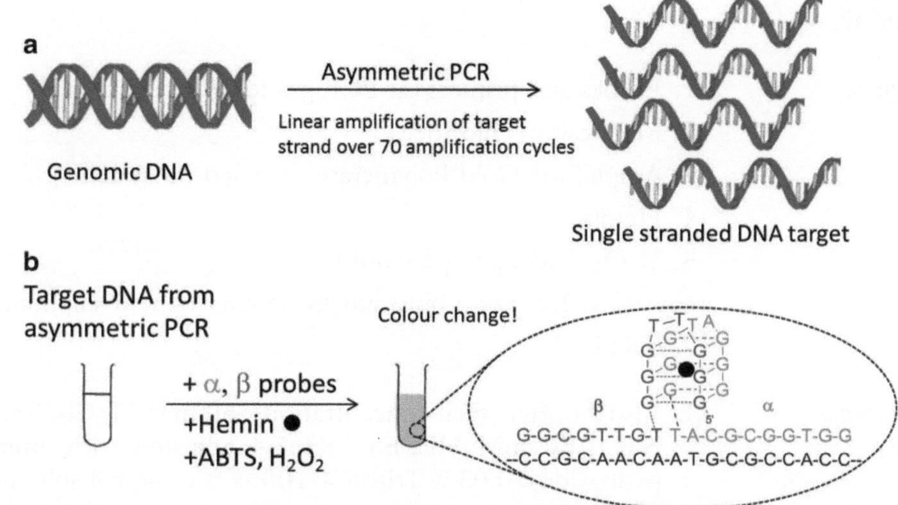

Fig. 2 Visual DNA detection assay strategy using split DNAzymes. (**a**) Asymmetric PCR to amplify a single-stranded DNA target. (**b**) Visual detection with the addition of specific α and β probes, Hemin, ABTS and H_2O_2, leading to a green product visible by eye. Sequences shown are for Mycobacterium

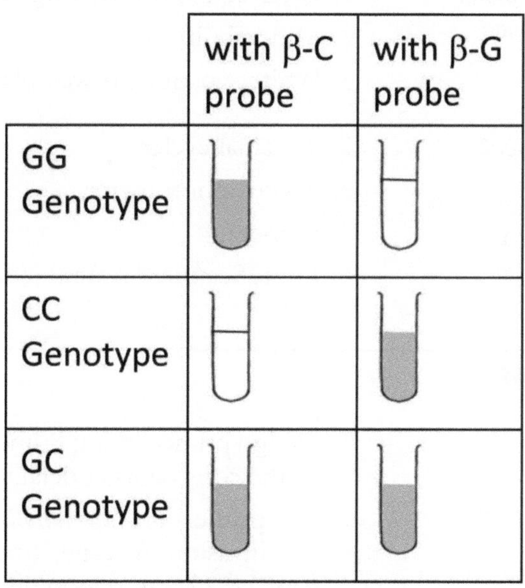

Fig. 3 The expected results for visual genotyping using our optimized protocol. The solution will appear green only when perfectly complementary with the β probe, for example individuals with the GG genotype will produce a positive green signal in the presence of β-C probe but not β-G probe and vice versa for the CC genotype. For heterozygotes, a positive signal will be produced with both β-C and β-G probes

2 Materials

2.1 Reagents

1. Probes and primers (at PCR grade).
2. Magnesium chloride.
3. AmpliTaq® Gold Polymerase (Applied Biosystems).
4. Hemin.
5. H_2O_2 (hydrogen peroxide).
6. ABTS [2,2′-azino-bis(3-ethylbenzothiazoline-6-sulphonic acid)].
7. dNTPs.

2.2 Preparation of Buffers

1. Visual buffer (final concentration): 50 mM $MgCl_2$, 53.33 mM NaCl, 50 mM HEPES (pH 7.4 adjusted with ammonium hydroxide), 0.03 % Triton X-100, 1 % dimethyl sulfoxide, and 150 mM ammonium acetate. Prepare in autoclaved MilliQ water (*see* **Note 1**).
2. 10× in-house PCR buffer: 100 mM Tris–HCl, pH 8.3, 400 mM NaCl (does not contain potassium salts, which can stabilize G-quadruplexes, resulting in false positives).

2.3 LightScanner® Accessories and Reagents

1. LCGreen® Plus Melting Dye.
2. Mineral oil.
3. White, opaque 96-well plates.

2.4 Equipment

1. Thermal cycler.
2. Spectrophotometer.
3. Camera.
4. Light Scanner® system (Idaho Technology, Inc.).

3 Methods

3.1 Design of Primers for Asymmetric PCR

1. Design primers using Primer3 (http://frodo.wi.mit.edu/), with all the constraints at default values.
2. The primers for *Salmonella enterica* flank a region of 119 bp in the invasion A gene, primers for *Mycobacterium smegmatis* flank a 72 bp region in the rRNA gene, and primers for rs2304682 flank a region of the 50 bp in the Emilin1 gene.

   The primers for these regions are listed in Table 1.

3.2 Design of Probes

1. Design binary probes, α and β, to recognize a stretch of the target sequence (*see* **Note 2**). Each probe is made up of two parts: the target recognizing sequences and the split DNAzyme sequence linked through triethylene glycol liners ($-(OCH_2CH_2)_3OPO_3-$).

Table 1
Oligonucleotide sequences used in this study

	Identity	Sequences
Positive control	peroxidase DNAzyme	5' – *GGG TAG GG*C *GGG TTG GG* – 3'
For Salmonella invA target	Left primer	5' – TCG TCA TTC CAT TAC CTA CC – 3'
	Right primer	5' – AAA CGT TGA AAA ACT GAG GA – 3'
	α probe	5' – ACC AAT ATC G-$(OCH_2CH_2)_3OPO_3$-*G GGT AGG G* – 3'
	β probe	5' – *GGG TTG GG*-$(OCH_2CH_2)_3OPO_3$-C CAG TAC G – 3'
	Synthetic 34nt	5' – CTG AAT ATC GTA CTG GCG ATA TTG GTG TTT ATG G – 3'
	Synthetic 50nt	5'– TGA TCG CAC TGA ATA TCG TAC TGG CGA TAT TGG TGT TTA TGG GGT CGT TC – 3'
	Synthetic 80nt	5' – ATC TGG TTG ATT TCC TGA TCG CAC TGA ATA TCG TAC TGG CGA TAT TGG TGT TTA TGG GGT CGT TCT ATA TTG ACA GAA TC – 3'
	Normal probe	5' – ACC AAT ATC GCC AGT ACG – 3'
For Mycobacterium rRNA gene target	Left primer	5' – GCA TCT AGT TCG TAA GAG TGT GG – 3'
	Right primer	5' – ACA TCA ATT TGT CCG CAA CA – 3'
	α probe	5' – GGT GGC GCA T-$(OCH_2CH_2)_3OPO_3$-*G GGT AGG G* – 3'
	β probe	5' – *GGG TTG GG*-$(OCH_2CH_2)_3OPO_3$-T GTT GCG G – 3'
	Synthetic 80nt	5' – TTA TCA GAA GAA AAA TTG CAC ATC AAT TTG TCC GCA ACA ATG CGC CAC CCA ACC TCA AAG GCC GGC AGC CAC ACT CTT AC – 3'
	Normal Probe	5' – GGT GGC GCA TTG TTG CGG – 3'
For Human Emilin1 SNP rs2304682 target	Left primer	5' – TCT GCT GAG GCT CTC CTG TT – 3'
	Right primer	5' – CTG CTT TGA AGT CCA CGT AGC – 3'
	α probe	5' – CTC TCC TGT TG-$(OCH_2CH_2)_3OPO_3$-*G GGT AGG G* – 3'
	β-C probe	5' – *GGG TTG GG*-$(OCH_2CH_2)_3OPO_3$-T GCT cGG T – 3'
	β-G probe	5' – *GGG TTG GG*-$(OCH_2CH_2)_3OPO_3$-T GCT gGG T – 3'
	Synthetic 50nt-c	5' – GAA GTC CAC GTA GCA CCc AGC ACA ACA GGA GAG CCT CAG CAG ATG TTT GC – 3'
	Synthetic 50nt-g	5' – GAA GTC CAC GTA GCA CCg AGC ACA ACA GGA GAG CCT CAG CAG ATG TTT GC – 3'
	Normal Probe	5' – CTG TTG TGC TcG GTG CTA CGT GGA – 3'

Shaded italics – Segments that form G-quadruplex; nt – nucleotides; lowercase – SNP site

2. Ensure that the α probe consists of the longer analyte binding arm (around 11 nucleotides) and forms the 5′ end of the probe; couple the 3′ end of this sequence to the split DNAzyme sequence 5′-GGGTAGGG-3′.
3. Design the β-probe with the shorter analyte binding arm (around 7 nucleotides) and forms the 3′ end of the probe; its 5′ end is coupled to the split DNAzyme 5′-GGGTTGGG-3′.
4. Design the Salmonella probes to target a contiguous 18 bp region of the invasion A (inA) gene and similarly, for Mycobacterium's probes, they target a segment of the 16S rRNA gene sequence.
5. For genotyping of rs2304682, design the β probe to have the SNP site centrally placed, with Tm of 26 °C. As there are two alleles present for this particular SNP, two β probes are designed to enable coverage for both the genotypes (*see* **Note 3**).
6. Check that the α and β probes form negligible hairpins and are noncomplementary to one another.

The probe sequences designed are listed in Table 1.

3.3 Determination of the Optimal Target Length Using Salmonella Synthetic Targets

1. Select three different lengths of Salmonella synthetic single-stranded targets covering (e.g., 34, 50, and 80 nt) with region of interest centrally positioned. The sequences of the synthetic targets are listed in Table 1.
2. Prepare 1 μM of the single-stranded targets in 50 μL of the 1× in-house PCR buffer (*see* **Note 4**).
3. Prepare a negative control containing 50 μL of 1× in-house PCR buffer.
4. Add 10 μL of the visual buffer to all the tubes.
5. Add 1 μM of α probe and 1 μM of the β probes only to the tubes.
6. Spin down the tubes followed by 10 min bench top incubation.
7. Add hemin to a final concentration of 125 nM to all the tubes, spin down the tubes followed by 20 min of incubation.
8. Add H_2O_2 and ABTS to all the tubes, to a final concentration each of 1 mM. Spin down and vortex gently to mix (*see* **Note 5**).
9. Compare the visual intensities of the different tubes to determine the optimal target length (Fig. 4) (*see* **Note 6**).

3.4 Asymmetric PCR

1. Optimize the asymmetric PCR by varying the primer ratios (attempted at 1:5, 1:10, and 1:20, with the larger primer concentration to amplify the desired strand), $MgCl_2$ concentration (1.5, 2.0, and 2.5 mM) and annealing temperature (T_a = 54–69 °C).
2. Prepare other reagents consisting of 10× in-house PCR buffer, 0.5 mM dNTPs, 2.5 U AmpliTaq® Gold Polymerase with

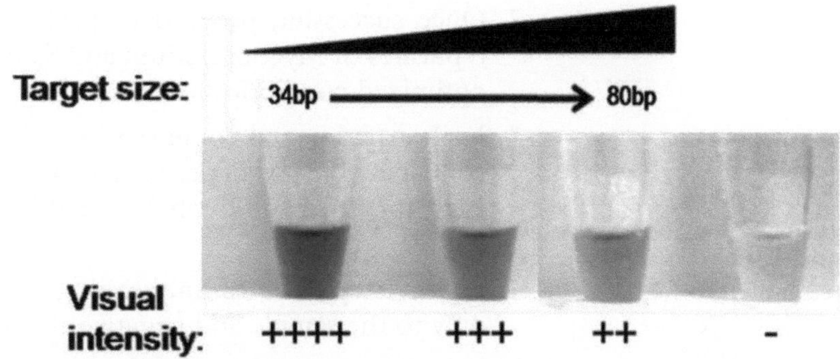

Fig. 4 The visual intensity decreases with increasing length of Salmonella synthetic single-stranded targets. Reproduced from Darius et al. [6] with permission from the Royal Society of Chemistry

Table 2
Asymmetric PCR conditions

	$MgCl_2$ (mM)	Left primer (μM)	Right primer (μM)	T_a (°C)
Salmonella	1.5	0.8	0.053	53.1
Mycobacterium	2.0	0.08	0.8	56.3
Human genomic	2.0	0.1	1.0	63.0

x ng of genomic DNA ($x=45$ for Salmonella, $x=50$ for Mycobacterium and Human genomic DNA) to a total reaction volume of 50 μL.

3. Perform thermocycling with conditions: 95 °C for 10 min followed by 70 cycles of 95 °C for 1 min, T_a for 1 min, 72 °C for 1 min (*see* **Note 7**).
4. Add 0.5 μL of the normal probes (10 μM) into 3 μL of post-PCR sample on a 96-well plate, overlaid with 20 μL of mineral oil.
5. Centrifuge at 161 × *g* for 1 min (*see* **Note 8**).
6. Load the plate into the LightScanner and select a start temperature of 55 °C and end temperature of 98 °C.
7. The optimized condition should not contain undesired product, has to the highest initial fluorescence and highest fluorescence change from beginning of melt to baseline.
8. The optimized conditions are as shown in Table 2.

3.5 Optimizing the Probe Ratio for Mycobacterium and Salmonella

1. Ensure that the α and β probes work by testing them against synthetic targets of 80 nt for Salmonella and Mycobacterium and following Subheading 3.3, **steps 2–8**. These are clean single-stranded sequences which must work to produce the desired color change, if not this could point to issues/problems with probe design.

2. Once successful, proceed to perform asymmetric PCR with replicates of Mycobacterium and Salmonella samples using the optimized conditions in Subheading 3.4.
3. Each experimental configuration consists of a positive control containing 0.3 μM of peroxidase DNAzyme and a blank control, both in 50 μL of 1× in-house PCR buffer and a post-PCR sample.
4. Follow **steps 4–8** of Subheading 3.3, adding α and β probes only to the sample and negative control (*see* **Note 9**).
5. Vary the α and β probes concentrations (between 0.2 and 1 μM) such that the visual intensity of the sample is the highest and that of the negative control remains low.
6. The optimized α and β probes concentrations are 1 μM each for Salmonella and 0.5 μM each for Mycobacterium, respectively.

3.6 Detecting for Salmonella and Mycobacterium from Genomic DNA

1. Perform asymmetric PCR of Mycobacterium, Salmonella, and human genomic (as non-target DNA) samples using the optimized conditions in Subheading 3.4.
2. Each experimental configuration consists of a positive control containing 0.3 μM of peroxidase DNAzyme and a blank control, both in 50 μL of 1× in-house PCR buffer, post-PCR samples of target Mycobacterium or Salmonella and non-target human DNA.
3. Follow **steps 4–9** of Subheading 3.3, adding α and β probes only to the post-PCR samples and negative control. The α and β probes concentrations are 1 μM each for Salmonella and 0.5 μM each for Mycobacterium.
4. Measure the absorbance readings at 419 nm using a spectrophotometer.
5. If necessary, capture the results using a camera (Fig. 5).

3.7 Optimizing the Probe Ratio for SNP Genotyping of rs2304682

1. Ensure that the α and β probes work by testing with synthetic targets of 50 nt and following Subheading 3.3, **steps 2–8**, adding β-C probe and β-G probe to Synthetic 50 nt-g and Synthetic 50 nt-c, respectively.
2. Select samples from each of the genotype group (GG, CC, and GC) that have been previously verified.
3. Perform asymmetric PCR with replicates of the samples using the optimized conditions in Subheading 3.4.
4. Each experimental configuration consists of a positive control containing 0.15 μM of peroxidase DNAzyme and a blank control, both in 50 μL of 1× in-house PCR buffer, and duplicates of the post-PCR products from each of the genotype group.

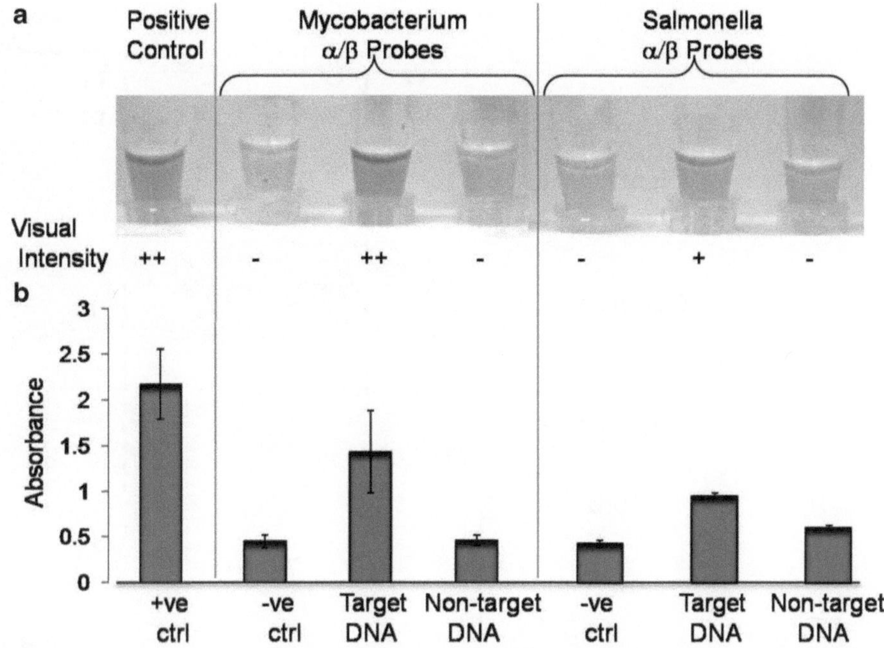

Fig. 5 (**a**) The visual readouts for detection of Salmonella and Mycobacterium DNA. –ve ctrl: contains no DNA, non-target DNA: asymmetric PCR of an Emilin1 SNP rs2304682. (**b**) Absorbance readings at 419 nm with error bars depicting results from three independent samples. Reproduced from Darius et al. [6] with permission from the Royal Society of Chemistry

5. Follow **steps 4–9** of Subheading 3.3 adding probes only to the samples and blank control. Add either β-C probe or β-G probe to each of the duplicates.
6. Vary the α and β probes concentrations (between 0.2 and 1 μM) such that the visual intensities of the homozygous samples with complementary probes are the highest compared with non-complementary probes. The visual intensities of the heterozygous sample for both probes should be similar (*see* **Note 10**) and that of the negative control remains low.
7. The optimized concentrations of the probes are 0.84 μM for α, 0.4 μM for β-G, and 0.67 μM for β-C.

3.8 Performing Visual Genotyping for SNP Genotyping of rs2304682 with Unknown Samples

1. With an unknown human genomic DNA, follow **steps 3–5** from Subheading 3.7, adding the probe concentrations as optimized previously.
2. Compare the visual intensity of the different duplicates across the sample to discern the genotype.
3. Measure the absorbance readings at 419 nm using a spectrophotometer.
4. If necessary, capture the results using a camera (Fig. 6).

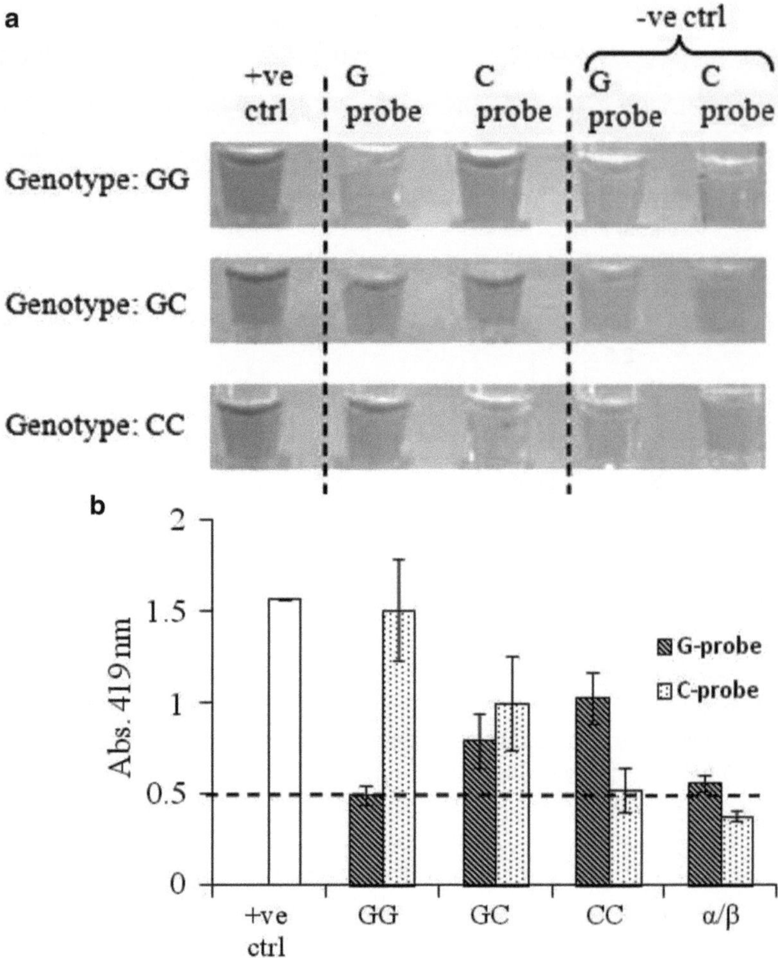

Fig. 6 (**a**) Visual genotyping of SNP rs2304682 for the three different genotypes: GG, GC, and CC. (**b**) Absorbance readings at 419 nm of five independent samples per genotype taken on a spectrophotometer. Reproduced from Neo et al. [7] with permission from the Royal Society of Chemistry

4 Notes

1. Always prepare fresh working stocks of hemin, hydrogen peroxide, and ABTS at the start of each day's experiment as these reagents are susceptible to atmospheric oxidation. The visual genotyping buffer can be stored at room temperature for a few months, mix the buffer well before use.
2. Check to ensure that there is no other polymorphism within the target sequence that is amplified, as this may compromise probe binding.
3. If the SNP of interest contains two alleles, i.e., G/C, design two β probes corresponding to β-G and β-C. Following convention, β-G probe binds to the CC genotype, β-C probe

binds to the GG genotype, and both the β probes will bind to the GC genotypes, changing the solution green.

4. 1 μM of the synthetic single-stranded targets are prepared in 50 μL of the 1× in-house PCR buffer to mimic post-PCR conditions.
5. Add drops of H_2O_2 and ABTS separately, at different sides of the Eppendorf tubes before spinning them down. This is to minimize any undesirable reactions between the two reagents.
6. Generally, the visual intensities vary inversely with the target length. The amplicon region is designed to be as short as possible without compromising the primers specificity for successful amplification of target strand during PCR.
7. Take note of the strand which should be amplified in excess and complementary to the probes designed.
8. Centrifuge in centrifuge at 161 × *g* for another minute to clear bubbles, if present.
9. The positive control contains intact peroxidase DNAzyme which is able to catalyze the reaction by itself without the need for probes. This is used to test that all reagents and buffers are working as expected.
10. The intensity of α and complementary β probes of the heterozygous genotype should yield signals with equal intensity. This will be ideal to prevent any inaccurate calling of the genotypes.

Acknowledgment

This work was supported by the Defence Science and Technology Agency, Grant No. 20080054-R6.

References

1. Li Y, Geyer CR, Sen D (1996) Recognition of anionic porphyrins by DNA aptamers. Biochemistry 35:6911–6922
2. Travascio P, Li Y, Sen D (1998) DNA-enhanced peroxidase activity of a DNA-aptamer-hemin complex. Chem Biol 5:505–517
3. Gellert M, Lipsett MN, Davies DR (1962) Helix formation by guanylic acid. Proc Natl Acad Sci U S A 48:2013–2018
4. Neo JL, Kamaladasan K, Uttamchandani M (2012) G-quadruplex based probes for visual detection and sensing. Curr Pharm Des 18:2048–2057
5. Kolpashchikov DM (2008) Split DNA enzyme for visual single nucleotide polymorphism typing. J Am Chem Soc 130:2934–2935
6. Darius AK, Ling NJ, Mahesh U (2010) Visual detection of DNA from salmonella and mycobacterium using split DNAzymes. Mol Biosyst 6:792–794
7. Neo JL, Aw KD, Uttamchandani M (2011) Visual SNP genotyping using asymmetric PCR and split DNA enzymes. Analyst 136:1569–1572
8. Zacchigna L, Vecchione C, Notte A, Cordenonsi M, Dupont S, Maretto S et al (2006) Emilin1 links TGF-beta maturation to blood pressure homeostasis. Cell 124:929–942
9. Shimodaira M, Nakayama T, Sato N, Saito K, Morita A, Sato I et al (2010) Association study of the elastin microfibril interfacer 1 (EMILIN1) gene in essential hypertension. Am J Hypertens 23:547–555

Chapter 13

Detection of Single-Stranded Nucleic Acids via Colorimetric Means, Using G-Quadruplex Probes

Herman O. Sintim and Shizuka Nakayama

Abstract

Many molecular biology experiments and clinical diagnostics rely on the detection or confirmation of specific nucleic acid sequences. Most DNA or RNA detection assays utilize radioactive or fluorescence labeling but although these tags are sensitive, safety issues (in the case of radiolabeling) or the need for expensive instrumentation (such as a fluorimeter or radiometric detector and the associated image analyzer softwares) can sometimes become impediments for some laboratories to use these detection tags. G-quadruplexes have emerged as efficient DNA-based peroxidases that can convert colorless compounds, such as 2,2′-azino-bis(3-ethylbenzothiazoline-6-sulfonic acid) (ABTS), into colored products and therefore are excellent platforms to use to detect nucleic acids via colorimetric means. Here, we describe the detection of a single-stranded DNA template using a split G-quadruplex probe that is catalytically non-proficient but becomes active after being reconstituted upon binding to the target template.

Key words Detection, Single-stranded, G-quadruplex, RNA, DNA, Peroxidase, ABTS

1 Introduction

Colorimetric detection of nucleic acids is highly desirable because the instrumentation needs for such assays are not high. Usually, a simple and cheap UV spectrophotometer is all that is required for detection. For cases where the concentration of the target is abundant or could be amplified by methods, such as PCR, it could even be possible to detect analytes by the naked eye. In the late 1990s, Sen reported that G-quadruplexes could bind to hemin and that the G-quadruplex/hemin complex had a peroxidation activity [1]. Since this important discovery, the use of G-quadruplexes to detect various analytes has increased [2–15].

We have used two guanine-rich probes, called split G-quadruplex probes [2, 12–15] to detect single-stranded nucleic acids via colorimetric means. In this approach, the two probes contain regions that can bind to a target in a juxtapose manner and also extra G-rich sequences (one probe contains three GGG units

Dmitry M. Kolpashchikov and Yulia V. Gerasimova (eds.), *Nucleic Acid Detection: Methods and Protocols*, Methods in Molecular Biology, vol. 1039, DOI 10.1007/978-1-62703-535-4_13,

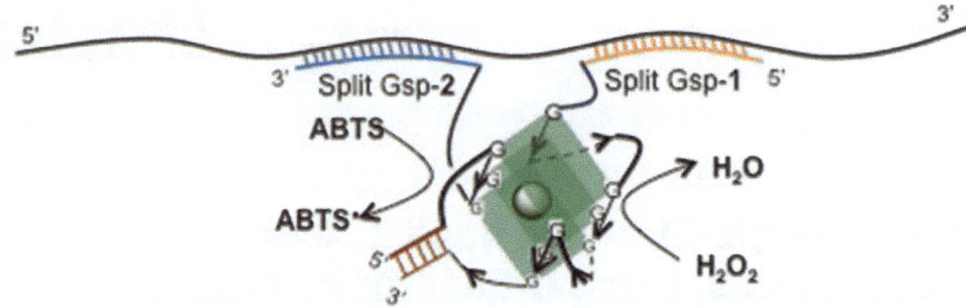

Fig. 1 Template enhanced reconstitution of a G-quadruplex DNAzyme and peroxidation reaction, using H_2O_2 as oxidant. Reprinted (adapted) with permission from Nakayama, S. and Sintim, H.O. (2009) J Am Chem Soc. 131, 10320–10333. Copyright 2009 American Chemical Society

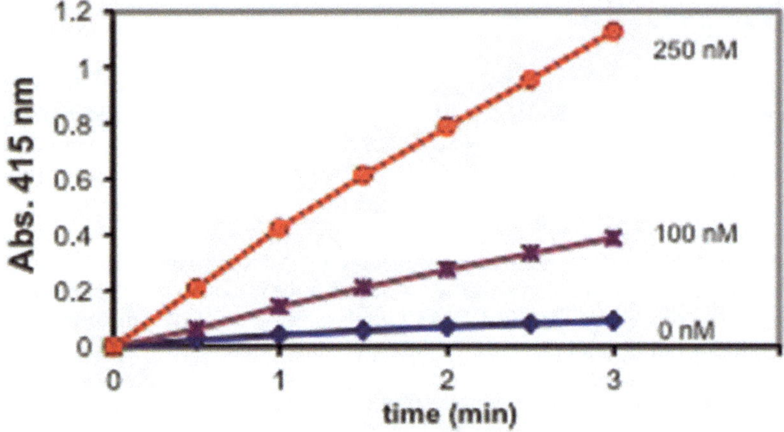

Fig. 2 Detection of the target template (0, 100, and 250 nM) using Split G-quadruplex probes and ABTS

and the other probe contains only one GGG unit; these probes are therefore called asymmetric split G-quadruplex probes). The two probes also have a limited region of complementarity, but because this region only involves 5 bp, the propensity of the two probes to associate together in the absence of a template is low [2]. Because neither probe contains four guanine repeats (a requirement for G-tetrad formation), the separate probes do not form intramolecular G-quadruplex on their own and hence each probe does not catalyze peroxidation reaction efficiently. However, in the presence of a target template that brings the two probes together, the G-quadruplex DNAzyme is efficiently reconstituted and the reconstituted enzyme can catalyze the oxidation of 2,2′-azino-bis(3-ethylbenzothiazoline-6-sulfonic acid) (ABTS) into ABTS radical cation (Fig. 1). This reaction can be followed colorimetrically, using a UV spectrophotometer and hence provides a simple means to detect nucleic acids. In this report, we detect a 70-nucleotide single-stranded DNA using the split G-quadruplex strategy (Fig. 2). In Subheading 3, we provide design principles for the

DNA probes that are used for the reconstitution of the G-quadruplex peroxidase and describe the procedure for the preparation of reaction buffers and details of the peroxidation reaction using the reconstituted split G-quadruplex probes (DNA).

2 Materials

2.1 Equipment

1. pH meter to adjust pH during the preparation of buffer solution (*see* **Note 1**).
2. UV–Vis spectrophotometer to monitor the reaction.
3. Cuvette with a 1-cm path length (total volume 100 μL).

2.2 Stock Solutions

Prepare all solutions using autoclaved distilled deionized water (ddwater). Store stock solutions at −20 °C.

1. Reaction Buffer (×5): 250 mM Tris–HCl, pH 7.9. Add 757 mg of Tris to 20 mL of ddwater in a volumetric flask. Adjust the pH to 7.9 with aqueous HCl (2.8 M) using a pH meter (*see* **Note 1**). Add more ddwater to the buffer solution to make a final volume of 25 mL. Check the pH again to ensure that the solution has a pH of 7.9.
2. ABTS stock solution: 20 mM ABTS in water. To prepare 300 μL of the stock solution, dissolve 3.3 mg of ABTS (*see* **Note 2**) in 300 μL of ddwater (*see* **Notes 3** and **4**).
3. Hemin stock solution: 25 μM hemin in 100 % DMSO. To prepare hemin stock solution, add 3 mg of hemin to 1.84 mL of DMSO to first make 2.5 mM hemin stock solution. Add 2 μL of the 2.5 mM hemin stock solution to 198 μL of DMSO to give a final concentration of 25 μM hemin stock solution (*see* **Note 5**). The final concentration of hemin in the stock solution can be verified by measuring of the solution's absorbance at 405 nm and calculated using hemin extinction coefficient ($\varepsilon_{405} = 180{,}000\ M^{-1}\ cm^{-1}$) (*see* **Notes 3**, **4**, and **6–8**).
4. H_2O_2 stock solution: 40 mM H_2O_2 in ddwater. To prepare H_2O_2 stock solution, add 1 μL of 30 % H_2O_2 to 221 μL of ddwater (*see* **Note 5**).
5. 2.5 M NH_4Cl.
6. 25 μM Split G-quadruplex probe-**1**.
7. 50 μM Split G-quadruplex probe-**2**.
8. Stock template solution: 200-fold more concentrated than in the final reaction mixture (*see* **Note 9**).

3 Methods

3.1 Design of the Split G-Quadruplex DNAzyme Probe

Sequences for template and probes used in this study are as follows:

Template: 5′-TTTTAGATTTCGTTATTTCATAAGTAGCAGCACGTAAATATTGGCGTTTGAATTCTTTTAATCTTTGTTT-3′ (70 bp).

Split G-quadruplex probe-**1** (Split Gsp-**1**): 5′-TTCAAACGCCAATATT TTTGGGTAGGGCGGGT*TTGAT*-3′.

Split G-quadruplex probe-**2** (Split Gsp-**2**): 5′-*ATCAA*TGGGTTTTTACGTGCTGCTACTTA-3′.

Split Gsp-**1** and Gsp-**2** were designed to contain segments that could hybridize with the template in a juxtapose manner (16 mer in each probe and color-coded as orange and blue, respectively). Additionally, split Gsp-**1** and Gsp-**2** contained five bases at 3′- and 5′-end, respectively (color-coded as brown), which were complimentary to each other to form a duplex. The intervening bases between the template-binding region and the duplex-forming region contained sequences that could form G-quadruplex. Although split Gsp-**1** and Gsp-**2** contain regions of complementarity, because a duplex of only 5 bp would form, which is not stable, the G-quadruplex peroxidase would reconstitute from the split probes only in the presence of a template. Intervening sequences of containing Ts were placed between the duplex-forming region of the probes and the G-quadruplex-forming regions of the probes to ensure that the two structural units were well separated to avoid the duplex from potentially interfering with the binding of peroxidase substrates to the G-quadruplex unit.

3.2 Peroxidation of ABTS Reaction, Catalyzed by Reconstituted G-Quadruplex DNAzyme

1. Prepare the reaction mixture. For this purpose, to 36.5 μL of ddwater (*see* **Note 10**), add 20 μL of Tris Buffer (×5) (*see* **Note 1**), 6 μL of 2.5 M NH_4Cl (to give a final concentration of 150 mM) (*see* **Note 11**), 10 μL of 25 μM Split G-quadruplex probe-**1** (to give a final concentration of 2.5 μM), 10 μL of 50 μM Split G-quadruplex probe-**2** (to give a final concentration of 5 μM) and 0.5 μL of stock template solution.
2. Heat the solution to 95 °C and keep at 95 °C for 5 min (*see* **Note 12**).
3. Leave the solution on the bench for 15 min to cool down to room temperature (*see* **Note 13**).
4. Put the solution on ice for 30 min.
5. Add 2 μL of Hemin stock solution (to give final hemin concentration of 0.5 μM. The final solution will also contain 2 % DMSO) (*see* **Notes 3**, **4**, and **6–8**). Leave the solution on the bench at room temperature for 30 min (*see* **Note 14**).

6. Add 10 μL of ABTS stock solution (to give a final concentration of 2 mM) at room temperature (no incubation).
7. Add 5 μL of H_2O_2 stock solution (to give a final concentration of 2 mM) and start to monitor the reaction by measuring the absorbance of the reaction mixture at 415 nm at 20 °C (*see* **Note 15**).

4 Notes

1. The reaction is sensitive to pH. It is therefore important to keep the pH of buffers used accurate and use a pH meter, and not a pH paper to determine the pH of the buffers. It is also important to calibrate the pH meter with standards.
2. ABTS purchased from Sigma was used without purification. However, because ABTS decomposes over time, it is not advisable to use an old ABTS sample.
3. It is advisable to use fresh ABTS, Hemin, H_2O_2 stock solutions each time. In our hands, using stock solutions that were prepared only a few hours before the assay gave more reproducible results than using old stocks.
4. Hemin and ABTS solutions were covered with aluminum foil to avoid light exposure.
5. Directly making about 1 mL of 25 μM of hemin would require measuring less than 1 mg of hemin. Measuring this small amount of hemin could introduce error so it is better to make a higher concentration, using more than 1 mg of hemin and then diluting the solution. In addition, it is better to use the 25 μM hemin stock solution for the assay because using the original 2.5 mM hemin stock solution would require adding only 0.02 μL of this stock solution to the peroxidation reaction (total volume of the reaction mixture is 100 μL). Measuring such a small volume would introduce errors.
6. 25 μM Hemin stock solution is in DMSO (The solubility of hemin in water is poor so it is difficult to prepare using only water). Although the assay is ultimately done in water, we discovered that using dry DMSO for the preparation of the hemin stock solution resulted in more reproducible results.
7. Using hemin at higher final concentration (>1 μM) gives higher background.
8. Since DMSO inhibits the peroxidation reaction, less than 2 % DMSO as final concentration is recommended.
9. For example, to detect a final template concentration of 250 nM, 0.5 μL of a template stock solution of 50 μM is added to the reaction mixture.

10. Presence of divalent or transition metals would inhibit the peroxidation reaction. Therefore water used should be deionized and distilled.
11. Potassium cation also works, but in general, ammonium cations are more superior than potassium [16]. Sodium gave reduced product and Li^+ does not work.
12. Heating is important to melt secondary structures in both template and probes before hybridization of probes to template.
13. In our experiments, the room temperature was 20 °C.
14. Incubation of all the oligonucleotides with hemin for 30 min allows forming a complex between G-quadruplex and hemin.
15. Temperature was controlled on a peltier cell.

Acknowledgment

This work was supported by the University of Maryland, College Park and Camille Dreyfus fellowship to H.O.S.

References

1. Travascio P, Li YF, Sen D (1998) DNA-enhanced peroxidase activity of a DNA aptamer-hemin complex. Chem Biol 5:505–517
2. Nakayama S, Sintim HO (2009) Colorimetric split G-quadruplex probes for nucleic acid sensing: improving reconstituted DNAzyme's catalytic efficiency via probe remodeling. J Am Chem Soc 131:10320–10333
3. Teller C, Shimron S, Willner I (2009) Aptamer-DNAzyme hairpins for amplified biosensing. Anal Chem 81:9114–9119
4. Zhao X-H, Kong R-M, Zhang X-B, Meng H-M, Liu W-N, Tan W et al (2011) Graphene-DNAzyme based biosensor for amplified fluorescence "turn-on" detection of Pb(2+) with a high selectivity. Anal Chem 83:5062–5066
5. Poon LCH, Methot SP, Morabi-Pazooki W, Pio F, Bennet AJ et al (2011) Guanine-rich RNAs and DNAs that bind heme robustly catalyze oxygen transfer reactions. J Am Chem Soc 133:1877–1884
6. Neo JL, Kamaladasan K, Uttamchandani M (2012) G-quadruplex based probes for visual detection and sensing. Curr Pharm Des 18:2048–2057
7. Yang X, Fang C, Mei H, Chang T, Cao Z, Shangguan D (2011) Characterization of G-quadruplex/hemin peroxidase: substrate specificity and inactivation kinetics. Chemistry 17:14475–14484
8. Bo H, Wang C, Gao Q, Qi H, Zhang C (2013) Selective, colorimetric assay of glucose in urine using G-quadruplex-based DNAzymes and 10-acetyl-3,7-dihydroxy phenoxazine. Talanta 108:131–135
9. Zhou X-H, Kong D-M, Shen H-X (2010) Ag^+ and cysteine quantitation based on G-quadruplex-hemin DNAzymes disruption by Ag^+. Anal Chem 82:789–793
10. Kolpashchikov DM (2010) Binary probes for nucleic acid analysis. Chem Rev 110:4709–4723
11. Zhang X-B, Wang Z, Xing H, Xiang Y, Lu Y (2010) Catalytic and molecular beacons for amplified detection of metal ions and organic molecules with high sensitivity. Anal Chem 82:5005–5011
12. Deng M, Zhang D, Zhou Y, Zhou X (2008) Highly effective colorimetric and visual detection of nucleic acids using an asymmetrically split peroxidase DNAzyme. J Am Chem Soc 130:13095–13102
13. Kong D-M, Wang N, Guo X-X, Shen H-X (2010) 'Turn-on' detection of Hg^{2+} ion using a peroxidase-like split G-quadruplex-hemin DNAzyme. Analyst 135:545–549

14. Xu J, Kong D-M (2012) Specific Hg^{2+} quantitation using intramolecular split G-quadruplex DNAzyme. Chin J Anal Chem 40:347–353

15. Kolpashchikov DM (2008) Split DNA enzyme for visual single nucleotide polymorphism typing. J Am Chem Soc 130:2934–2935

16. Nakayama S, Sintim HO (2012) Investigating the interactions between cations, peroxidation substrates and G-quadruplex topology in DNAzyme peroxidation reactions using statistical testing. Anal Chim Acta 747:1–6

Chapter 14

Lateral Flow Biosensors for the Detection of Nucleic Acid

Lingwen Zeng, Puchang Lie, Zhiyuan Fang, and Zhuo Xiao

Abstract

The detection of nucleic acid is of central importance for the diagnosis of genetic diseases, infectious agents, and biowarfare agents. Traditional strategies and technologies for nucleic acid detection are time-consuming and labor-intensive. Recently, isothermal strand-displacement reaction-based lateral flow biosensors have attracted a great deal of research interest because they are sensitive, simple, fast, and easy to use. Here, we describe a lateral flow biosensor based on isothermal strand-displacement polymerase reaction and gold nanoparticles for the visual detection of nucleic acid.

Key words Lateral flow biosensor, Nucleic acid, Gold nanoparticle

1 Introduction

The detection of nucleic acid is of central importance for the diagnosis of genetic diseases, infectious agents [1], and biowarfare agents [2].

Various strategies and technologies have been developed for the detection of nucleic acid such as Southern blot, agarose gel electrophoresis, and polyacrylamide gel electrophoresis. However, such conventional methods are time-consuming and labor-intensive, which have been hampering their implementation for wider and more versatile applications. The advent of new methods and technologies including real-time polymerase chain reaction (RT-PCR) [3], DNA microarrays [4], surface plasmon resonance (SPR) [5, 6], and isothermal strand-displacement polymerase reaction (ISDPR)-based fluorescence method [7, 8] or chemical method [9] offers fast and sensitive tools to detect nucleic acid. However, complicated instrumentation and highly trained personal are needed, which prevent their use in many laboratories. So, continuing efforts have been made to seek ideal tools for fast, sensitive, low-cost, and easy-to-use detection of nucleic acid.

Dmitry M. Kolpashchikov and Yulia V. Gerasimova (eds.), *Nucleic Acid Detection: Methods and Protocols*,
Methods in Molecular Biology, vol. 1039, DOI 10.1007/978-1-62703-535-4_14,

Lateral flow biosensor is a strip-type biosensor. It is commonly composed of three parts: sample pad, nitrocellulose membrane, and absorbent pad. Through the aggregation of indicator such as functionalized gold nanoparticle, the analytical result can be read from observing the color change of test zone on the strip.

Variety of lateral flow biosensors have been developed for the detection of different analytes, such as proteins [10, 11], cancer cells [12], and copper [13]. This analytical assay has four advantages: (1) It allows visual detection of analytes, requiring no expensive or complicated instruments; (2) It is easy to use. It does not involve multiple incubation and washing steps commonly performed in most other assays; (3) This assay can be performed in a short period of time; (4) Without use of expensive instruments, it is cost effective. Thus this assay is suitable for in field and point-of-care diagnosis, having great potential for the detection of genetic diseases, cancer-related mutations, and infectious agents.

Here, we describe a lateral flow biosensor based on ISDPR and gold nanoparticles for the visual detection of nucleic acid.

2 Materials

Ultrapure water (prepared by purifying deionized water to attain a resistivity of 18MΩ.cm at 25 °C) and analytical grade reagents are used to prepare all solutions. All reagents are prepared and stored at room temperature (unless indicated otherwise). All waste disposal regulations are diligently followed when disposing waste materials.

2.1 Colloidal Gold

1. 1 % $HAuCl_4$ solution. Weigh 1 g of $HAuCl_4$ to a beaker. Add water to a volume of 90 mL. Mix gently and adjust volume to 100 mL. The solution can be stored at 4 °C for 3 months.
2. 1 % Trisodium citrate solution. Weigh 1 g of Trisodium citrate to a beaker. Add water to a volume of 90 mL. Mix gently and adjust volume to 100 mL. Prepare this solution every time before use (*see* **Note 1**).
3. Bovine serum albumin (BSA).

2.2 AuNP–DNA Conjugates Preparation

1. Streptavidin (SA): Weigh 1 mg of SA in a 1.5-mL Eppendorf tube. Add ultrapure water to a volume of 1 mL. Store at 4 °C.
2. Biotinylated thiol-modified hairpin probe (probe 1, Biotin-5′-TCTTGGACACAACTAACGCCATGGCTAGACTGTGTCCAAGA-3′-thiol).
3. Target DNA. GTCTAGCCATGGCGTTAGTTGTGTCTTT.
4. Non-target DNA with random sequences (random sequence 1, GGTAGAAGGGAGGGCTAGTTGTGTCTTT; random sequence 2, GTCAGTTCCCCTTGGTTCTGTGTCTTT;

and random sequence 3, AGTGTGGATTCGCACTCCTT GTGTCTTT).

5. Rinsing buffer: 20 mM Na_3PO_4, 5 % BSA, 0.25 % Tween-20, 10 % sucrose, 0.1 % NaN_3.
6. Suspension solution: 100 g/L BSA, 0.5 g/L NaN_3, 38 mM $Na_2B_4O_7$. Weigh 10 g of BSA, 0.05 g of NaN_3, and 0.762 g of $Na_2B_4O_7$ to the beaker. Add water to a volume of 90 mL. Mix gently and adjust volume to 100 mL. Store at 4 °C after filtration by 0.22 μm Millipore membrane.
7. AuNPs solution.
8. Aging solution: 100 mM phosphate buffer, pH 7.0, 1 % sodium dodecyl sulfate, 1.5 M NaCl.

2.3 Lateral Flow Biosensor Construction

1. Nitrocellulose membrane, 25 mm × 30 cm, capillary rate: 140 s, thickness: 145 μm.
2. Sample pad buffer: 1 % Triton, 1 % BSA, 2 % glucose, 50 mM boric acid, pH 8.0.
3. Fiberglass, 16 mm in width. Fiberglass was treated by soaking in a sample pad buffer, dried, and stored in low-humidity at room temperature.
4. Absorbent paper, 17 mm in width.
5. Lateral flow dispenser (Shanghai Kinbio, Shanghai, China).
6. Paper cutter (Programmed high speed cutter, Shanghai Kinbio, Shanghai, China).

2.4 Isothermal Strand-Displacement Reaction (ISDR)

1. Polymerase Klenow fragment exo-, 5 U/μL.
2. Deoxynucleoside triphosphates (dNTPs), 2.5 mM each.
3. Reaction solution: 50 mM Tris–HCl, pH 8.0, 50 nM tag-primer (probe 2), 3 U polymerase Klenow fragment exo-, 50 μM dNTPs, 6 % DMSO, 0.1 % BSA, 1 mM DTT, 5 mM $MgCl_2$, and 1× enzyme reaction buffer.

2.5 Detection of Nucleic Acid by Lateral Flow Biosensor

1. 4× SSC: 0.6 M NaCl, 50 mM trisodium citrate, pH 7.0. Adjust to pH 7.0 with HCl.
2. Hand-held strip reader (Shanghai Kinbio, Shanghai, China).

3 Methods

Carry out all procedures at room temperature unless otherwise specified.

3.1 Preparation of AuNPs and AuNP–DNA Conjugates

1. Mix 4 mL of 1 % $HAuCl_4$ solution to 400 mL of ultrapure water in a 500 mL round bottom flask (*see* **Note 2**). Heat and stir to boil for 5 min (*see* **Note 3**).

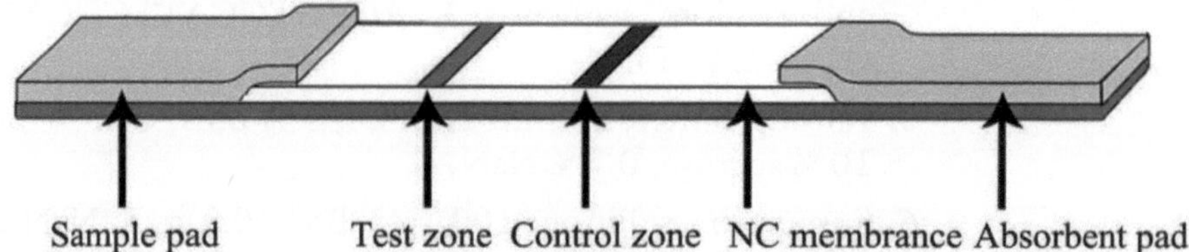

Fig. 1 Construction of lateral flow biosensor. The biosensor composed of three parts: sample pad, nitrocellulose membrane, and absorbent pad, two lines are dispensed onto the nitrocellulose membrane to form a test zone and a control zone

2. Add 6.8 mL of 1 % trisodium citrate solution rapidly to that boiling $HAuCl_4$ solution (0.01 %). After the solution turns red, boil it for 10 additional min, then cool to room temperature with gentle stirring.
3. Store the resulting AuNPs solution at 4 °C (*see* **Note 4**) and use it for preparation of AuNP–DNA conjugates.
4. To prepare AuNP–DNA conjugate, add 100 μL of 100 μM 5′-biotinated 3′-thiolated DNA hairpin probe to 500 μL of fourfold concentrated AuNPs solution, and shake the mixture gently at 4 °C for 24 h.
5. Subject 500 μL of the DNA-coated AuNPs to "aging" by adding 55 μL of aging solution. Keep the solution at 4 °C for 12 h.
6. Centrifuge the aged DNA-coated AuNPs at 13400×g for 20 min and rinse them three times with rinsing buffer to remove any unbound DNA. Resuspend the red pellet in 500 μL of rinsing buffer and store the suspension in a refrigerator at 4 °C until use.

3.2 Construction of Lateral Flow Biosensor

1. Dispense 30 μL of 100 μM 15-mer DNA probe (probe 4 GAAAGATAGAAAGAT) and 30 μL of 1 mg/mL streptavidin (SA) onto the nitrocellulose membrane simultaneously to form a test zone and a control zone, respectively, with a lateral flow dispenser (Fig. 1).
2. Dry the membrane at room temperature for 12 h.
3. Attach a strip of sample pad, nitrocellulose membrane, and absorbent pad along the long axis of an adhesive plate with an overlap of 2–3 mm, according to the layout shown in Fig. 1. Then cut the plate into 0.4-cm wide strips using a paper cutter (*see* **Note 5**).

3.3 Detection of Nucleic Acid by Lateral Flow Biosensor

1. To detect nucleic acid using the lateral flow biosensor, mix 2 μL of hairpin probe (probe 1)-conjugated AuNPs with different amounts of synthetic target DNA, non-target DNA with random sequences, extracted human genome DNA, hepatitis C

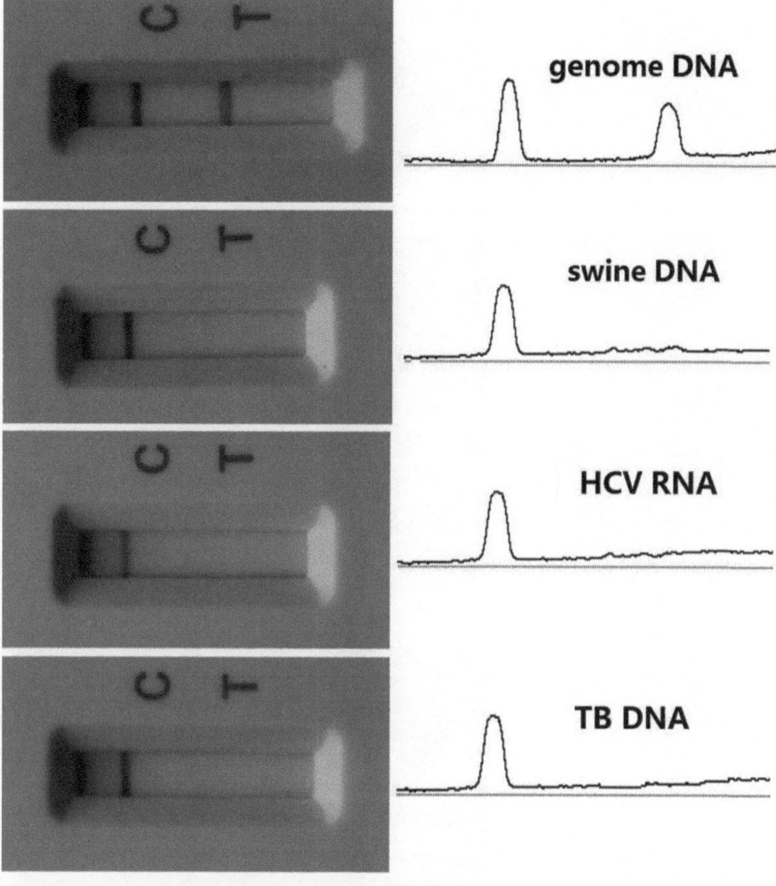

Fig. 2 Typical photo images (*left*) and corresponding intensities (*right*) of lateral flow biosensor for detection of human genome DNA, swine DNA, HCV RNA, and TB DNA were used as negative control

virus (HCV) RNA, swine DNA, or tuberculosis (TB) DNA in 20 μL reaction solution (*see* **Note 6**) (Figs. 2 and 3).

2. Incubate the mixture at 42 °C for 30 min (*see* **Note 7**).
3. Load 20 μL of ISDPR product onto the sample pad along with 20 μL of 4× SSC. Scan the biosensor using a hand-held strip reader 10 min later.

4 Notes

1. We find that it is best to prepare this fresh each time.
2. The round bottom flask must be very clean to guarantee the pureness, avoid any floating and sedimentation of the sythesized colloidal gold solution.

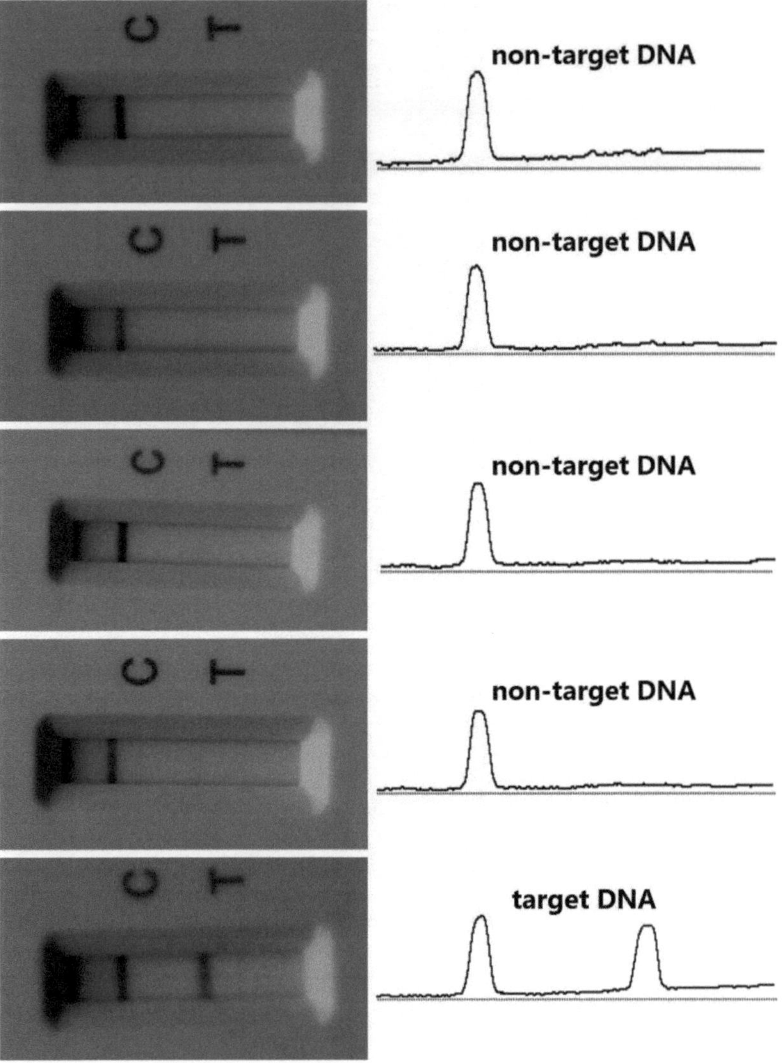

Fig. 3 Photo images (*left*) and corresponding intensities (*right*) of the lateral flow biosensor for specificity analysis with target DNA and non-target DNA (DNA with random sequences)

3. In order to make the size of sythesized colloidal gold uniform, the agitating speed is set to be 600 rpm. So the intensity field of the magnetic stirrer is.
4. We find that sythesized colloidal gold can be stored at 4 °C for at least 1 year.
5. The environment for construction of lateral flow biosensor must be strictly controlled, temperature 25–27 °C, humidity is lower than 30 %.

6. To reduce the operation error for multiple sample detection, we: prepare a mix of reaction buffer first and then subdivide into each sample.
7. We find that 2 μL of AuNP–DNA conjugate is the optimum concentration and 30 min is the optimum reaction time.

Acknowledgments

This work was financially supported by The Foundation of Special Research Program of Science and Technology Bureau of Guangzhou (No. 2008Z1-E581, No. 2009Ss-P030), Chinese Academy of Science (KSCX2-YW-R-164), and Department of Education of Guangdong Province (GXZD0901).

References

1. Sampath R, Hall TA, Massire C, Li F, Blyn LB, Eshoo MW et al (2007) Rapid identification of emerging infectious agents using PCR and electrospray ionization mass spectrometry. Ann N Y Acad Sci 1102:109–120
2. Park SJ, Taton TA, Mirkin CA (2002) Array-based electrical detection of DNA with nanoparticle probes. Science 295:1503–1506
3. Elkady A, Tanaka Y, Kurbanov F, Sugauchi F, Sugiyama M, Khan A et al (2010) Performance of two real-time RT-PCR assays for quantitation of hepatitis C virus RNA: evaluation on HCV genotypes 1–4. J Med Virol 82:1878–1888
4. Liu RH, Yang JN, Lenigk R, Bonanno J, Grodzinski P (2004) Self-contained, fully integrated biochip for sample preparation, polymerase chain reaction amplification, and DNA microarray detection. Anal Chem 76:1824–1831
5. He L, Musick MD, Nicewarner SR, Salinas FG, Benkovic SJ, Natan MJ et al (2000) Colloidal Au-enhanced surface plasmon resonance for ultrasensitive detection of DNA hybridization. J Am Chem Soc 122:9071–9077
6. Sipova H, Zhang SL, Dudley AM, Galas D, Wang K, Homola J (2010) Surface plasmon resonance biosensor for rapid label-free detection of microribonucleic acid at subfemtomole level. Anal Chem 82:10110–10115
7. Guo QP, Yang XH, Wang KM, Tan W, Li W, Tang H et al (2009) Sensitive fluorescence detection of nucleic acids based on isothermal circular strand-displacement polymerization reaction. Nucleic Acids Res 37:e30
8. Shi C, Ge Y, Gu H, Ma C (2011) Highly sensitive chemiluminescent point mutation detection by circular strand-displacement amplification reaction. Biosens Bioelectron 26: 4697–4701
9. He YQ, Zeng K, Zhang XB, Gurung AS, Baloda M, Xu H et al (2010) Ultrasensitive electrochemical detection of nucleic acid based on the isothermal strand-displacement polymerase reaction and enzyme dual amplification. Electrochem Commun 12:985–988
10. Fang ZY, Ge CC, Zhang WJ, Lie PC, Zeng LW (2011) A lateral flow biosensor for rapid detection of DNA-binding protein c-jun. Biosens Bioelectron 27:192–196
11. Chen J, Fang Z, Lie P, Zeng L (2012) Computational lateral flow biosensor for proteins and small molecules: a new class of strip logic gates. Anal Chem 84:6321–6325
12. Liu GD, Mao X, Phillips JA, Xu H, Tan WH, Zeng LW (2009) Aptamer-nanoparticle strip biosensor for sensitive detection of cancer cells. Anal Chem 81:10013–10018
13. Fang Z, Huang J, Lie P, Xiao Z, Ouyang C, Wu Q et al (2010) Lateral flow nucleic acid biosensor for Cu^{2+} detection in aqueous solution with high sensitivity and selectivity. Chem Commun 46:9043–9045

Chapter 15

Electrochemiluminescence Detection of c-Myc mRNA in Breast Cancer Cells on a Wireless Bipolar Electrode

Mei-Sheng Wu, Guang-sheng Qian, Jing-Juan Xu, and Hong-Yuan Chen

Abstract

Electrochemiluminescence (ECL) on bipolar electrode (BPE) is a sensitive, portable, and low-cost approach which has been employed to detect DNA and proteins. Here, we develop an ultrasensitive method for intracellular mRNA assay based on mRNA-mediated reporter DNA liberation and $Ru(bpy)_3^{2+}$-conjugated silica nanoparticles ($RuSi@Ru(bpy)_3^{2+}$) tag-based signal amplification.

Key words Bipolar electrode (BPE), Electrochemiluminescence (ECL), Tumor cells, mRNA, Quantum dots, Signal amplification

1 Introduction

Highly sensitive quantification of specific ribonucleic acid (RNA) transcripts from tumor cells is an essential tool for the study of gene transcription and identification of the subtypes of disease. Fluorescence [1–3], electrochemical [4–6] and electrochemiluminescence [7–9] (ECL) are the most commonly used method in mRNA expression assay. However, all current techniques for quantifying specific RNA require either calibration with known concentrations of pure mRNA or the correction of fluorescence intensity that are time-consuming and labor-intensive [10].

Among these technologies, ECL method for the detection of mRNA exhibits many merits. For example, high sensitivity, good stability, fast, and a wide linear dynamic range [11]. The electrodes can be embedded into microchannels for constructing portable and high throughput detector. Recently advances in wireless ECL biosensor on bipolar electrodes (BPEs) open up new opportunities for DNA and protein measurement in which faradic reactions occur at the two poles of BPE simultaneously [12–14]. Importantly, there is no need to connect BPE with external voltage, which simplifies the design of ECL-microfluidic chip. When a sufficient high

Dmitry M. Kolpashchikov and Yulia V. Gerasimova (eds.), *Nucleic Acid Detection: Methods and Protocols*, Methods in Molecular Biology, vol. 1039, DOI 10.1007/978-1-62703-535-4_15, © Springer Science+Business Media New York 2013

electric field is applied across the microchannel, faradic reactions occur at the two ends of BPE.

Here we present a novel RNA sensing platform based on bipolar ECL technology using MCF-7 cells as a model, where the ECL signal is significantly amplified by RuSi@Ru$(bpy)_3^{2+}$ probes. Studies demonstrate that c-Myc mRNA is overexpressed in many kinds of human cancers including breast cancer. In order to protect mRNA from degradation in ECL detection procedure, we employ in situ hybridization within living cells and measure the released reporter DNA. CdSe@ZnS-antisense DNA/reporter DNA conjugates are transfected into cells. After the hybridization of intracellular mRNA and antisense DNA, reporter DNA is released from CdSe@ZnS quantum dots and captured by a wireless ECL biosensor. The biosensor consists of a U-shape ITO BPE sensor surface, which is embedded into a microchannel. The hybridization of reporter DNA with antisense DNA/RuSi@Ru$(bpy)_3^{2+}$ tags on BPE caused the separation of the ECL probe away from electrode surface, thus resulting in a decrease of ECL signal [15].

2 Materials

All solutions are prepared using Millipore (model milli-Q) purified water and stored at 4 °C in a refrigerator. All the other chemicals are of analytical grade.

2.1 DNA Probes

1. Antisense DNA probe: 5′-NH_2-$(CH_2)_6$-CTC GGT T**GT CCT GGA TGA TGA TGT TTT** CAA CC-$(CH_2)_6$-NH_2-3′.
2. Reporter DNA probe: 5′-**AAA ACA TCA TCA TCC AGG A**-3′.

2.2 PBS Solution Components

1. PBS solution: 0.1 M Phosphate-buffered saline, pH 7.4. Weigh 13.61 g KH_2PO_4 and transfer to a 1-L volumetric flask. Make up to 1 L with water. Weigh 22.82 g $K_2HPO_4 \cdot 3H_2O$ and transfer to another 1-L volumetric flask. Add ultrapure water to a volume of 1 L. Slowly add KH_2PO_4 solution into 500 mL K_2HPO_4 solution till the solution pH of 7.4.
2. High salt PBS solution: 0.1 M PBS buffer solution, pH 7.4, 1 M NaCl, 1 mM tripropylamine (TPA). Weight 5.85 g of NaCl to a 125 mL reagent bottle and add as-prepared 100 mL of PBS buffer solution into it.

2.3 Reagents for Nonspecific Determination

1. 1×10^{-7} M Single-stranded DNA binding protein (SSB) in PBS.
2. 1×10^{-7} M DNase I in PBS.

2.4 Cell Culture

1. 96-well Cell Culture Plate.
2. CO_2 incubator.

3. High-glucose Dulbecco's Modified Eagle's Medium (DMEM): To prepare 1 L of high-glucose DMEM add 900 mL of water and a packet of DMEM into a sterile glass beaker. Make sure to rinse out the packet at least twice. Once the powder has dissolved, add 100 mL of water into it and adjust pH to 7.2 with $NaHCO_3$. Filter sterilizes with a 0.22 μm filter. Aliquot and store at −20 °C.
4. Fetal calf serum (FBS): Sera should be aliquoted and stored at −20 °C.
5. Penicillin–streptomycin solution (100× stock solution): 1 % penicillin and 1 % streptomycin. Weigh 1 g of penicillin and 1 g of streptomycin to a sterile glass beaker. Add 100 mL water into it. Store at −20 °C.
6. Sterile PBS: Dissolve 0.2 g of KCl, 0.2 g of $KHPO_4$, 8.0 g of NaCl, and 2.16 g of $Na_2HPO_4{\cdot}7H_2O$ in 1 L of water; adjust pH to about 7.2 with $NaHCO_3$. Sterilize with a 0.22 μm filter and store at 4 °C.
7. T/E solution: 0.25 % Trypsin, 0.5 mM EDTA. Store the sterile solution at 4 °C.
8. Cell culture solution: High-glucose DMEM, 10 % FBS, 100 μg/mL penicillin, 100 μg/mL streptomycin.
9. 75 % Ethanol solution.
10. A set of scissor and pipets autoclaved at 121 °C for 30 min.
11. MCF-7 cells (breast cancer cell line) and LO2 cells (normal liver cell).
12. Culture flasks with vented caps, sterile: T-75 flask.
13. Cell lysis solution: 1 % (w/v) SDS. Dissolve 1 g sodium dodecyl sulfate (SDS) in water.

2.5 CdSe@ZnS–dsDNA Conjugates

1. Solution A: 2.4 mM Na_2SO_3, 94 mg/L Se powder. Mix 0.015 g of Na_2SO_3 and 50 mL water in a 100 mL three-necked bottle. Add 0.0047 g of Se powder into the above mixture and react by heating for 2 h at 100 °C.
2. Solution B: 5 mM $CdCl_2$, 12 mM 3-Mercaptopropionic acid (MPA), pH 9.0. Dissolve 0.055 g of $CdCl_2{\cdot}2.5H_2O$ and 52 μL of MPA in 50 mL of water. Adjust pH to 9.0 using 1 M $NH_3{\cdot}H_2O$.
3. Sodium sulfide solution: 2.4 mM Na_2S in water.
4. $Zn(OOCH_2CH_3)_2$ solution: 3.2 mM $Zn(OOCH_2CH_3)_2$ in water.

2.6 RuSi@Ru(bpy)$_3^{2+}$ Probe

1. 40 mM $Ru(bpy)_3^{2+}$.
2. 10 % APTES alcoholic solution: 10 % (v/v) aminopropyl triethoxysilane (APTES) in ethanol. Mix 1 mL of APTES and 9 mL of ethanol.

3. Ru(bpy)$_3^{2+}$-NHS solution: 10 g/L Ru(bpy)$_3^{2+}$-NHS in DMSO. Dissolve 1 mg of Ru(bpy)$_3^{2+}$-NHS in 100 μL DMSO.
4. GA solution: 2.5 % (v/v) glutaraldehyde (GA) in water. Mix 1 mL of 25 % (v/v) GA and 9 mL water.

2.7 Microfluidic Channel and ITO Microelectrode

1. Indium tin oxide (ITO)-coated (thickness, ~100 nm; resistance, ~10 Ω/square) aluminosilicate glass slides. Cut the ITO glass electrode into small slices (4 cm×4 cm).
2. 2 M KOH in 2-propanol.
3. SG-2506 borosilicate glass (with 145 nm thick chrome film and 570 nm thick positive S-1805 type photoresist).
4. 0.5 % NaOH: Dissolve 0.5 g of NaOH in 100 mL of water.
5. Ammonium cerium (IV) nitrate solution: 0.2 M $Ce(NH_4)_2(NO_3)_6$ in 3.5 % (v/v) acetic acid. Mix 100 g of $Ce(NH_4)_2(NO_3)_6$, 17.5 mL of CH_3COOH, and 500 mL of H_2O in a plastic bottle.
6. HF–NH_4F solution: 1 M HF–NH_4F. Add 460.6 mL of water in a 600-mL plastic bottle; dissolve 9.23 g of ammonium fluoride into it; add 17.3 mL of concentrated nitric acid and 22.1 mL of hydrofluoric acid into the plastic bottle.
7. Sylgard 184 (including PDMS monomer and curing agent).
8. The as-prepared U-shape PDMS microchannel (arms 500 μm width, spacing 1.5 cm).
9. 1 M NaOH: Dissolve 4 g of NaOH in 100 mL of water.
10. ITO glass etching solution. Add 100 mL of water in a glass beaker; add 50 mL of concentrated hydrochloric acid gradually into the glass beaker under constant stirring until it becomes cold; then add 50 mL of concentrated nitric acid with the same procedure; and dissolve 3.24 g of $FeCl_3$ into it.
11. 5 % APTES alcoholic solution: 5 % (v/v) aminopropyl triethoxysilane (APTES) in ethanol. Mix 0.5 mL of APTES and 9.5 mL of ethanol.
12. A PDMS slice with straight microchannel (3.5 cm long, 500 μm wide, and 45 μm high).

2.8 Electrochemiluminescence Detection

1. Two driving electrodes (Pt wires).
2. 660A electrochemical workstation (CH Instruments, China).
3. ECL detection solution: Add 10 μL of TPA high salt PBS solution to make 100 mL ECL detection solution and store at 4 °C.

3 Methods

3.1 Preparation of CdSe@ZnS–dsDNA Conjugates

1. Prepare solution A and solution B (50 mL of each solution).
2. Add solution A into solution B and continue stirring at 100 °C about 2 h. Cool the reaction solution to room temperature.
3. Add 3 mL of sodium sulfide solution and 3 mL of $Zn(OOCH_2CH_3)_2$ solution to the above mixture under constant stirring at 60 °C for 5 h.
4. Separate the precipitate of quantum dots (QDs) from the solution by centrifuging, washing three times with ethanol and ultrapure water.
5. Redisperse the synthesized QDs in 5 mL of ultrapure water. Store in 4 °C. The core-shell system can be characterized by fluorescence spectra (*see* **Note 1**).
6. Centrifuge the prepared CdSe@ZnS QDs and vacuum-dry overnight.
7. Dissolve 1 mg of dried CdSe@ZnS QDs in 1 mL of PBS buffer solution. Activate the surface of CdSe@ZnS QDs by adding 50 mg of *N*-(3-dimethylaminopropyl)-*N*′-ethyl-carbodiimide hydrochloride (EDC) and 25 mg of *N*-hydroxysuccinimide (NHS) for 30 min under gentle stirring.
8. Add 50 μL of 1×10^{-7} M amine-terminated antisense DNA into the above solution under gentle mixing for 4 h.
9. Centrifuge the antisense DNA functionalized QDs, wash the pellet with PBS buffer solution.
10. Add 50 μL of 1×10^{-7} M reporter DNA and hybridize for 3 h at 37 °C to form CdSe@ZnS–dsDNA conjugates.
11. Characterize the prepared CdSe@ZnS–dsDNA conjugates by UV–vis absorption spectra.
12. Add 50 μL of 2 % BSA into CdSe@ZnS–dsDNA conjugates and incubate for 30 min to block the unoccupied site.

3.2 RuSi@Ru(bpy)$_3^{2+}$ Probe Preparation

1. Mix 1.77 mL of Triton X-100, 7.5 mL of cyclohexane, 1.8 mL of *n*-hexanol, and 340 μL of 40 mM $Ru(bpy)_3^{2+}$ in a glass beaker under constant magnetic stirring to form the water-in-oil microemulsion.
2. Add 100 μL of tetraethyl orthosilicate (TEOS) and 60 μL of NH_4OH, the reaction is allowed to continue for 24 h.
3. Add acetone to destroy the emulsion. Collect the RuSi nanoparticles (NPs) by centrifuging and washing with ethanol and water (*see* **Note 2**).
4. Treat the RuSi NPs with 10 % APTES for 2 h, then rinse twice with ethanol to remove loose bound APTES. Add 2 mL of PBS buffer solution to suspend RuSi NPs.

5. Mix 500 μL of $Ru(bpy)_3^{2+}$-NHS solution with 500 μL of RuSi NPs solution under constant stirring at room temperature for 24 h (*see* **Note 3**).
6. Collect RuSi@$Ru(bpy)_3^{2+}$ by centrifuging and washing several times with PBS buffer solution to remove excess $Ru(bpy)_3^{2+}$.
7. Redisperse the obtained nanoparticles in 500 μL of PBS buffer solution.
8. Add 2.5 % GA into RuSi@$Ru(bpy)_3^{2+}$ solution and incubate for 2 h. Wash the nanoparticles with water and redisperse in 500 μL water.

3.3 Preparation of Microfluidic Channel and ITO Microelectrode

1. Cut the ITO glass electrode into small slices (4 cm×4 cm). Immerse these ITO slices in a boiling solution of 2 M KOH in 2-propanol for 20 min, then wash with milli-Q water.
2. The electrode and microchannel micropattern (Straight line or U-shape curve) are predesigned on a mask.
3. Place the mask on a SG-2506 borosilicate glass (with 145 nm thick chrome film and 570 nm thick positive S-1805 type photoresist) and exposed to UV radiation for 2 min.
4. Immerse the glass plate into 0.5 % NaOH solution for 1 min and wash with water.
5. Immerse the glass plate into 0.2 M ammonium cerium (IV) nitrate solution. Wash the glass plate with water.
6. Etch this glass plate in a water bath with 1 M HF–NH_4F solution (40 °C, 1.1 μm/min) for 45 min. Wash the glass plate with water and dry.
7. Weigh the Sylgard 184 (including PDMS monomer and curing agent) in a 10:1 (base:curing agent) ratio. Then degassing PDMS was cast on the glass mold for 30 min in 80 °C. After cooling, the PDMS slice is peeled off from the mold.
8. Punch two holes (~1.5 mm) at the two ends of the PDMS microchannel as reservoirs.
9. Reversibly attach the as-prepared U-shape PDMS microchannel (arms 500 μm width, spacing 1.5 cm) to the ITO surface (4 cm×4 cm).
10. Introduce 1 M NaOH into the microchannel for 1 h (*see* **Note 4**). Rinse the channel with water and dry at 80 °C for 30 min.
11. Fill the channel with carbon ink and dry in thermostat at 30 °C for 30 min (*see* **Note 5**). After that, peel off the PDMS slice leaving corresponding patterns on ITO substrate (*see* **Note 6**) and incubate it for an additional 30 min.
12. Etch the ITO glass with the prepared ITO glass etching solution for 45 min at room temperature (*see* **Note 7**). Remove the

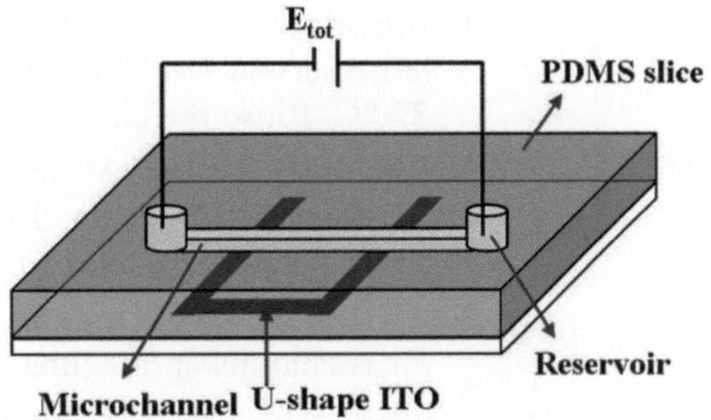

Fig. 1 Layout of the microchip with a wireless bipolar electrode

carbon ink with ethanol. Wash the obtained U-shape ITO electrode with water and dry.

13. Place the as-prepared U-shape ITO microelectrode in a plastic box and add 50 mL of 5 % APTES into it. Seal the plastic box and incubate for 12 h (*see* **Note 8**).
14. Carefully rinse the ITO microelectrode with ethanol to remove loosely bound molecules and heat it at 80 °C for 10 min.
15. Embed the prepared U-shape ITO electrode into a PDMS slice with straight microchannel (3.5 cm long, 500 μm wide, and 45 μm high). The direction of the microchannel is perpendicular to the U-shape ITO microelectrode (Fig. 1).

3.4 Trypsin–EDTA Treatment for Preparation of Cell Suspension

1. MCF-7 cells and LO2 cells are adherent cells lines that are maintained in T-75 culture flasks.
2. Remove the culture medium and rinse twice with 2 mL of sterile PBS.
3. Slowly add 1.5 mL of T/E solution and incubate at room temperature for 1 min.
4. Remove the T/E solution and add 10 mL of cell culture medium.

3.5 Cell Transfection and Lysing

1. Seed 150 μL of cell suspensions in 96-well plates and grow for 2 days. In this study, LO2 cell line is used as a negative control, whereas MCF-7 cell line is used as a positive control.
2. Culture the cells in 100 μL of fresh medium and incubate with 50 μL of CdSe@ZnS–dsDNA conjugates at 37 °C for 12 h.
3. Discard the cell culture medium and wash twice with sterile PBS.
4. Add 50 μL of cell lysis solution into cells solution and incubate for 30 min at 37 °C to obtain cell lysate samples.

3.6 Electrochemiluminescence Detection on a Wireless Bipolar Electrode

1. Introduce 2.5 % GA solution into the microchannel from the left reservoir (*see* **Note 9**) and incubate the channel for 2 h at 37 °C. Rinse the channel with PBS buffer solution for twice.
2. Introduce 1×10^{-7} M antisense DNA into the microchannel. Incubate at 37 °C for 3 h. Rinse the channel with ultrapure water for 30 s (*see* **Note 10**).
3. Add 20 μL of GA modified RuSi@Ru$(bpy)_3^{2+}$ probe into the microchannel and incubate for 2 h at 37 °C. After that, introduce 20 μL of 2 % BSA into the microchannel and incubate for 30 min (*see* **Note 11**).
4. Rinse the microchannel with PBS buffer solution twice. Introduce high salt PBS solution into microchannel and incubate for 1 h (*see* **Note 12**).
5. Place two driving electrodes (Pt wires) at the two ends of the microchannel (*see* **Note 13**) and connect the electrodes with 660A electrochemical workstation.
6. Introduce different concentration of reporter DNA into the microchannel and incubate at 37 °C for 40 min. Rinse the channel twice with PBS buffer solution.
7. Fill the channel with ECL detection solution. Place the microfluidic chip on the optical window of an MPI-E multifunctional electrochemical and chemiluminescent analytical system.
8. Apply a linear voltage from 1 to 3.5 V on a driving electrode (*see* **Note 14**) and then collect the ECL signal. Draw the calibration curve (abscissa: reporter DNA concentration, ordinate: maximum ECL intensity).
9. Introduce cell lysate samples into microchannel and incubate at 37 °C for 40 min. Rinse the channel twice with PBS buffer solution. Then repeat **steps** 7 and **8** and measure the ECL intensity of cell lysate samples. Read the concentrations on the calibration curve.
10. Mix 10 μL of 1×10^{-7} M SSB and 1 mL of 1×10^{-7} M reporter DNA and then introduce the mixture into the microchannel. Incubate for 40 min at 37 °C. Rinse the channel twice with PBS buffer solution. Repeat **steps** 7 and **8**. Compare the changes of ECL signal before and after adding SSB. This is used to estimate the nonspecific reaction between SSB and reporter DNA.
11. Mix 10 μL of 1×10^{-7} M DNase I and 1 mL of 1×10^{-7} M reporter DNA and then introduce the mixture into the microchannel. Incubate for 40 min at 37 °C. Rinse the channel twice with PBS buffer solution. Repeat **steps** 7 and **8**. Compare the changes of ECL signal before and after adding DNase I. This is employed to estimate the nucleases digestion between DNase I and reporter DNA.

12. Rinse the microchannel with ultrapure water for 30 s at room temperature to regenerate DNA biosensing surface. Then the biosensor is regenerated and can be used for the next determination of reporter DNA by repeating **steps 4–8**.

4 Notes

1. CdSe nanoparticles should be totally encapsulated by ZnS in order to decrease the cytotoxicity of nanoparticles. The emission peak of CdSe disappears and a strong new emission peak emerges at 470 nm from the fluorescence spectra indicating the formation of CdSe@ZnS core-shell nanoparticles.
2. The diameter of silica nanoparticles is mainly affected by the amount of water. After the formation of RuSi nanoparticles, it needs a high rotating speed to separate the nanoparticles due to its good solubility in water.
3. The color of the nanoparticles changes into red gradually indicating the conjugation of $Ru(bpy)_3^{2+}$ on RuSi NPs surface.
4. Prior to form carbon ink micropattern on ITO glass surface, it is very important to ensure that hydroxyl groups only be modified on the surface of the ITO layer. If the U-shape ITO electrode is immersed in 1 M NaOH solution to fabricate OH surface, the hydroxyl groups will be functionalized on both ITO electrode and glass surface, resulting in the immobilization of antisense DNA/ECL probe on both hydroxyl-functionalized ITO and glass surface. It seriously affects the precision of measurement. To solve this problem, the patterned OH surface should be prepared by attaching U-shape microchannel on ITO glass substrate. Then carbon ink is added into microchannel to form U-shape electrode micropattern and protect the patterned hydroxyl groups on ITO substrate.
5. The incubation time during thermostatic period should be precise controlled. Long incubation time will increase the solvent evaporation which decreases the carbon ink volume and forms air chamber in microchannel. As a consequence, it destroys the carbon ink micropattern on ITO surface.
6. The PDMS slice should be peeled off promptly from ITO surface, otherwise the wet carbon ink will flow along the interface between microchannel and ITO surface.
7. Do not agitate the solution while waiting for etching the uncovered ITO glass substrate. Besides, temperature plays major role in etch rate of ITO electrode. Studies demonstrate that when the temperature is 10 °C, it needs about 45 min to

etch ITO layer on glass substrate. While it increases to 30 °C, the etch process could be achieved within 13 min.

8. The plastic box should be sealed to avoid the evaporation of ethanol solvent.
9. When GA solution flows through the left arm of U-shape microelectrode, ultrapure water is dropped on the right reservoir. It forms an air chamber in microchannel which avoids the modification of GA on the right arm of U-shape microelectrode. It allows that antisense DNA and ECL probes are conjugated on one arm of the U-shape ITO microelectrode in the next experiments.
10. It is important to rinse the channel with ultrapure water to get rid of salts which ensures the conversion of stem-loop antisense DNA to linear ssDNA, facilitating the modification of RuSi@ $Ru(bpy)_3^{2+}$ probes on antisense DNA.
11. Blocking the unoccupied sites of GA is the key element to avoid false-positive signal.
12. Since DNA has a net negative charge at the phosphate groups on the DNA backbone, it may cause electrostatic interaction between DNA and RuSi@$Ru(bpy)_3^{2+}$. As a result, RuSi@ $Ru(bpy)_3^{2+}$ will bind to DNA backbone via nonspecifically electrostatic immobilization. High salt PBS solution decreases the nonspecific interaction between RuSi@$Ru(bpy)_3^{2+}$ and DNA. Importantly, it reduces the electrostatic repulsion between the negative charges on the backbone which brings RuSi@$Ru(bpy)_3^{2+}$ close to the electrode surface. After hybridization with reporter DNA, the rigid rod-like structure of dsDNA makes RuSi@$Ru(bpy)_3^{2+}$ be away from electrode leading to a decrease of ECL signal.
13. External voltage is applied on the two driving electrodes in the two reservoirs to produce electric field across the microchannel. It causes a potential difference on the two arms of U-shape ITO electrode in this microchannel (ΔE_{elec}). The value of ΔE_{elec} varies according to the electric field applied across the channel. It should exceed a critical value to driving the occurrence of faradic reactions on both arms of U-shape ITO electrode. In this experiments, $Ru(bpy)_3^{2+}$ and TPA are oxidized at the left arm of U-shape ITO electrode, while oxygen is reduced at the another arm.
14. Check the electric field direction across the channel. Since ECL probe is modified on the left arm of U-shape ITO electrode, the higher voltage should be applied on the right reservoir of the microchannel.

Acknowledgment

This work was supported by the 973 Program (Grant 2012CB932600), the National Natural Science Foundation (Grants 21025522 and 20890020) of China.

References

1. Kauraniemi P, Bärlund M, Monni O, Kallioniemi A (2001) New amplified and highly expressed genes discovered in the ERBB2 amplicon in breast cancer by cDNA microarrays. Cancer Res 61:8235–8240
2. Paige JS, Nguyen-Duc T, Song WJ, Jaffrey SR (2012) Fluorescence imaging of cellular metabolites with RNA. Science 335:1194
3. Yamada H, Maruo R, Watanabe M, Hidaka Y, Iwatani Y, Takano T (2010) Messenger RNA quantification after fluorescence-activated cell sorting using in situ hybridization. Cytometry A 77A:1032–1037
4. Yeung SS, Lee TM, Hsing IM (2008) Electrochemistry-based real-time PCR on a microchip. Anal Chem 80:363–368
5. Lee AC, Dai ZY, Chen BW, Wu H, Wang J, Zhang A et al (2008) Electrochemical branched-DNA assay for polymerase chain reaction-free detection and quantification of oncogenes in messenger RNA. Anal Chem 80:9402–9410
6. Walter A, Wu J, Flechsig GU, Haake DA, Wang J (2011) Redox cycling amplified electrochemical detection of DNA hybridization: application to pathogen E. coil bacterial RNA. Anal Chim Acta 689:29–33
7. Vandevyver C, Motmans K, Raus J (1995) Quantification of cytokine mRNA expression by RT-PCR and electrochemiluminescence. Genome Res 5:195–201
8. Baeumner AJ, Humiston MC, Montagna RA, Durst RA (2001) Detection of viable oocysts of Cryptosporidium parvum following nucleic acid sequence Based amplification. Anal Chem 73:1176–1180
9. Lo WY, Baeumner AJ (2007) RNA internal standard synthesis by nucleic acid sequence-based amplification for competitive quantitative amplification reactions. Anal Chem 79:1548–1554
10. Medley CD, Drake TJ, Tomasini JM, Rogers RJ, Tan W (2005) Simultaneous monitoring of the expression of multiple genes inside of single breast carcinoma cells. Anal Chem 77:4713–4718
11. Li J, Yang L, Luo S, Chen B, Li J, Lin H et al (2010) Polycyclic aromatic hydrocarbon detection by electrochemiluminescence generating Ag/TiO2 nanotubes. Anal Chem 82:7357–7361
12. Chow KF, Chang BY, Zaccheo BA, Mavré F, Crooks RM (2010) A sensing platform based on electrodissolution of a Ag bipolar electrode. J Am Chem Soc 132:9228–9229
13. Zhan W, Alvarez J, Crooks RM (2002) Electrochemical sensing in microfluidic systems using electrogenerated chemiluminescence as a photonic reporter of Redox reactions. J Am Chem Soc 124:13265–13270
14. Wu MS, Xu BY, Shi HW, Xu JJ, Chen HY (2011) Electrochemiluminescence analysis of folate receptors on cell membrane with on-chip bipolar electrode. Lab Chip 11:2720–2724
15. Wu MS, Qian GS, Xu JJ, Chen HY (2012) Sensitive electrochemiluminescence detection of c-Myc mRNA in breast cancer cells on a wireless bipolar electrode. Anal Chem 12:5407–5414

Chapter 16

High-Throughput Label-Free Detection of DNA Hybridization and Mismatch Discrimination Using Interferometric Reflectance Imaging Sensor

Sunmin Ahn, David S. Freedman, Xirui Zhang, and M.Selim Ünlü

Abstract

Optical label-free biosensors have demonstrated advantages over fluorescent-based detection methods by allowing accurate quantification while also being capable of measuring dynamic bimolecular interactions. A simple, high-throughput, solid-phase, and label-free technique, interferometric reflectance imaging sensor (IRIS), can quantify the mass density of DNA with pg/mm^2 sensitivity by measuring the optical path difference. We present the design of the IRIS instrument and complementary microarrays that can be used to perform a quantitative analysis of DNA microarrays. Finally, we present methods to accurately calculate the hybridization efficiency and identify SNPs from dynamic measurements, as well as supporting software algorithms needed for robust data processing.

Key words DNA microarrays, Label-free detection, Optical interferometry, Hybridization, SNP, High-throughput analysis, Quantitative analysis, Spot finding

1 Introduction

In the past two decades, advancements in biotechnology have provided us with unprecedented amounts of genomic data. In particular, DNA microarrays have been widely used in biological and medical research. The high-throughput capacity of DNA microarrays provides an enormous economic advantage in genomic analysis. Label-free biosensors are creating additional advancements in biotechnology, and interest in label-free sensing has grown exponentially [1]. This growth is driven by several advantages over traditional fluorescence-based sensing. These advantages include reduced cost and preparation time with the elimination of the labeling step, more accurate observation of molecular interaction in their native states, and the ability to provide quantitative measurements. Quantitative analysis is difficult with fluorescent techniques because of the heterogeneity in labeling efficiency and noncontrolled photobleaching. An additional advantage of

Dmitry M. Kolpashchikov and Yulia V. Gerasimova (eds.), *Nucleic Acid Detection: Methods and Protocols*, Methods in Molecular Biology, vol. 1039, DOI 10.1007/978-1-62703-535-4_16, © Springer Science+Business Media New York 2013

label-free sensing is the ability to dynamically monitor molecular interactions, consequently, providing information on binding kinetics. Real-time kinetic measurements are not available through conventional labeling approach because the detection must occur after the non-reacted reporting molecules have washed off the sensor. Thus, developing applications for high-throughput label-free detection of DNA microarrays are of great interest.

One such application is the detection of single-nucleotide polymorphisms (SNPs). Genetic variations caused by SNPs occur as often as every few hundred base pairs in genomic DNA [2, 3]. Many SNPs have been identified as important biomarkers [4, 5]. Thus, successful detection of SNPs will expedite advancements in personalized medicines as well as the diagnosis and treatment of hereditary diseases. A common method for genotyping of SNPs uses MALDI-TOF MS (matrix assisted laser desorption/ionization time of flight mass spectrometry), but samples are usually labeled with mass tags [6], and mass spectrometry is generally only capable of very limited throughput. While there have been efforts to increase throughput for SNP detection for MALDI-TOF MS-based methods, the throughput offered has yet to achieve the multiplexing capability of microarrays [7]. Very few SNP genotyping solutions offer high-throughput capabilities by using microarrays [8, 9], but these techniques require labeling the targets. In general, these systems can be very expensive, bulky, and complex, and their utilization requires complicated sample preparation. Most importantly, they do not offer quantitative expression analysis [5]. A few label-free sensors have demonstrated SNP detection, but these systems are rather difficult to implement. For example, they require enzymatic reactions on sensors, demand precise control of the environment (i.e., temperature), and present limited throughput [10–12]. Thus, there is an increasing demand for SNP genotyping and gene expressions technology that is simple, cost-effective, quantitative, and amenable to automated high-throughput analysis.

Recently, we introduced a biosensor, interferometric reflectance imaging sensor (IRIS), for high-throughput label-free detection of biomolecular interactions on a glass surface with a buried reference plane [13, 14]. The detection principle of IRIS is summarized in Fig. 1. We have demonstrated label-free detections of

Fig. 1 (continued) changes of five different probe oligonucleotides are plotted. The data shown is the average of 15 spots per probe type (T_1P: positive control, complementary to T_1, T_1N: negative control, noncomplementary to T_1, T_2PM: perfect match probe, perfectly complementary to T_2, T_2MM: mismatch probe, single point mutation to T_2PM, T_2DM: double mismatch probe, two point mutations to T_2PM). A brief overview of the experiment is given. Target 1 (T_1, 40mer ssDNA) is introduced (t = 40 min), and the wash buffer (t = 100 min) washes away weakly interacting duplexes. Only the probes that are complementary to T_1 remain hybridized as indicated by ~2.5 ng/mm^2 mass increase on T_1P. When deionized water (ddH$_2$O) is introduced, all duplexes denature, and the surface is regenerated. Target 2 (T_2, 20mer ssDNA) is introduced (t = 190 min), and probes that are complementary to T_2 (T_2PM, T_2MM, T_2DM) show similar mass increase regardless of the presences of mutations. Probes with double mutations start denaturing with the introduction of the wash buffer (t = 260 min). When a buffer with much weaker ionic strength (0.5 mM [Na$^+$]) is introduced, all probes start denaturing, and different probe types display different denaturation kinetics. This figure was adapted from ref. [14] with permission from Elsevier

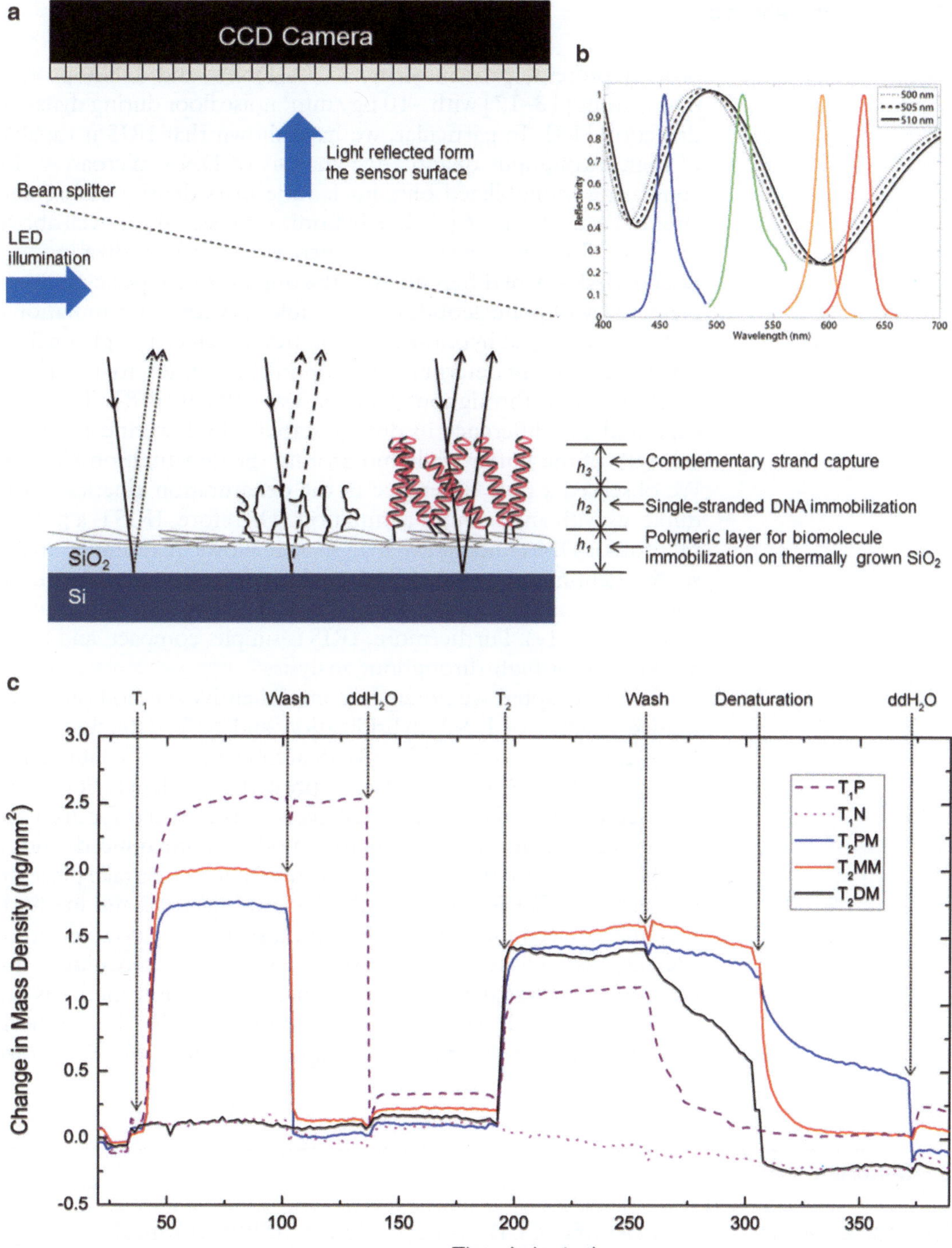

Fig. 1 Schematic representation of label-free detection of DNA microarray. (**a**) Detection principle of IRIS is shown. DNA microarray on a SiO_2–Si-layered substrate is illuminated with multiple wavelengths, and microarray images are acquired with a CCD camera. The interference of reflected light from the interfaces of the layered substrate, air–SiO_2 (or buffer–SiO_2) and SiO_2–Si, results in varied intensity as a function of wavelength, as illustrated in (**b**). A functional polymeric surface is used to specifically immobilize DNA on the surface (h_1). Accumulation of biomolecules on the surface changes the optical path length (h_2 and h_3), causing a quantifiable optical phase shift in the reflected intensity. The spectral shift of the reflectance is measured allowing one to calculate the optical path difference assuming that the refractive index of the DNA is the same as SiO_2 [13]. (**c**) A typical result obtained by dynamic detection of DNA hybridization and denaturation is shown. The surface mass density

protein–protein, protein–virus, DNA–DNA, and DNA–protein interactions [13–17] with ~10 pg/mm^2 noise floor during dynamic detection [13]. In particular, we have shown that IRIS is capable of high-throughput quantitative analysis of DNA microarrays by comparing immobilized oligonucleotide mass density before and after hybridization [16]. While hybridization signal was well above the noise floor, the level of mass increase upon hybridization was not an ideal way of differentiating the duplexes with perfectly complementary oligonucleotides from duplexes with point mutations. Fish et al. explored hybridization kinetics to detect a single nucleotide mutation, but determining an optimal condition to investigate SNP in a high-throughput manner was difficult [18]. Thus, we exploited the difference in duplex stability by lowering the ionic strength of the buffer while monitoring the denaturation kinetics. We observed a clear difference in the denaturation kinetics of the duplexes with and without a mutation. Therefore, IRIS is a powerful tool for DNA microarray analysis as it allows quantitative analysis for genomic expression and enables SNP detections without additional complex biochemical reactions or precise temperature control (Fig. 1c). Furthermore, IRIS is simple, compact, and easily automated for high-throughput analysis.

In this chapter, we present a comprehensive method on quantitative analysis of DNA hybridization and SNP detection using IRIS. Instructions on building IRIS are given, and the fabrication processes of DNA microarrays are presented in detail. For label-free detection methods, data analysis is a critical step for accurate quantification. An interferometric model is introduced and a method to analyze the biomolecular interactions that take place on the sensor surface is described. Particularly, materials and methods that are required for dynamic detection, such as the flow cell components, as well as data processing and analysis techniques are described in great detail because dynamic detection methods are more complicated than static detection methods. Finally, methods to calculate the hybridization signal and identify SNPs are presented.

2 Materials

2.1 Interferometric Reflectance Imaging Sensor

2.1.1 Instrument

1. Scientific CCD camera. Retiga-2000R (QImaging, Corp., Surrey, BC, Canada).
2. LED illumination source. ACULED VHL, ACL01-MC-RGYB (Excelitas Technologies, Pfaffenhofen, Germany) (*see* **Note 1**).
3. Microscope objective. 2× Plan Achromatic, 0.06 NA.
4. Beam splitter. Pellicle coated for the visible spectrum.

5. Lenses. Achromatic doublets for the visible spectrum.
6. Diffuser. N-BK7 Ground Glass Diffuser, 220 Grit.
7. Iris diaphragm.
8. Stage. 1/2″ XYZ translation stage.
9. Mechanical components for optical setup. Cage system and lens holders.
10. Data acquisition card.
11. Computer. Acquisition and processing PC with Firewire card and Windows XP operating system.

2.1.2 Substrate

1. Silicon substrate. 100 mm double side polished with 500 nm of thermally grown oxide (Silicon Valley Microelectronics, Inc., Santa Clara, CA, USA).
2. Positive photoresist.
3. Spinner.
4. Mask aligner.
5. Hot plate.
6. Developer.
7. Plasma asher.
8. Dicer.
9. Buffered oxide etch (BOE). 6:1 volume ratio of 40 % NH_4F in water to 49 % HF in water.
10. Photomask. 5-in. green soda-lime substrate with iron oxide.
11. Acetone.

2.2 Surface Chemistry

1. Copoly (DMA-NAS-MAPS) (Lucidant Polymers, Sunnyvale, CA, USA) (*see* **Note 2**).
2. Ammonium sulfate solution: aqueous solution of $(NH_4)_2SO_4$ at 40 % saturation level. To prepare a stock solution, add 242 g of ammonium sulfate to 1 L of deionized water.
3. Acetone and methanol.
4. Plastic Petri dish.
5. Plasma asher.
6. Vacuum drying oven.

2.3 DNA Microarray Preparation and Hybridization

Prepare spotting buffer and hybridization buffer in 2× concentration. Dilute the buffers to 1× concentration for spotting and hybridization. For buffers, use deionized water filtered with Barnstead Nanopure Diamond water purification system with 18.2 MΩ resistivity setting and a 0.2 μm particle filter.

1. Spotting buffer (2×): 300 mM sodium phosphate buffer, pH 8.5. Prepare 300 mM Na_2HPO_4 and adjust the pH to 8.5

with 300 mM NaH_2PO_4. The concentration of the final spotting buffer (1×) is 150 mM. Store at room temperature.

2. MES stock buffer (12×): 1.22 M MES, 0.89 M NaCl. To prepare 100 mL of 12× MES stock buffer, add 6.46 g of MES hydrate, 19.3 g of MES sodium salt to 80 mL of deionized water and mix. Adjust pH to 6.5–6.7 with HCl and/or NaOH and bring the volume up to 100 mL. Store at 4 °C.
3. Hybridization buffer (2×): 200 mM MES, 2 M NaCl, 40 mM EDTA. To prepare 50 mL of hybridization buffer (2×), mix 8.3 mL of MES stock buffer (12×), 17.7 mL of 5 M NaCl, 4.0 mL of 0.5 M EDTA, and 20 mL of deionized water. The content of hybridization buffer (1×) is 100 mM MES, 1 M NaCl, 20 mM EDTA. Store at 4 °C.
4. Wash buffer: SSPE solution (6×), 0.01 % Tween-20. Store at room temperature.
5. Blocking solution: 50 mM ethanolamine, 50 mM Tris–HCl, pH 8.5, 150 mM NaCl. Store at room temperature.
6. Microarray spotter.
7. Multipurpose rotating shaker.

2.4 Flow Components for Dynamic Detection

Here, we describe the custom made flow cell, as shown in Fig. 2, and all other necessary components for executing a real-time binding experiment.

1. Flow cell.
   (a) Laser grade window (BK7) with less than 5 arc minutes wedge and custom anti-reflection coatings (CVI Melles Griot, Albuquerque, NM) (*see* **Note 3**).
   (b) Gasket: 250 μm thick silicone rubber sheet (*see* **Note 4**).
   (c) The chip holder (*see* **Note 5**).
   (d) Two polyoxymethylene (POM) blocks (*see* **Note 6**).
   (e) Four thumb screws.
2. Millex-GS filter unit with 0.22 μm pore size.
3. Peristaltic pump.
4. Platinum cured silicone tubing (*see* **Note 7**).
5. Lens cleaning tissue.
6. Ethanol and isopropanol.

2.5 Software

Label-free biosensors require new methods to process and analyze raw data. Physical artifacts such as dirt and salt, which are difficult to detect by fluorescence-based sensors, are easily detected by label-free biosensors, and they can skew the result. This makes the software used with IRIS a critical component of the entire system. A custom application written in the MATLAB computing

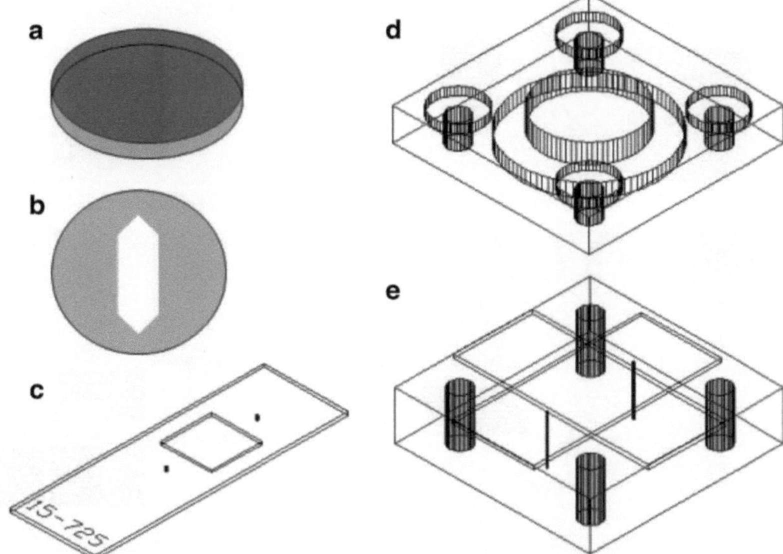

Fig. 2 Five different components of the flow cell for IRIS are presented. (**a**) BK-7 window with anti-reflection coating on both sides. (**b**) Silicone rubber sheet gasket. (**c**) A custom designed chip holder to keep the IRIS chip in place. (**d**) The top block of the flow cell. It has indents for the windows and thru holes for the screws. (**e**) The bottom block of the flow cell. It has threaded holes for the screws to secure the assembly, and two small thru holes for tubing. The bottom block also has two different layers of indents crossing at a 90° angle. The top indent was designed to fit a chip holder which will place the IRIS chip at the same location for different experiments. The bottom indent was designed to fit a resistive heater to control the temperature of the IRIS chip. Using a resistive heater is optional. In this protocol, we do not present temperature-dependant dynamic detection methods

environment was created to acquire, process, and analyze label-free IRIS images.

1. MATLAB (MathWorks, Natick, MA, USA) (*see* **Note 8**).
2. Curve fitting tool box for MATLAB (MathWorks, Natick, MA, USA) (*see* **Note 9**).

3 Methods

3.1 Interferometric Reflectance Imaging Sensor

Here we describe the optical setup of IRIS. The schematic representation is shown in Fig. 3a.

1. Create an optical path with the illuminating LEDs by randomizing the light with a diffuser and then collect the light with a lens. Ensure the diaphragm before the lens is maximized.

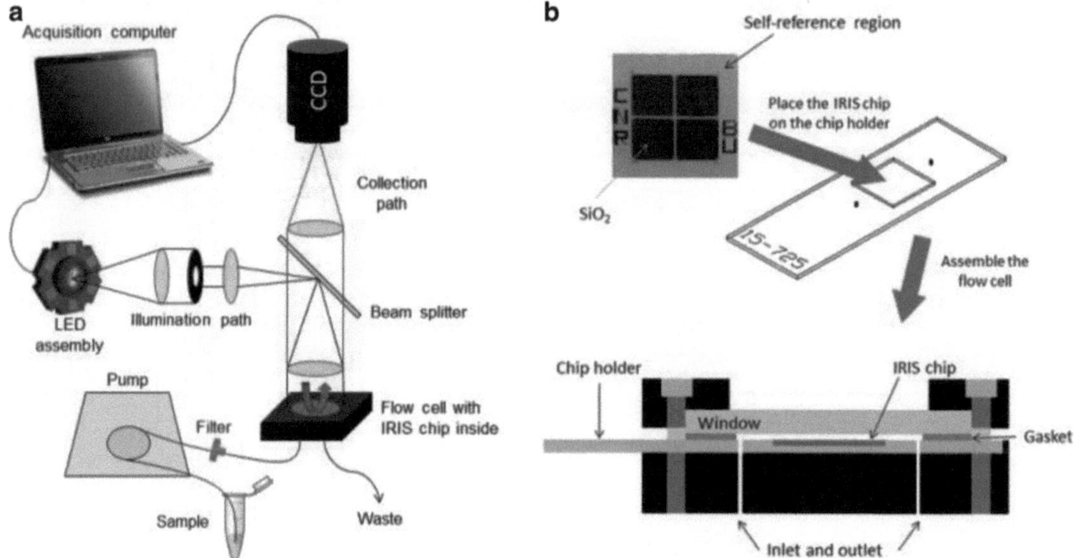

Fig. 3 The experimental setup of IRIS for dynamic detection. (**a**) The optical setup for IRIS and the related flow components are shown. (**b**) Flow cell assembly is presented

2. With a beam splitter, focus the light beam so that it is directed down to the sample stage and the reflected off the IRIS sensor surface.
3. Use a microscope objective to focus the light beam onto the sensor surface.
4. Collect reflected light from the sample after it has been passed through the objective and then transmitted through the beam splitter.
5. Use a tube lens after the beam splitter to focus the sample image onto a CCD sensor.
6. Iterate through the multiple LEDs via a USB DAQ.

3.2 Oligonucleotide Design

1. Design the oligonucleotide probe to be perfectly complementary to a region of the investigating target while minimizing secondary structure formation (*see* **Note 10**).
2. For detecting SNPs, design the oligonucleotide probe to avoid that the mutation is at a position close to the 5′ or 3′ termini of the probe.
3. Introduce an NH_2-modification at the 5′ end of the probes for immobilization on the copoly (DMA-NAS-MAPS) functionalized surface.

The sequences of oligonucleotides used for SNP detection in this work are listed in Table 1.

Table 1
Oligonucleotide sequences used for SNP detection

Name	Sequence
Target	5′-TGC AGA CGA CCA GCG GAA AT-3′
PM	5′-NH_2 ATT TCC GCT GGT CGT CTG CA-3′
MM	5′-NH_2 ATT TCC GCT $\underline{C}$GT CGT CTG CA-3′
DM	5′-NH_2 ATT TCC $\underline{C}$CT GGT C$\underline{C}$T CTG CA-3′

The underlined nucleotide is the mutation

3.3 DNA Microarray Fabrication

3.3.1 IRIS Substrate Fabrication

1. Apply S1813 photoresist to a silicon/silicon dioxide wafer by spinning at 2,000 rpm to obtain an approximately 2 μm photo-patternable surface layer. Perform a 90 s softbake at 90 °C on a hot plate.
2. Transfer self-reference region pattern with positive hard-contact UV lithography using the MA6 mask aligner with a 30 s exposure.
3. Develop the photoresist in developer for approximately 40 s.
4. Hard bake the photoresist at 120 ° C for 10 min on a hot plate.
5. Etch to the self-reference region using BOE for 7–8 min to expose the silicon surface. The BOE etch rate is approximately 80 nm per minute for thermally grown oxide.
6. Strip photoresist with acetone and sonication (5 min). Rinse in deionized water and then dry. Strip remaining photoresist with oxygen plasma at 300 sccm and 300 W for 2 min.
7. Protect wafer with photoresist, but only perform soft and hard bakes, no exposure or development.
8. Dice wafer into individual dies (chips) using dicing saw. Typical dimensions of chips are 15 mm × 15 mm.
9. Strip photoresist with acetone and sonication (10 min). Rinse in deionized water and then dry. Strip remaining photoresist with oxygen plasma at 300 sccm and 500 W for 5 min. Patterned IRIS chips can be stored indefinitely in a clean environment until use.

3.3.2 Surface Functionalization

1. Clean patterned IRIS chips by sonicating in acetone for 5 min, rinsing in methanol, and then rinsing in deionized water. Dry chips with nitrogen gas.
2. Plasma ash the clean IRIS chips with oxygen plasma at 300 sccm and 500 W for 10 min.
3. Prepare 1 % (w/v) copoly (DMA-NAS-MAPS) in a mixture of water and 40 % ammonium sulfate solution (1:1).

4. Place the IRIS chips in the 1 % (w/v) copoly (DMA-NAS-MAPS) solution in a plastic Petri dish and incubate at room temperature for 30 min with shaking (*see* **Note 11**).
5. Rinse the IRIS chips with deionized water, and then wash three times in deionized water for 3 min each on a shaker.
6. Dry the IRIS chips with argon gas, and dry the chips in a vacuum oven at 80 °C for 15 min.
7. Use the chips immediately for microarray printing (*see* **Note 12**).

3.3.3 Oligonucleotide Spotting and Immobilization

DNA microarrays are fabricated by spotting oligonucleotides on the functionalized IRIS chips with a robotic arrayer (*see* **Note 13**).

1. Design a microarray pattern with at least three replicate spots per oligonucleotide sequence.
2. Set the humidity inside the spotter to 55 % and the temperature of the plate to 20 °C.
3. Spot the oligonucleotide probes at 25 μM concentration in the spotting buffer (1×) on copoly (DMA-NAS-MAPS) functionalized IRIS chips.
4. Upon completion of spotting, keep the DNA microarrays at 70–75 % relative humidity overnight at room temperature.

3.3.4 Washing and Blocking

1. Wash the IRIS chips in the wash buffer four times for 10 min each in a petri dish on a rotating shaker.
2. Rinse the IRIS chips briefly with deionized water.
3. Incubate the slides with the blocking buffer at room temperature for 1 h.
4. Rinse the chips with deionized water.
5. Dry the chips with argon gas.

3.4 Dynamic Label-Free Detection of DNA Hybridization and Denaturation

3.4.1 Mirror Image Acquisition

A mirror image is acquired to correct for any spatial nonuniformity of illumination on the surface.

1. Place a silicon mirror (15 mm × 15 mm) in the chip holder and assemble the flow cell as shown in Fig. 3b.
2. Connect all tubing including the filter and place the flow cell on the stage of IRIS for image acquisition, as shown in Fig. 3a.
3. Flow hybridization buffer (1×) through the flow cell.
4. Open acquisition software and turn-on camera.
5. Place a silicon mirror in the field-of-view, then check focus.
6. Set the number of images to be averaged for every illumination source (*see* **Note 14**).
7. Select the maximum exposure time to be used on the camera and ensure that the camera is not being saturated for every illumination source (*see* **Note 15**).

8. Acquire a mirror scan to be used for correcting any spatial nonuniformity.
9. After acquiring the mirror image, disassemble the flow cell and rinse with deionized water and dry.
10. Wash the tubing by flowing 70 % ethanol and then drying (*see* **Note 16**).
11. Clean the window by rinsing with water then carefully wiping with isopropanol and lens cleaning tissue.

3.4.2 Dynamic Image Acquisition for Detecting Hybridization and SNP

1. Place the IRIS chip in the chip holder and repeat **steps 1–6** described in Subheading 3.4.1. Mirror Image Acquisition.
2. Set the exposure time to be the same as the mirror scan.
3. Set data acquisition software to continuously acquire images.
4. Acquire images for 30 min while hybridization buffer (1×) is flowing to obtain a baseline signal.
5. Introduce DNA targets in hybridization buffer (1×).
6. After hybridization reaches equilibrium, flow diluted hybridization buffer with [Na^+] = 50 mM as a wash buffer for 20 min. Use deionized water to dilute the hybridization buffer to reduce the [Na^+] concentration.
7. Introduce diluted hybridization buffer with [Na^+] at a concentration of less than 1 mM. Such ionic strength will start denaturing duplexes at room temperature (*see* **Note 17**).
8. To denature all duplexes and regenerate the microarray, wash it with deionized water.
9. After the experiment is completed, disassemble the flow cell and rinse the flow cell and gasket with deionized water and dry.
10. Clean the window by rinsing with water and then carefully wiping with isopropanol and lens cleaning tissue.
11. Discard the tubing and filter.

3.5 Data Processing

By collecting multispectral data from IRIS chips, it is possible to individually fit each pixel to a reflection curve. This is done by using a nonlinear fitting algorithm available in MATLAB to fit the data to an interferometric model.

1. Data is normalized to the nonuniform spatial illumination as captured by the mirror scan. This is done by dividing the chip image file by the mirror scan for every spectral image collected.
2. To account for temporal variations in illumination power, the data processing software automatically identifies self-reference regions in every image and then divides the intensity measurements by this value, for all wavelengths, to normalize the image [19].

3. Each pixel is then fit to the reflection function that is approximated with the following equation:

$$R = |r|^2 = \frac{r_1^2 + r_2^2 + 2r_1r_2\cos(2\phi)}{r_1^2r_2^2 + 2r_1r_2\cos(2\phi)}, \tag{1}$$

where r_1 and r_2 are the Fresnel reflection coefficients of the air–SiO_2 (or buffer–SiO_2) and Si–SiO_2 interfaces, respectively. The reflection coefficients can be calculated by:

$$r_1 = \frac{n_{ox} - n_1}{n_{ox} + n_1} \text{ and } r_2 = \frac{n_{Si} - n_{ox}}{n_{Si} + n_{ox}}, \tag{2}$$

where n_1, n_{ox}, and n_{Si} are the refractive indices of air (or buffer solution), SiO_2, and Si, respectively. The optical path difference is described by the phase difference, ϕ, from Eq. 1, which is given by:

$$\phi = \frac{2\pi d}{\lambda} n_{ox} \cos\theta \tag{3}$$

Here, d is the thickness of the layer (SiO_2 or SiO_2 plus the biomolecule layer), n_{ox} is the refractive index of SiO_2, λ is the wavelength of the incident light, and θ is the angle of incidence. The system uses low angles of illumination and collection, therefore the angle of incidence can be assumed to be near zero. The thickness, d, is determined by minimizing the error when solving Eqs. 1 and 3 with constants calculated from Eq. 2. Accumulation of biomaterial is measured as an increase in the thickness of this layer.

4. The processing software then generates output files that contain an image of calculated thickness from the previous step as well as fitting error for every pixel.

3.6 Data Analysis

3.6.1 Spot Finding/Tracking

Oligonucleotide spots on the surface must be located before they can be analyzed. During the analysis stage, users can manually locate the spots and select the areas used to determine the signal, but manual spot finding may introduce error in the calculation due to user bias. Automating the spot finding algorithm eliminates these effects. Zoiray Technologies developed a software that allows automatic spot finding from a single image. While the software is robust and easy to use, it does not allow automatic spot finding from a stack of images, hence analyzing data by finding spots one image at a time becomes extremely laborious. To avoid this lengthy process, one could identify individual spots from the first image and analyze the same areas for the rest of the images acquired during a single experiment. However, mechanical movements of a microarray during a dynamic, or real-time, data acquisition caused by the flow system or ambient vibrations can result in fluctuations in the spatial locations of individual spots with respect to the imaging camera. Thus, post-processing software is needed to account for micro-motion of the sensor.

There are two possible strategies to track spots from a stack of images acquired during a real-time experiment. First, image registration techniques can be used. They allow the analysis of the same areas identified from the first image throughout the rest of the experiment by aligning the subsequent images to the first one. Alignment marks on the sensor surface are used to register the images by rotation and lateral translation. While the images can be automatically aligned with 1 pixel resolution, we found that registration performs better if the user manually steps through the translation in sub-pixel increments. Second, an automated spot finding algorithm that can locate every spot for each image throughout the experiment can be used. The automated spot finding algorithm has been shown to be robust to these micro-motions and enables real-time experiments to be analyzed.

While both techniques are capable of tracking individual spots in an automated manner, the spot finding algorithm provides greater accuracy than previous image registration techniques (*see* **Note 18**). Thus, data analysis protocol using the spot finding algorithm is given below [20]. The following analysis is done for the specified field-of-view which can be selected as the entire image if desired. Automatically identified self-reference regions are ignored during analysis.

1. Identify the background by looking at a histogram and identifying the maximum histogram level. This is assumed to be the background level.
2. Specify a level greater than the background. Then perform binary thresholding against the background. The default level is set up as 1 nm.
3. Identify all spots and then eliminate the spots that do not satisfy some basic user settings. These include the minimum and maximum spot sizes as well as symmetry constraint. Any spots that do not satisfy these demands are no longer processed.
4. Calculate the signal as the average value of the background compared to an average value of the center of the spot. These parameters are user specified and correspond to the radii used to find these areas. The spot value, which is defaulted to 80 %, calculates a radius that is smaller than the found spot and then finds the average height in this area. The area is reduced from the edge of the spot to reduce the effect of coffee-ring and other edge behavior. The background area is found by taking an annular ring outside of the spot, as shown in Fig. 4a. The ring values are user specified and default to 120 and 140 %.

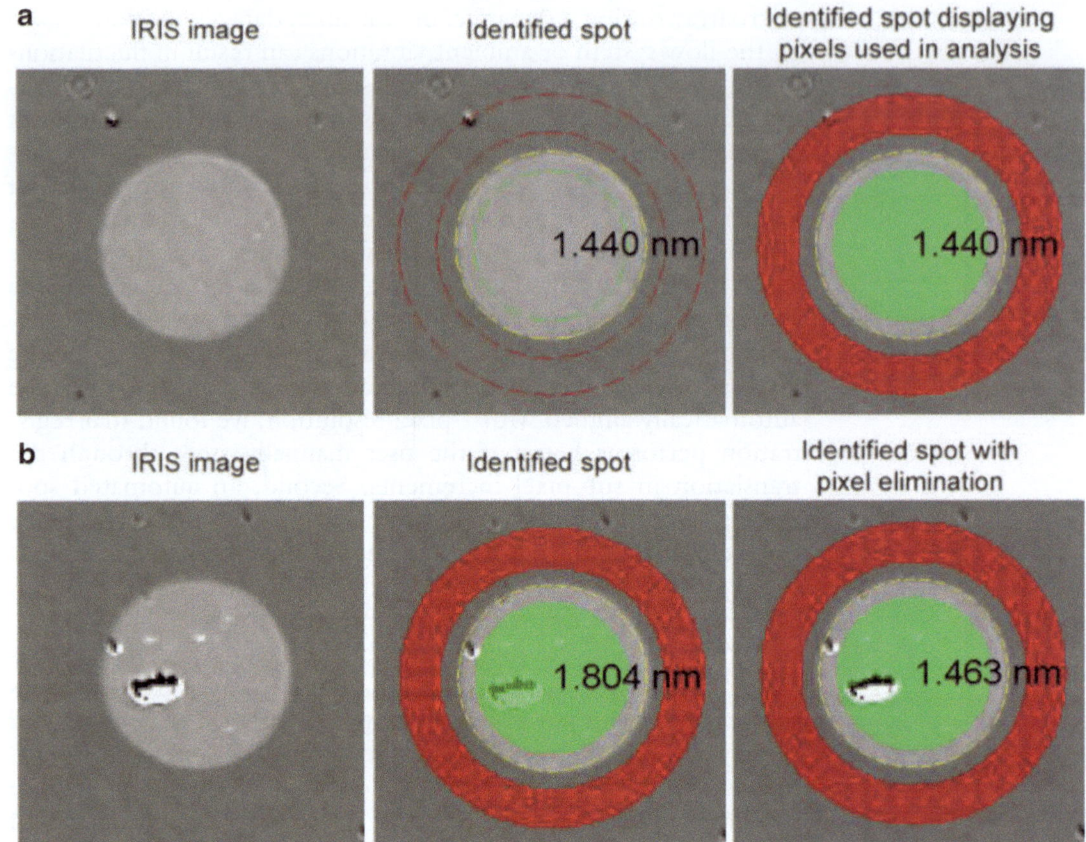

Fig. 4 Spot finding algorithm. (**a**) A spot, or a sensor, is easily identified by accumulating the vertical and horizontal lines throughout the image. The center and the radius of the sensor are identified as indicated by the *yellow dotted circle*. User specifies the radius inside sensor (*green dotted circle*) as well as the area outside the sensor (*red dotted circles*) to be used as the local background. In this example, a spot height of 1.440 nm was obtained by measuring the differential optical path length of the area defined by 80 % of the sensor radius (*green pixels*) and the area defined by 120 and 160 % of the sensor radius (*red pixels*). (**b**) A spot with a small area artifact, most likely caused by a piece of dirt is analyzed. The *dirt pixels* must be eliminated for accurate quantitative analysis. When only specifying the three radii for the spot finding algorithm, the obtained spot height is 1.804 nm. However, we filtered out all pixels that had fitting error of more than one standard deviation of the entire image. After eliminating the pixels with large error, the obtained spot height is decreased by 19 % to 1.463 nm, a value close to the spot height of a spot without any artifact (the *right image* from (**a**)). As seen in the *right image*, the artifact is not included, and only the *green pixels* are used in calculating the spot height

5. Eliminate undesired pixels using available advanced options. This allows the user to eliminate pixels that fall outside of up to three standard deviations from the median fitting error. In this way, pixels that do not fit well are not included in the calculation of spot or background heights. This often occurs with scratches, dirt, salt, or other undesired artifacts. An example spot using pixel elimination is shown in Fig. 4b.

3.6.2 Quantitative Analysis of Hybridization

The spot finding software outputs quantified observations of the sensor, which is the optical path difference of the spot to its background (i.e., spot height). Spot height information can easily be converted into the surface mass density (ρ_m) or molecular surface density (ρ_N) for quantitative analysis.

1. Obtain surface mass density using the following equation:

$$\rho_m = 0.8\,h, \tag{4}$$

where ρ_m is the surface mass density in ng/mm^2, h is the output signal in nm, and 0.8 is the optical density to mass density conversion factor for DNA, which was obtained previously based on using a fixed refractive index model [21].

2. Calculate hybridization efficiency by understanding the molecular surface density of the probe before the hybridization (ρ_{Np}) and the molecular surface density of the captured target after hybridization (ρ_{Nt}). Molecular surface densities are given by the following equations with the units in number of molecules/mm^2.

$$\rho_{Np} = \frac{0.8\times10^{-9}\bullet h_i N_A}{MW_p} \tag{5}$$

$$\rho_{Nt} = \frac{0.8\times10^{-9}\bullet(h'_f - h_i)N_A}{MW_t}, \tag{6}$$

where N_A is Avogadro's number and MW_p and MW_t are molecular weights of the probe and the target, respectively. h_i is the initial spot height (nm) before hybridization and h_f' is corrected height (nm) for refractive index difference between ssDNA and dsDNA, which is given by:

$$h'_f = 0.941\bullet h_f, \tag{7}$$

where h_f is the measured spot height (nm) after hybridization. The correction factor, 0.941, is the ratio of the refractive index of ssDNA ($n \sim 1.45$) to that of dsDNA ($n \sim 1.54$) [22]. Hybridization efficiency can be calculated by simply taking the ratio of the surface molecular densities of probe and target (divide Eq. 5 by Eq. 6):

$$\%\ \text{Hybridization} = \frac{\rho_{Np}}{\rho_{Nt}}\times100 \tag{8}$$

3.6.3 SNP Detection

1. Characterize the denaturation kinetics of different spots by fitting an exponential decay function:

$$f = a_1 e^{(-a_2 t)} + a_3 \tag{9}$$

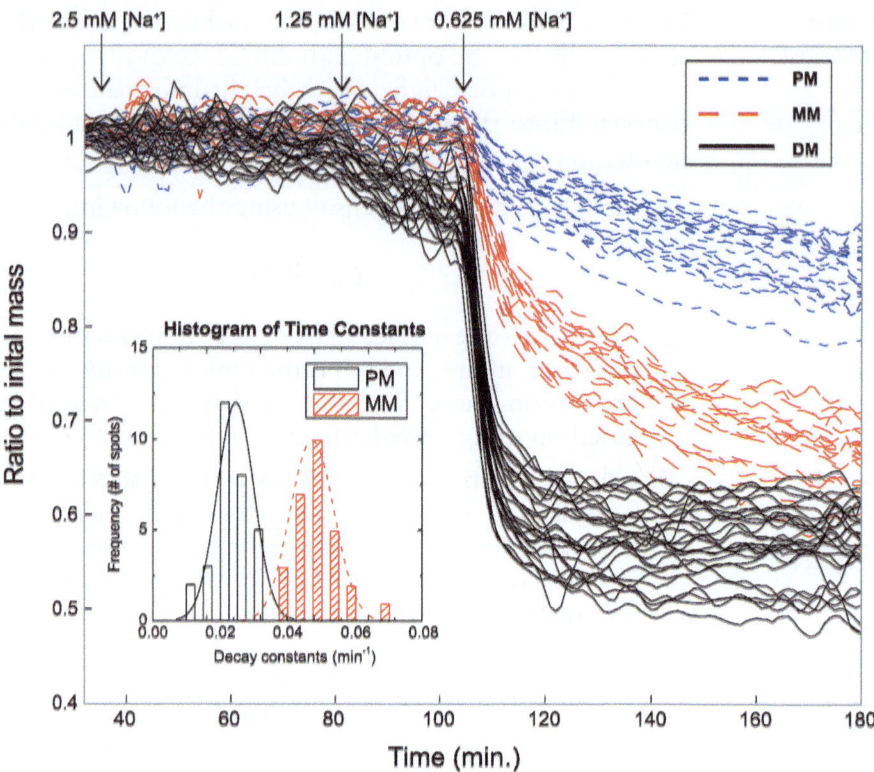

Fig. 5 Denaturation kinetics of DNA duplexes with and without mutation. Mass change ratio upon the introduction of buffer with low ionic strength is shown. Different duplexes display very different denaturation kinetics (PM-*blue dotted line*, perfect match; MM-*red dash line*, single mismatch; DM-*black solid line*, double mismatch). Single exponential decay functions are fit to mass change ratio of each spot, and the decay constants are shown as a histogram (*inset*). Higher decay constants represent faster denaturation kinetics. The free energy difference (ΔG) between PM and MM is 2.2 kcal/mol, and we can distinguish the two population with 97.2 % CI. The decay constants from the DM probes are too high to be shown in this histogram. This figure was adapted from ref. 16 with permission from Elsevier

While a_1 and a_3 do not have any experimental significance, they improve the quality of the fitting and do not affect the values found for the decay constant a_2. An example of DNA denaturation kinetics is shown in Fig. 5.

2. Plot a histogram of the decay constants. If a mutation is present, the population of the decay constants will have a different distribution from the ones obtained from perfectly complementary duplexes, as shown as an inset in Fig. 5.

4 Notes

1. The ACULED VHL LED array is used because the spatial locations of the four different color LEDs are as close as possible. The distance between the LEDs affects the illumination uniformity.

2. Full chemical name of copoly (DMA-NAS-MAPS) is *N*,*N*-dimethylacrylamide(DMA), *N*,*N*-acryloyloxysuccinimide (NAS), and [3-(methacryloyl-oxy) propyl] trimethoxysilyl (MAPS). It can be synthesized as described in refs. [23, 24].
3. One surface of the window has a standard anti-reflection coating for visible and near IR light. The other surface has anti-reflection coating for the same wavelength range, but it is designed for glass–water interface. In addition, the anti-reflection coating is made of a material with very low water solubility to survive aqueous flow experiments. The exact composition of the anti-reflection coating is not disclosed by the manufacturer.
4. The rubber sheet is cut out to make an approximately 50 μL large flow chamber. For the flow cell assembly, the gasket is sandwiched between the IRIS chip and the glass window forming the side walls for the flow chamber.
5. The chip holder is machined out of aluminum to hold IRIS chip in place in the assembled flow cell. The chip holder has the same dimension as a conventional glass slide for easy handling of the IRIS chip. The dimension also allows IRIS chip to be scanned by a conventional microarray fluorescence scanner. Aluminum is chosen as the material of the chip holder because of its good thermal conductivity. It is designed to be placed on top of a resistive heater in an assembled flow cell for temperature control.
6. The top block and the bottom block serve as the main structure. Both pieces are made out of polyoxymethylene (POM), or commercially known as Delrin. POM is the choice of material as it is easy to machine and nonconductive. It also has high heat resistance and low water absorption.
7. The diameter of the tubing determines the flow rate. The inner diameter of the tubing used for the peristaltic pump is 0.04 in., and the inner diameter of the tubing leading to and from the flow cell is 0.03 in.
8. MATLAB is selected as the programming interface because of the familiarity to the developers and because it allows one to easily use many image processing and fitting algorithms. Additionally, all initial modeling is done with MATLAB, allowing rapid testing of new substrate and illuminations designs.
9. Processing, referred to as fitting, can be done using the nonlinear fitting algorithm available in MATLAB's curve fitting toolbox. This process can be computationally demanding, but it is rapidly parallelizable as each pixel can be processed independently. Multi-core processors such as modern graphic processing units (GPU) are ideal for fitting IRIS images. Zoiray Technologies developed fitting software that used GPUs but was difficult to modify. To keep up with the development of

new fitting models, computation time was traded for a reduction in software developer time. To this end, a cluster of computers is used for fitting large numbers of datasets with MATLAB.

10. Possible hairpin structures as well as homodimers and heterodimers of the oligonucleotide probes that can decrease the hybridization efficiency can be calculated with OligoAnalyzer provided by Integrated DNA Technologies (Coralville, IA, USA). The URL for this free software is http://www.idtdna.com/analyzer/Applications/OligoAnalyzer/.
11. Be sure to use a plastic Petri dish for copoly (DMA-NAS-MAPS) coating. A glass container will compete with the IRIS chip for functionalization.
12. The functionalized IRIS chips can be stored in the dessicator for up to 3 weeks. Bake the stored functionalized IRIS chips in a vacuum oven at 80 °C for 15 min prior to spotting.
13. In our work, BioOdyssey Calligrapher MiniArrayer (Bio-Rad, Hercules, CA, USA) was used for oligonucleotide spotting. Alternative arrayers can be used to produce comparable results.
14. Averaging images reduces noise, hence increases the sensitivity of the sensor. However, there is a practical limitation to consider. Averaging 200 images with 0.03 s exposure at every wavelength takes ~30 s to acquire a single data point and result in temporal resolution of greater than 30 s. On the contrary, averaging 20 images with 0.03 s exposure results in temporal resolution of less than 10 s, although the noise will be greater by threefold. The trade-off between the sensitivity and the kinetic characterization capability needs to be considered. In the method we present, we averaged 50 images per wavelength.
15. An ideal target is 80 % of full-well capacity.
16. Make sure the tubing is completely dry to avoid trapping bubbles during the flow experiment.
17. Free energy of the duplexes can be calculated with OligoAnalyzer, as described in **Note 10**, to check the duplex stabilities in different ionic strength buffers.
18. We evaluated the performance of dynamic data processing techniques by analyzing the change in mass density of a protein microarray under a flow for 37 h. The % change in mass density was obtained by using (1) automatic image registration, (2) manual image registration, and (3) spot finding algorithm. The results are compared to the mass density obtained by identifying individual sensors using the software developed by Zoiray Technology. The differences in % change in mass density from using different techniques are presented in Fig. 6. The spot finding algorithm gave a much closer result to the

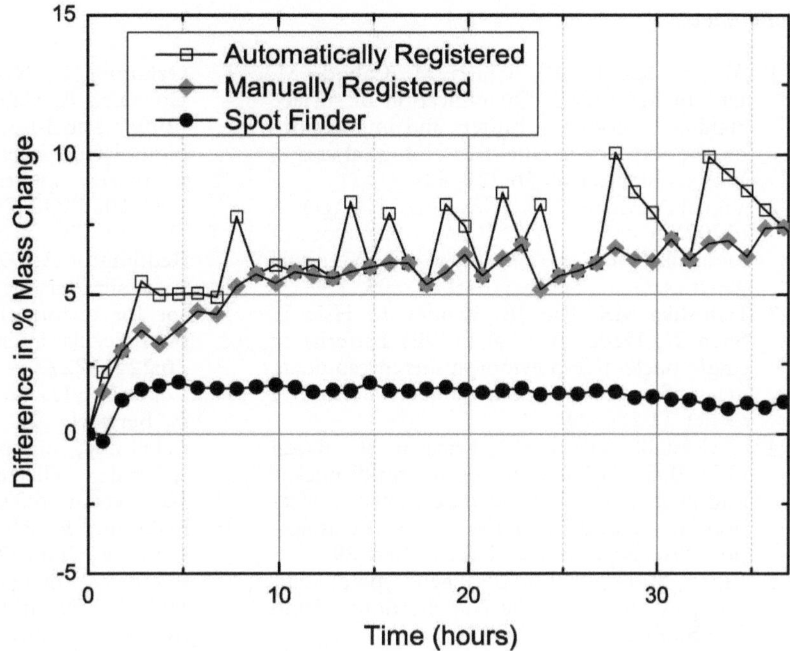

Fig. 6 Comparison of different methods for dynamic data processing

result obtained using Zoiray Technology's software than image registration techniques did. The average percent difference between the results obtained via the spot finding algorithm from Zoiray's software was 1.4 ± 2.5 % with 95 % confidence interval. The average percent difference between the automatic registration and manual registration techniques from Zoiray's software was 6.5 ± 4.2 % and 5.5 ± 4.3 %, respectively, with 95 % confidence interval. The spot finding algorithm also resulted in the least amount of signal fluctuations throughout the experiment at 0.4 %, whereas the fluctuations from automatic image registration and manual image registration were 2.0 % and 1.5 %, respectively.

Acknowledgments

We thank Dr. Emre Özkumur for his contributions in developing IRIS and the presented protocol, Professor Marcella Chiari for providing her expertise in surface chemistry, Professor Mario Cabodi for his advice on flow cell design, and Alex Reddington for his help in chip fabrication.

References

1. Ahn S, Spuhler PS, Chiari M, Cabodi M, Unlum MS (2012) Quantification of surface etching by common buffers and implications on the accuracy of label-free biological assays. Biosens Bioelectron 36:222–229
2. Vignal A, Milan D, SanCristobal M, Eggen A (2002) A review on SNP and other types of molecular markers and their use in animal genetics. Genet Sel Evol 34:275–305
3. Halushka MK, Fan JB, Bentley K, Hsie L, Shen N, Weder A et al (1999) Patterns of single-nucleotide polymorphisms in candidate genes for blood-pressure homeostasis. Nat Genet 22:239–247
4. Gaylord BS, Massie MR, Feinstein SC, Bazan GC (2005) SNP detection using peptide nucleic acid probes and conjugated polymers: applications in neurodegenerative disease identification. Proc Natl Acad Sci USA 102:34–39
5. Kim S, Misra A (2007) SNP genotyping: technologies and biomedical applications. Annu Rev Biomed Eng 9:289–320
6. Peyret N, Seneviratne PA, Allawi HT, SantaLucia J Jr (1999) Nearest-neighbor thermodynamics and NMR of DNA sequences with internal A center dot A, C center dot C, G center dot G, and T center dot T mismatches. Biochemistry 38:3468–3477
7. Fei ZD, Ono T, Smith LM (1998) MALDI-TOF mass spectrometric typing of single nucleotide polymorphisms with mass-tagged ddNTPs. Nucleic Acids Res 26:2827–2828
8. Nicolae DL, Wen XQ, Voight BF, Cox NJ (2006) Coverage and characteristics of the affymetrix GeneChip human mapping 100K SNP set. PLoS Genet 2:665–671
9. Shen R, Fan JB, Campbell D, Chang W, Chen J, Doucet D et al (2005) High-throughput SNP genotyping on universal bead arrays. Mutat Res 573:70–82
10. Zhang SB, Wu ZS, Shen GL, Yu R (2009) A label-free strategy for SNP detection with high fidelity and sensitivity based on ligation-rolling circle amplification and intercalating of methylene blue. Biosens Bioelectron 24:3201–3207
11. Ingebrandt S, Han Y, Nakamura F, Poghossian A, Schoning MJ, Offenhausser A (2007) Label-free detection of single nucleotide polymorphisms utilizing the differential transfer function of field-effect transistors. Biosens Bioelectron 22:2834–2840
12. Rant U, Arinaga K, Scherer S, Pringsheim E, Fujita S, Yokoyama N et al (2007) Switchable DNA interfaces for the highly sensitive detection of label-free DNA targets. Proc Natl Acad Sci USA 104:17364–17369
13. Ozkumur E, Needham JW, Bergstein DA, Gonzalez R, Cabodi M, Gershoni JM et al (2008) Label-free and dynamic detection of biomolecular interactions for high-throughput microarray applications. Proc Natl Acad Sci USA 105:7988–7992
14. Daaboul GG, Vedula RS, Ahn S, Lopez CA, Reddington A, Ozkumur E et al (2011) LED-based interferometric reflectance imaging sensor for quantitative dynamic monitoring of biomolecular interactions. Biosens Bioelectron 26:2221–2227
15. Lopez CA, Daaboul GG, Vedula RS, Ozkumur E, Bergstein DA, Geisbert TW et al (2011) Label-free, multiplexed virus detection using spectral reflectance imaging. Biosens Bioelectron 26:3432–3437
16. Ozkumur E, Ahn S, Yalcin A, Lopez CA, Cevik E, Irani RJ et al (2010) Label-free microarray imaging for direct detection of DNA hybridization and single-nucleotide mismatches. Biosens Bioelectron 25:1789–1795
17. Ahn S, Huang C, Ozkumur E, Zhang X, Chinnala J, Yalcin A et al (2012) TATA binding proteins can recognize nontraditional DNA sequences. Biophys J 103:1510–1517
18. Fish DJ, Horne MT, Searles RP, Brewood GP, Benight AS (2007) Multiplex SNP discrimination. Biophys J 92:L89–L91
19. Vedula R, Daaboul G, Reddington A, Ozkumur E, Bergstein DA, Ünlü MS (2010) Self-referencing substrates for optical interferometric biosensors. J Mod Opt 57: 1564–1569
20. Ahn S, Freedman DS, Massari P, Cabodi M, Ünlü MS (2013) A mass-tagging approach for enhanced sensitivity of dynamic cytokine detection using a label-free biosensor. Langmuir 29:5369–5376
21. Ozkumur E, Yalcin A, Cretich M, Lopez CA, Bregstein DA, Goldberg BB et al (2009) Quantification of DNA and protein adsorption by optical phase shift. Biosens Bioelectron 25:167–172
22. Elhadj S, Singh G, Saraf RF (2004) Optical properties of an immobilized DNA monolayer from 255 to 700 nm. Langmuir 20: 5539–5543
23. Pirri G, Damin F, Chiari M, Bontempi E, Depero LE (2004) Characterization of a polymeric adsorbed coating for DNA microarray glass slides. Anal Chem 76:1352–1358
24. Cretich M, Longhi R, Corti A et al (2009) Epitope mapping of human chromogranin A by peptide microarrays. Methods Mol Biol 570:221–232

Chapter 17

Graphene–PAMAM Dendrimer–Gold Nanoparticle Composite for Electrochemical DNA Hybridization Detection

Kumarasamy Jayakumar, Rajendiran Rajesh, Venkataraman Dharuman, and Rangarajan Venkatesan

Abstract

Graphene oxide is chemically functionalized using planar structured first generation polyamidoamine dendrimer (G1PAMAM) to form graphene core GG1PAMAM. The monolayer of GG1PAMAM is anchored on the 3-mercapto propionic acid monolayer pre-immobilized onto a gold transducer. The GG1PAMAM is decorated using gold nanoparticles for the covalent attachment of single-stranded DNA through simple gold-thiol chemistry. The single- and double-stranded DNAs are discriminated electrochemically in the presence of redox probe $K_3[Fe(CN)_6]$. Double-stranded-specific intercalator methylene blue is used to enhance the lower detection limit. The use of linear and planar G1PAMAM along with the graphene core has enhanced the detection limit 100 times higher than the G1PAMAM with the conventional ethylene core. This chapter presents the details of GG1PAMAM preparation and application to DNA sensing by electrochemical methods.

Key words Graphene, PAMAM dendrimer, DNA, Electrochemical, Gold, Nanoparticle, Composites

1 Introduction

1.1 Electrochemical DNA Hybridization Sensing

DNA hybridization detection is made mainly using fluorescence and radioactive methods [1–4]. Although the radioactive technique is sensitive and resulting in lower background signal, the technique is currently outdated due to its harmful effects on health. The fluorescence technique is highly sensitive, selective, reliable, and readily available for medical and environmental applications. The recent technological developments in nanoelectronics and silicon technology enabled the development of miniaturized devices for fluorescence measurement. However, the high cost, complicated instrumentation, experimental protocol, and complex data analysis force the development of alternative instruments with

Dmitry M. Kolpashchikov and Yulia V. Gerasimova (eds.), *Nucleic Acid Detection: Methods and Protocols*, Methods in Molecular Biology, vol. 1039, DOI 10.1007/978-1-62703-535-4_17,

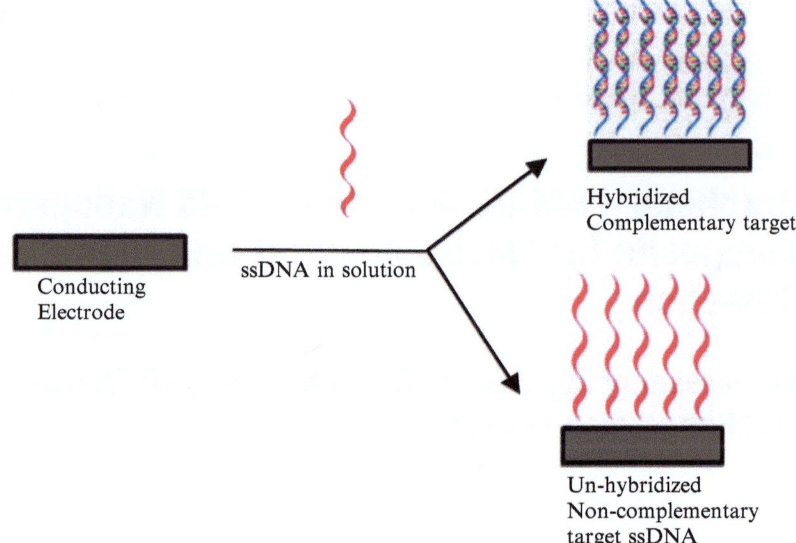

Scheme 1 General representation of DNA detection using electronic transducers

desirable properties of portable, low cost, and simple operation with direct readout similar to glucometers. Hence, the main driving force for employing the electrochemical DNA hybridization detection technology is its simplicity, sensitivity, low cost, and compatibility for mass fabrication of sensor elements and easy integration to develop miniaturized portable devices for onsite evaluation of clinical and environmental samples [5, 6]. Generally, the electrochemical DNA hybridization detection involves three steps. In the first step, the capture DNA probe (called, hereinafter, as recognition layer) is immobilized on a solid metal transducer (e.g., Au, carbon, metal oxide), Scheme 1. In the second step, hybridization of the target DNA with the recognition layer is made on the same transducer surface. Similar to the fluorescence measurements, both the complementary and noncomplementary targets (with respect to the recognition layer (capture probe) immobilized on the solid transducer) hybridization experiments should be done in parallel. Finally, the difference in the electrical signal of the transducer (either current or resistance) induced by the target hybridization indicates the occurrence of the hybridization and/or formation of the double-stranded DNA on the surface. For the noncomplementary target hybridization insignificant signal change occurs. Figure 1 shows the behaviors of the complementary (a and c) and noncomplementary (b and d) target hybridized gold surfaces monitored using cyclic voltammetry (Fig. 1a, b) and impedance techniques (Fig. 1c, d). However, the sensitivity, selectivity, data reliability, and repeatability parameters are greatly influenced by the type of transducer, methods of capture DNA

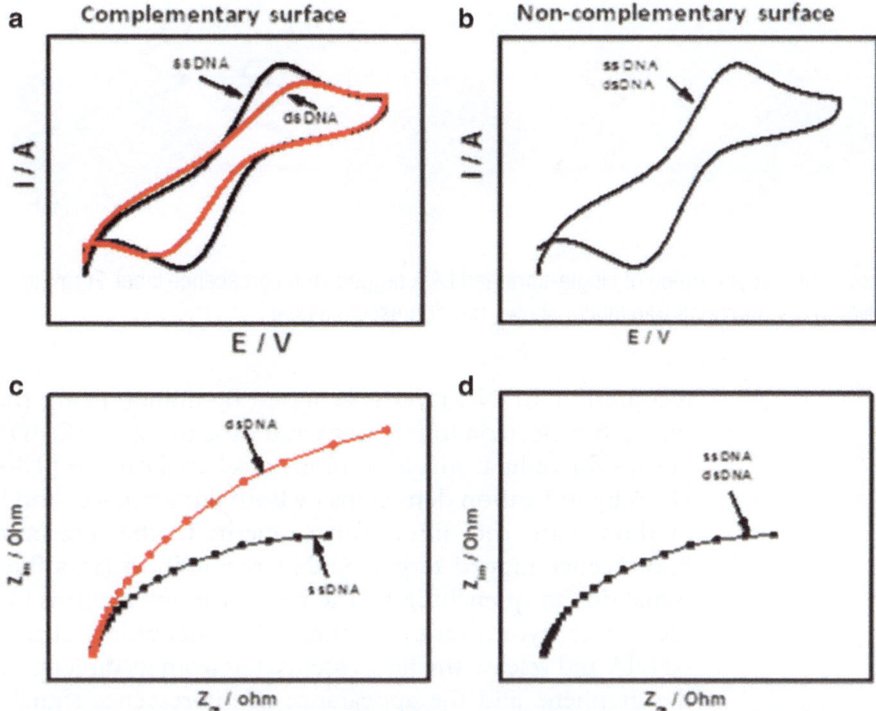

Fig. 1 Electrochemical behavior of complementary and noncomplementary target hybridized surface monitored by voltammetric and impedance techniques in the presence of $[Fe(CN)_6]^{3-/4-}$ redox probe

immobilization, electrolyte solution, electroactive molecules (e.g., $K_3[Fe(CN)_6]/K_4[Fe(CN)_6]$, $[Ru(NH_3)_6]^{2+}/[Ru(NH_3)_6]^{3+}$ and other metal complexes, DNA intercalators methylene blue, eosin blue, etc.), and electrochemical technique used to monitor the hybridization reactions. Another important parameter which limits the electrochemical sensor devices is the sluggish electron transfer between the biomolecules and transducer. Numerous papers have been published on various aspects/methods of electrochemical DNA sensing using different types of transducers [1–18]. The transducers such as conducting metals, semiconductors, conducting polymer wires, and allotropes of carbon were reported towards the development of portable electronic devices. In depth discussion is out of scope of this chapter and limited to DNA detection using graphene transducers.

1.2 DNA Sensing at Graphene Transducers

Graphene, discovered by Geim and Novoselov [7–12], has been investigated for a decade on different applications including electrocatalysis, energy storing supercapacitors, and biosensing due to its high abundance, unique physical, thermal, electrical, and mechanical properties originating from its two dimensional single atom thick planar structure consisting of SP^2 carbons. The intimate association between the graphene transducer and the target

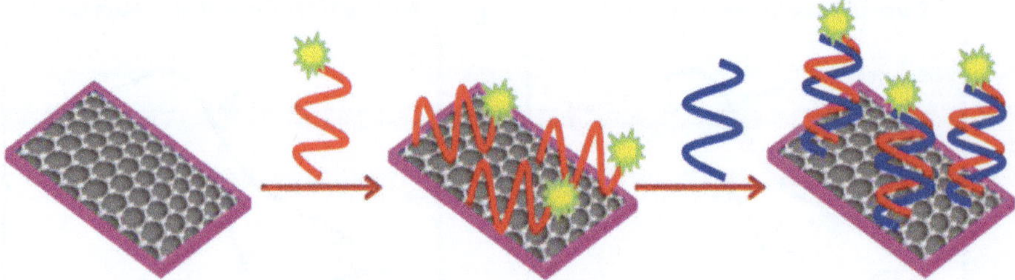

Fig. 2 Noncovalent immobilization of single-stranded DNA tagged with florescence label. Formation of double-stranded DNA on the graphene transducer shows the fluorescent signal

recognition ssDNA layer is achieved by immobilizing the ssDNA through π-electron interactions and used to form GO/DNA composites for cellular imaging, biomedical applications [13–15], and DNA hybridization detections by both fluorescence- and label-free methods. In the fluorescence method, the pre-immobilized fluorescence-tagged target ssDNA recognition lacks fluorescence signal due to quenching by the π-electron interactions [16], while the target hybridization induces the orientation change of the ssDNA and release the fluorescence tag from its direct contact with the graphene and the appearance of fluorescence signal indicates the hybridization, Fig. 2. Circular dichroism supports orientation change upon the duplex formation by showing the negative effect [17]. Both the linear and hairpin types of DNAs labeled with fluorescent tags have been used, however, physisorbed DNAs are prone to the nonspecific displacement. Although it is highly sensitive, shortcomings such as low efficiency fluorescence labeling of DNA, complex multistep reaction, contamination of the products, and high cost permit the development of cost effective and label-free DNA detection devices and simple protocols. In this category, the single-stranded DNA immobilized through π–π electron interactions on graphene and the presence of ssDNA is detected by the appearance of four different oxidation peaks corresponding to the guanine, adenine, cytosine, and thiamine bases simultaneously and efficiently which could not be achieved on other carbon allotropes, either glassy carbon or graphite [18, 19]. In contrast, the dsDNAs showed feeble oxidation signals for these bases on graphene. This is because the target hybridization with the ssDNA induces the orientation change so that the negatively charged phosphate backbones of the dsDNA are repelled electrostatically by the graphene and DNA bases are not in direct contact with the graphene. Alternatively, the electrostatic immobilization of the negatively charged DNA on the graphene with pre-grafted with *N,N*-bis-(1-aminopropyl-3-propylimidazol salt)-3,4,9,10-perylene tetracarboxylic acid diimide (PDI) showed improved target DNA detection [20]. However, this type of DNA sensing leads damaging of DNA strands and less sensitive to the low-quantity DNA

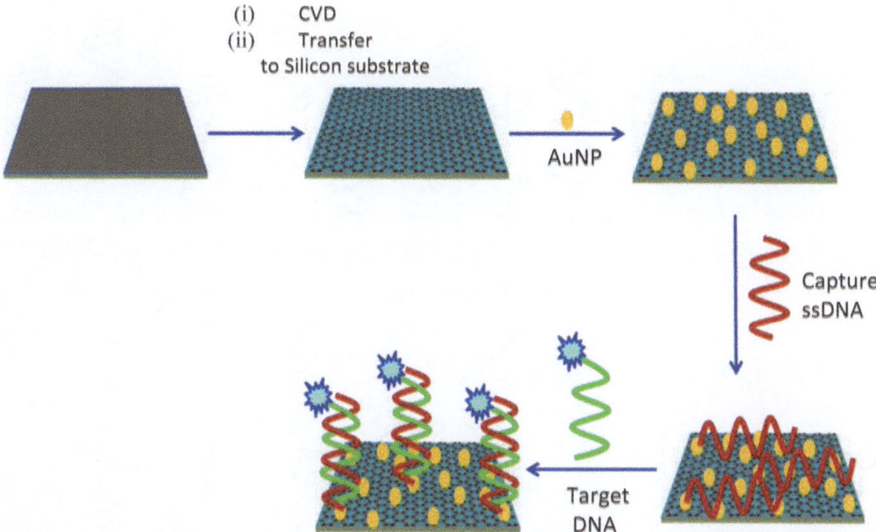

Fig. 3 The CVD prepared single layer graphene on copper plate is transferred to silicon substrate and AuNPs are decorated the layer for attaching thiol capped single-stranded DNA via gold-thiol linkage. This platform has been used for optical and electronic detection of target DNA

samples and thus covalent immobilization of the DNA is used for increasing the sensor specificity and reusability [21, 22]. The hydrophilic oxygen functional groups of the GO are used for the covalent attachment of bovine serum albumin (BSA) [23], DNA molecules [24, 25], and enzymes [26–28] via a carbodiimide coupling, biotin–streptavidin interactions, and metal functionalization, Fig. 3.

Poly(amidoamine) (PAMAM) dendrimers are three dimensional nanosized synthetic molecules with internal cavities and large number of -NH_2 active sites at their terminals for the covalent attachment of ssDNA via amide linkage [29] is excellent platform for enhancing the sensitivity of DNA hybridization sensing. It is well established that the 5′ end thiol functionalized single-stranded DNA to form covalent bond with gold surface (Au + HS-ssDNA → Au-S-ssDNA + H^+) and used in the constructions of DNA-based microarrays and the colloidal AuNPs are also widely used for DNA analysis [30]. In order to simplify the DNA attachment chemistry, to increase the conductivity and detection sensitivity, the Dendrimer–AuNP [29, 30] and graphene–AuNP composites [31–35] have been reported for the DNA detection. But, the detection sensitivity of the dendrimer–AuNP composites decreases with increasing dendrimer generations due to their capping effect on the AuNP by their circular structures.

The electronic properties of graphene depend on the number of layers; several methods have been developed for preparing single monolayer graphene [36–38], but producing single layer of

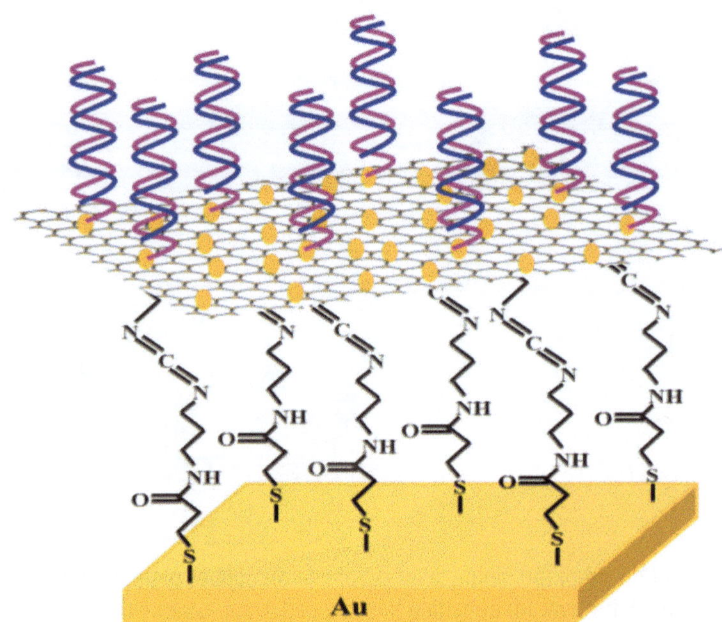

Fig. 4 Single layer of graphene core first generation PAMAM dendrimer attached on the monolayer of mercaptopropionic acid on the gold transducer and decorated with gold nanoparticle for DNA attachment via gold-thiol chemistry for electrochemical DNA detection

graphene requires high-tech expensive instruments. Although, the mechanical exfoliation of highly oriented pyrolytic graphite using scotch tape is simple and cost effective, the graphene synthesized by chemical oxidation and reduction is advantageous for functionalization of defective carbons [39]. In addition the electronic properties of single-layer graphene could be modified by the functionalization of its surface functional groups viz., epoxy, hydroxyl, and carboxyl groups using metals, metal oxides, polymers, and biomolecules. Recent advances focus on the use of functionalized graphene monolayer as the transducer for biomolecular sensing. Here, we report functionalization of monolayer of graphene using the planar first generation polyamidoamine dendrimer (G1PAMAM) at the defective carbons of the chemically exfoliated graphene oxide covalent attachment through amide coupling reactions on the pre-immobilized mercapto propionic acid on the gold transducer. To further simplify the immobilization protocol for anchoring the single-stranded DNA covalently on the GG1PAMAM via simple gold-thiol linkage, the AuNPs are electroless deposited on the GG1PAMAM and used for DNA hybridization sensing by electrochemical method, Fig. 4. In this chapter, the preparation of GG1PAMAM and its application for electrochemical DNA detection is presented.

2 Materials

2.1 Buffers

Phosphate saline buffer pH 7.4

The phosphate buffer prepared using the composition $[NaH_2PO_4]=100$ mM; $[NaCl]=10$ mM; $[KCl]=2.7$ mM and total ionic strength of PBS was 0.0577 M and of the chloride ion was 0.00635 M.

Sodium Saline Citrate (SSC) Buffer

20× Concentrate buffer was bought from Sigma. It has the composition of 0.3 M sodium citrate and 3 M NaCl and pH 7. It is diluted four times and used in target hybridization experiments at an ambient temperature. All solutions including distilled water and buffers are autoclaved before use.

2.2 DNA Sequences

The required DNA sequences are designed based on gene data bank, for instance, NCBI, discussed in later section. In the present work, following short-chain 27mers synthetic oligonucleotides synthesized by MWG biotech, Ebersberg, Germany, with HPLC purification were used.

Capture probe: 5′-HS-$(CH_2)_6$-CGA T CTG TTT TAT GT**A** GGG TTA GGT CA-3′ (I)

Complementary target to I: 5′-TG ACC TAA CCC TAC ATA AAA CAG-3′ (II)

Noncomplementary target to II: 5′-TAC CAT TCT CAT CTC TGA AAA CTT CCG-3′ (III)

Single base mismatch target to I: 5′-TGA CCT AAC CC**C** ACA TAA AAC AG-3′ (IV)

2.3 Equipments

Electrochemical potentiostat from CH Instrument, Inc., USA (model 650D, 440B). Other potentiostates from Auto lab, IVM (Netherlands), Gamrey potentiostat, PAR EG &G potentiostat can also be used. CH Instrument is used in the present work for all electrochemical characterization. All instruments used for surface characterization are from different scientific companies and widely used in literatures. The specifications are given only when specific parameters and/or care taken for the particular measurement.

3 Methods

3.1 Preparation of Graphene Core First Generation PAMAM Dendrimer

3.1.1 Graphene Synthesis

Graphene can be prepared in multi- and single-layered structures by selecting appropriate preparation method. It has been experimentally proved that the single layer graphene is more reactive than the double or triple or multilayer structures [40–42]. Monolayer and few layer of graphene were first synthesized by Boehm in 1962 by graphene oxide reduction [43] and intercalation effect on graphene sheets [44]. Other than these, liquid phase exfoliations, thermal decomposition of silicon carbide, and chemical vapor deposition (CVD) methods have been used for the synthesis of graphene monolayer. The mechanical exfoliation method, developed by Geim and Novoselov, involves peeling of graphene layer from carbon flakes using scotch tape and repeatedly transferred onto the SiO_2 substrate, named as scotch tape method. Indeed this forms a basis to fabricate small scale transistors and exciting alternative for the existing silicon technology for the development of innovative products; however, it has limitation in producing number of layers [45, 46]. In liquid phase extraction method, the graphite is dispersed in *N*-methylpyrrolidone (NMP) and exfoliation occurs due to strong attractive force between the solvent and carbon monolayer [47, 48]. Thermal decomposition of silicon carbide at 1,100–1,200 °C followed by sublimation leaves the graphene and the lattice plan controls the growth of the monolayer graphene. In spite of its high purity product, transfer of the monolayer to other substrates is difficult [49]. Similar to thermal decomposition, CVD method also involves heating the graphite flake at 1,000 °C and deposition on the Ni and Cu catalyst [50]. The CVD prepared single-layered graphene decorated with gold nanoparticle has been shown to be the promising candidate for biosensing, Fig. 3.

The graphite oxide reduction method developed by Hummers [51–53] involves production of graphene oxide by the oxidation of cyleon graphite using H_2SO_4 in the presence of sodium nitrate, $KMnO_4$, and hydrazine reduction. This GO contains the disruption planar sheet of sp^2-bonded carbon atoms arranged in a honey comp structure. These defects are further used for tuning its intrinsic electronic conduction properties. Hence, modified Hammer's method is used here to prepare graphene oxide from graphite flakes.

Preparation of Graphene Oxide [GR-COOH] from Graphite

Graphene oxide is prepared following the Hummer's and Offeman's [50–53] procedure. 1 g of graphite and 1 g of $NaNO_3$ were mixed in 50 mL of concentrated H_2SO_4 and stirred together in an ice bath for 30 min at 1,000 rpm. 3 g of crystalline $KMnO_4$ was slowly added to the above mixture in about 30 min. Since this is an exothermic reaction, the reaction temperature needs to be maintained below 20 °C. Hence, care was taken during the addition of $KMnO_4$ at a controlled rate of manual addition. The mixture was then transferred into a water bath which has temperature at 35 °C and stirred for 1 h. This results in the formation of a thick viscous brown color paste. To the same vessel, 50 mL of deionized water was added gradually and temperature is increased to 98 °C slowly. Since the mixture contained excess concentrated H_2SO_4, the temperature was found to increase initially to ~120 °C and finally attained a constant temperature of 98 °C after all water was added. The mixture was reddish yellow in color and was transferred over a boiling water bath and kept for 30 min and allowed to cool to room temperature. After 15 min, the mixture was treated with 150 mL deionized water followed by 10 mL of 30 % H_2O_2 solution.

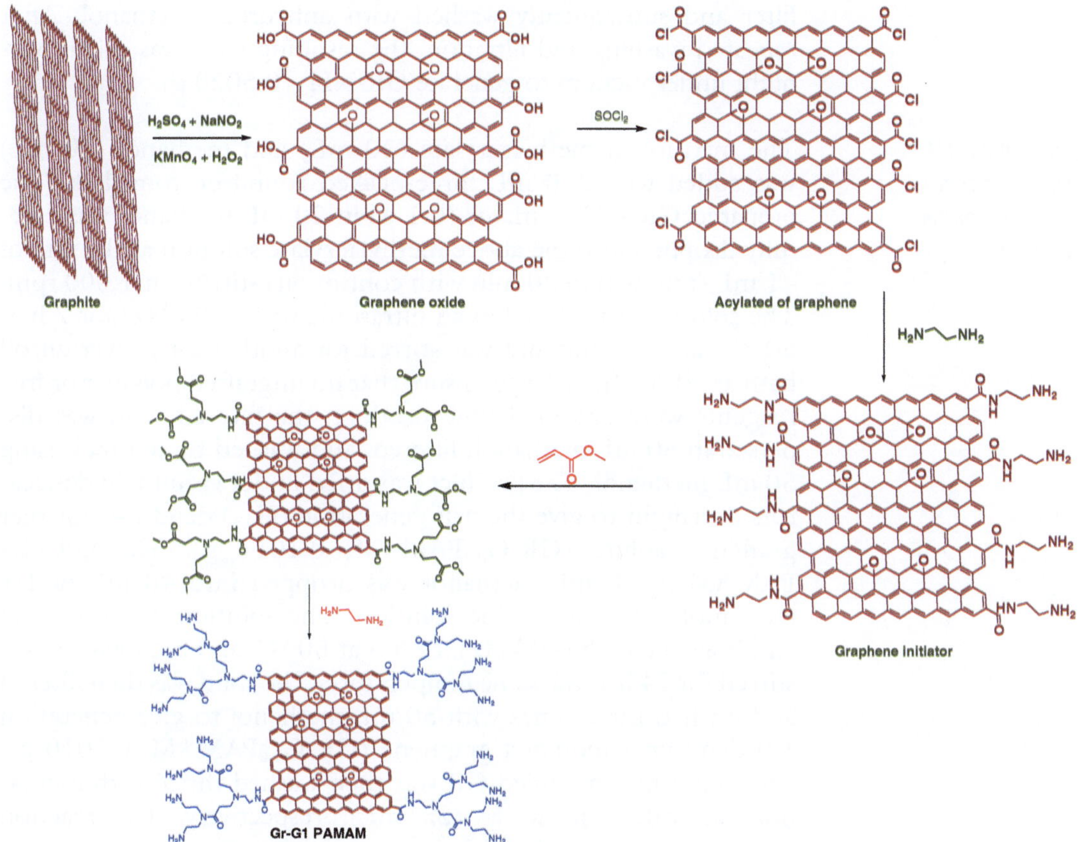

Scheme 2 Experimental sequence of graphene core first generation PAMAM dendrimer synthesis

The warm solution was filtered and washed with deionized water until the pH of the filtrate showed pH 7. The precipitate is separated and dried at 65 °C over a water bath under vacuum. The precipitate is the graphene oxide (GO) and also called as graphene acid, GR-COOH, Scheme 2.

Acylation of Graphene Acid

The GR-COOH (0.5 g) was suspended in $SOCl_2$ (30 mL) and stirred for 24 h at 70 °C. This solution was filtered, washed with anhydrous tetrahydrofuran (THF), and dried under vacuum for 24 h at room temperature, generating acylated graphene (GR-COCl) (0.4953 g). In this step, thionyl chloride acts as chlorinating agent and the hydroxyl group is replaced by chlorine following the SNi mechanism. The THF was used for the removal of unreacted thionyl chloride from the mixture.

Preparation of GR-NH_2 Initiator from GR-COCl

The GR-COCl (0.4953 g) prepared above was mixed with ethylenediamine (20 mL) and sonicated in an ultrasonic bath (40 kHz) for 5 h at 60 °C. This mixture is then stirred at the same temperature for another 24 h over an oil bath. The resulting solid was separated by vacuum filtration using 0.22 μm Millipore polycarbonate membrane filter and subsequently washed with anhydrous methanol. After repeated washing and filtration, the resulting solid was dried overnight under vacuum to generate GR-NH_2 (0.5020 g).

Growth of PAMAM Dendrimers on the GR Surface Initiated by GR-NH_2

The mixture of methylacrylate (20 mL) and methanol (50 mL) was added to a 250 mL three-necked round-bottom flask. The prepared GR-NH_2 initiator (0.1 g) in 20 mL methanol was carefully dropped into the above methylacrylate solution at the rate of ~1 mL/min within 20 min with continuous stirring at 1,000 rpm. The solution was placed in an ultrasonic bath (40 kHz) for 7 h at 50 °C, and the mixture was stirred for another 24 h over an oil bath at 50 °C. In order to ensure that no ungrafted polymer or free reagents were present in the product, the filtered solid was dispersed in 50 mL methanol, filtered, and washed three times using 50 mL methanol. The product was dried under vacuum in desiccators overnight to give the half generation ($G_{0.5}$) dendritic polymer grafted graphite (GR-$G_{0.5}$PAMAM, 0.1011 g). The GR-$G_{0.5}$ PAMAM in 20 mL methanol was dropped into 40 mL of 1:1 methanol/ethylenediamine solution. The solution was placed in an ultrasonic bath (40 kHz) for 5 h at 50 °C, and the mixture was stirred for 24 h at the same temperature. The solid was then filtered and washed three times with 50 mL methanol to give generation 1.0 dendrimer-modified graphene (GR-$G_{1.0}$PAMAM, 0.1010 g). The dendrimer-modified GR was then washed thrice with methanol (25 mL) and water (25 mL), respectively. The reaction sequence is presented in Scheme 2.

3.2 Bio Functionalization of Gold Transducer

3.2.1 GG1PAMAM Monolayer Immobilization on Gold Electrodes

The Au surface was modified by dipping into the ethanol containing 1 mM MPA for 2 h and washed with blank ethanol followed by phosphate buffer of pH 7.4. *See* **Note 1** for surface preparation of polycrystalline Au and behavior of MPA monolayer on the gold surface. The GG_1PAMAM layer was then attached onto the MPA monolayer by dropping a mixture of 1 μg GG_1PAMAM and 5 mM EDC in PBS (*see* **Note 2**: NHS not used in this case), Fig. 4. The attachment of monolayer of GG_1PAMAM is indicated clearly by the decreased CV peak currents compared to the unmodified electrode. The current decreases to a little extent, Fig. 5, curve a and b, due to the fact that the short-chain MPA forms phase segregated monolayer and the structure of GG1PAMAM appears like molecular domains on the Au chips [54].

3.2.2 DNA Probe Design Procedure

1. To design a DNA sequence, one must preselect the DNA sequence to be immobilized on the solid surface and the DNA sequence information can be obtained using the free access DNA data banks (database) in which almost all the DNA sequence encoding all biofunctions are stored. Some DNA data banks are as follows.

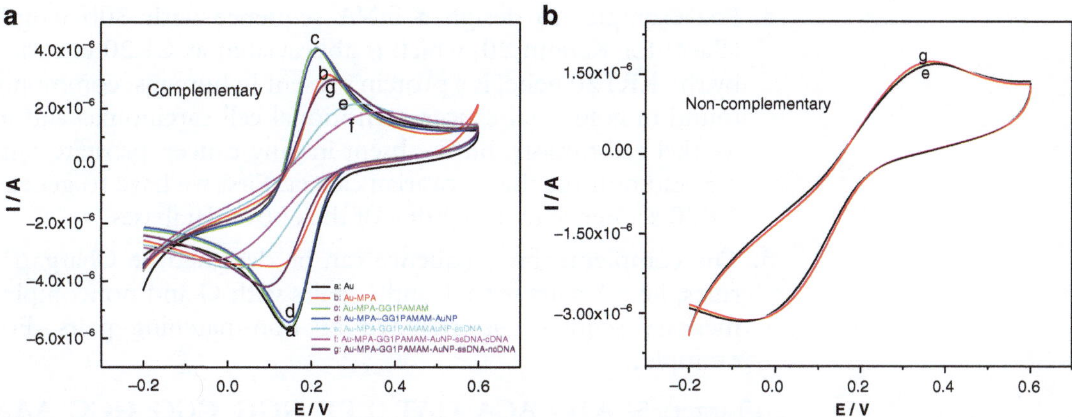

Fig. 5 (**a**) CV behavior of Au electrode at different modifications measured in the presence of 1 mM $[Fe(CN)_6]^{3-/4-}$ measured in PBS (pH 7.4) and scan rate 50 mV/s. Bare Au (*curve a*); Au-MPA (*curve b*); Au-MPA-GG1PAMAM (*curve c*); Au-MPA-GG1PAMAM-AuNP (*curve d*); Au-MPA-GG1PAMAM-AuNP-S-ssDNA (*curve e*); Au-MPA-GG1PAMAM-AuNP-S-dsDNA (*curve f*). Noncomplementary (*curve g*)

- NCBI: National Center for Biotechnology Information
- DDBJ: DNA DataBank of Japan
- EMBL: European Molecular Biology Laboratory

2. Using the internet search engine like Google.com, the above data banks home page can be easily accessed. In the Home page of the desired data bank, *type the gene name in the query box* and proceed. This step maps the DNA sequence with the existing DNA sequences using online tools like BLAST (Basic Local Alignment search tool) and FastA (Fast Alignment tool) which gives information on the source organism from which it is derived, biological function of that particular gene, length, etc.
3. Following the mapping, the exons containing both coding sequence and noncoding introns appear differentiated by color variation and only the number of coding bases will be given in each exon. The coding region (genatlas data bank) (Black color) is differentiated from the noncoding sequence (Blue color) by color and each exon have the number of coding bases near to the exon number, e.g., 108 bp, 165 bp. In that we have to identify the region which is responsible for the particular disease, if it is mutation of ACG to ATG of exon 20 in 790b means you have to get all the coding regions in the particular gene and start count from the exon1 coding region (NOTE: all exons have coding region in black color, but in exon1 ATG, a codon, is in red color and sequence starting the counting (in between the black color)) that codes for methionine. This condition is only for exon1, while you start counting the bases for all other exons we can count all the coding sequence which is black in color to till 790b.

4. For example, to design a DNA sequence with 30b length (Base) for Keratin 20, which is abbreviated as CK20 encoded by the KRT20 gene, is a protein present in humans, commonly found in colorectal cancer, transitional cell carcinomas and in Merkel cell tumors, but is absent in lung cancer, prostate cancer, and non-mucinous ovarian cancer. First we have to get the CK20 sequence by using one of the above databases.
5. The complementary sequence can be designed be Chargaff's rules, i.e., A pairs with T and C pairs with G and noncomplementary sequence is designed by non-matching pairs. For example,

Target: 5′-**ATC ACA GAT TTT GGG CGG GCC AAA CTG CTG**-3′

Complementary: 3′-**TAG TGT CTA AAA CCC GCC CGG TTT GAC GAC**-5′

Noncomplementary: 3′-**CGA CAC TCG GGG TTT ATT TAA CCC AGT AGT**-5′

Another important point we have to remember while designing DNA sequence is 3′ and 5′ end positioning. The 3′ always binds from the 5′ end group because the 5′ end have phosphate group and 3′ end have OH group. So each reaction of complementation occurs from 3′ position.

3.2.3 DNA Immobilization and Hybridization at GG1PAMAM Surfaces

The Au nanoparticles (*see* **Note 3** for preparation) were deposited onto the GG_1PAMAM by dipping the modified electrode for 2 h at 4 °C [55, 56]. 7 μL of 1 μM HS-ssDNA in 1 M NaCl is then drop casted on the GG1PAMAM–AuNP monolayer for 2.30 h. After extensive rinsing with the blank PBS buffer, monolayer surface thus formed is now allowed to hybridize with 7 μL of 1 μM (used in most of the experiments except in target concentration variation) complementary (II) and/or noncomplementary (III) targets in the hybridization buffer (4× SSC) for 2 h. Double-stranded DNA specific intercalator methylene blue was used further to enhance the discrimination effect (*see* **Note 4**). This is done by treating the hybridized surfaces with SDS (5 % in PBS pH 7.4) for 10–15 min followed by intercalation with methylene blue (1 mM in PBS pH 7.4) for 10–15 min and reassessing the surface behavior again in the presence of the $[Fe(CN)_6]^{3-}$.

3.3 Electrochemical Characterization of Immobilized HS-ssDNA and Target DNA Hybridization Sensing

In general, three electrode system consisting of working electrode (in the present study, gold is working electrode), Pt wire of length 2 cm and 1 mm diameter as counter electrode and Ag/AgCl reference electrode is used. The electrodes are dipped in the PBS containing 1 mM $[Fe(CN)_6]^{3-/4-}$ solution and connected to a potentiostat. The solution should be purged with inert N_2 gas to remove the dissolved O_2 from the measuring solution. For all type of electrochemical measurements (*see* **Note 5**) this configuration is constantly used. Generally, electrochemical techniques measure a change in either current or resistance of the electrode. In label-free method of DNA detection, change in the sensor signal influenced by the reaction between the recognition layer and the target DNA is monitored in the absence and presence of any reversible redox couple.

4 Notes

1. *Cleaning of gold surface*
   The gold undergoes slow aerial oxidation and absorbs organic materials and degrades from its original activity which requires surface cleaning. Polycrystalline gold electrode was initially cleaned in piranha solution (*70 %:30 % H_2SO_4:H_2O_2, caution: H_2SO_4 should be added slowly to the H_2O_2 to avoid the explosion due high heat generation and highly reactive towards organic matters*) by dipping the same for 2 h followed by mechanical polishing using ~1 μm diameter alumina slurry and sonicated in distilled water. Cleanses of the gold is verified by potential scanning in the window −0.5 to 1.6 V at a scan rate 20 mV/s in 0.2 M H_2SO_4 until the gold oxide formation region shows three distinct peaks with constant peak currents [57]. This method minimizes the surface roughness and produces more reproducible surfaces. In some reports, the dipping time is varied between 2 and 12 h which meant highly contaminated surfaces of gold. Further, the surface behavior is measured in the presence of 1 mM $[Fe(CN)_6]^{3-/4-}$ redox couple in 0.01 M PBS buffer. When we use this redox probe for assessing the surface state, the anodic peak-to-cathodic peak separation (ΔE_p) at low scan rates 5–50 mV/s (at least) should remain constant. When many electrodes are required, the change in the peak parameters between them should be maintained at ±0.3 μA. Currently, many labs are using the gold-coated silicon or mica chips in place of polycrystalline gold. In such cases, cleaning of the chips using piranha solution for 1–2 min are found to be sufficient for freshly prepared chips without polishing. For chips stored under inert (nitrogen or Argon) gas atmosphere for longer period cleaning more than 5–10 min is required to attain the ΔE_p in the range 60–70 mV in phosphate buffer.
   (a) *Immobilization of 3-mercaptopropionic acid (MPA) on gold surface*
   Since the alkane thiols form easily the covalent bond with gold to form monolayer via the $Au + HS\text{-}R \rightarrow Au\text{-}S\text{-}R + H^+$, they are extensively applied in electronics and sensor applications. The reaction can be simply done by incubating the gold

substrate in the respective alkane thiol solution. The formation of monolayer on the gold surface is confirmed by observing the decreased CV peak currents and increased ΔE_p than the unmodified Au surface. This behavior is consistently observed for alkane thiol molecules that are insulating. However, control of the monolayer property for the reproducible results is essential. In general, thiol monolayer formation on gold requires minimum of 1 h incubation. The variation in the peak current and ΔE_p between different electrodes should be maintained within the range of 5–10 % error. In the present work, MPA monolayer is formed by just dipping the pre-cleaned Au surface in 1 mM MPA in PBS. Some time the monolayer may not be formed perfectly on the gold surface and the current may remain closer to the bare electrode. In those cases, the electrode cleanness need to be rechecked. The freshly prepared piranha solution is essential for obtaining good activity of the gold electrode, described above. Ethanol solvent can be used in place of solvent PBS for preparing monolayer solution for better quality monolayers. For this, 75 % ethanol is best suited for monolayer formation. Generally, the MPA forms defective layer and care must be taken to form close packed layer which can be verified by observing maximum current decrease compared to the bare electrode and high intense hydrogen bonding band in the surface FTIR.

2. *Preparing Ethylene Diamine Carboamide*
   Generally, the ethylene diamine carboamide (EDC) and *N*-hydroxysuccinimide (NHS) are used together to activate amine terminal of the SAM monolayers for the attachment of ssDNA functionalized with acid group. The EDC and NHS are mixed in equal ratio and dropped on the monolayer surface. In this work, the carboxyl group of MPA is modified by the GG1PAMAM dendrimer by incubating the MPA monolayer in the solution containing the mixture of EDC (5 mM) and GG1PAMAM (10 mg) for 12 h at an ambient temperature. The NHS is avoided because the acid group is more reactive in the presence of EDC than using the mixture of EDC and NHS in phosphate buffer [58].

3. *AuNP preparation*
   Au nanoparticles were prepared by Frens method [59]. In this method required quantity of $AuCl_3 \cdot nH_2O$ (0.025 M stock solution, 5 mM) is taken in water and boiled. Sodium citrate (2 mg/2 mL) is added to the above boiling solution and boiling is continued until the solution turns from yellow to wine red color. The solution is transferred in eppendrof tube and stored at 4 °C until use. The formation of AuNP is confirmed by UV visible absorption peak at 520 nm and the average size

of the Au nanoparticle is 17.5 nm confirmed by transmission electron microscope (TEM).

4. *Immobilization of capture probe DNA on the pre-immobilized GG1PAMAM–AuNP layer*
   Thiolated DNA is immobilized on Au or AuNP by covalently by the reaction Au + HS-ssDNA → Au-S-ssDNA + H^+. Here, the HS-ssDNA is acting as an insulating layer and, hence, after their immobilization the CV peak current decreases and the charge transfer resistance (R_{CT} from impedance spectroscopy) increases. The relative change in these parameters depends again on the solvent used for immobilization. The extent of decrease of CV peak currents indicates the quality of the immobilized layer. Solvents such as water, ethanol, methanol, phosphate buffer, tris buffer, and salt solutions are being used. Among these, NaCl and $MgCl_2$ are used extensively as they stabilize the negatively charged ssDNA by neutralization and minimize the electrostatic repulsion. In our case, we have used 1 M NaCl for immobilizing the 1 μM HS-ssDNA on AuNP and it gives most dense ssDNA monolayer on the surface.

5. *Electrochemical characterization*
   A three electrode system consisting of working (graphene, graphene deposited on any conducting metal, or carbon electrodes), Ag/AgCl reference and Pt wire counter electrode is used in all electrochemical measurements. For electrochemical detection of DNA, several electrochemical techniques such as cyclic voltammetry, chronoamperometry, chronopotentiometry, and impedance are used to characterize the behavior of DNA films immobilized on the electrode surface.

Step 1: Characterization of Au in phosphate buffer in the presence of 1 mM $[Fe(CN)_6]^{3-/4-}$
The pre-cleaned Au electrode is examined for its behavior in the phosphate buffer in the first step. In Fig. 5 curve a represents the behavior of the unmodified Au measured in the presence of 1 mM $[Fe(CN)_6]^{3-/4-}$ at a scan rate 50 mV/s. The unmodified electrode showed ΔE_p 80 mV. The anodic and cathodic peak currents increases with scan rates and their respective potentials move towards higher potentials confirming the quasi reversible nature of the $[Fe(CN)_6]^{3-/4-}$ redox reaction. For further comparison only one scan rate is taken. For this purpose, it is recommended that the electrode need to be scanned for constant current behavior at the desired scan rate.

Step 2: Immobilization of Mercapto-Propionic Acid (MPA) on gold
The gold electrode is incubated in phosphate buffer containing 1 mM MPA at an ambient temperature for 1 h. The electrode is removed and washed with the blank solution. Now the behavior of the MPA layer is monitored in the presence of 1 mM $[Fe(CN)_6]^{3-/4-}$ redox couple in phosphate buffer and compared at the same scan

rate. The attachment of MPA on the gold surface blocks the active gold sites and also the MPA at pH 7.4 provides negative charge by its dissociation (HS–CH_2–CH_2–COO^-) and repel the $[Fe(CN)_6]^{3-/4-}$ direct diffusion through the molecular defects resulting in decreased peak current and increased ΔE_p. These parameters changes are highly dependent on the surface characteristics of the sensor and quality of the layer. The stepwise CV behavior of these layers in the presence of $[Fe(CN)_6]^{3-/4-}$ is presented in Fig. 5.

Step 3: Immobilization of GG1PAMAM on the MPA layer

The MPA layer is then incubated for 6 h in the phosphate buffer (pH 7.4) containing the coupling reagent EDC and GG1PAMAM at the ratio 9:1. The CV curve c represents the behavior of GG1PAMAM monolayer attached on the MPA layer. The layer showed increased peak currents and reduced ΔE_p equivalent to the unmodified bare Au electrode. This unexpected behavior is attributed to the following two reasons. (1) The lone pair on the NH groups of GG_1PAMAM catalyze the $[Fe(CN)_6]^{3-/4-}$ redox reaction as reported for the short-chain cysteamine [60]. (2) Another reason is that both the G_1PAMAM dendrimer and graphene are linear and planar; the negative charges on the COO^- head groups of the MPA layer could be minimized. These factors might reduce the negative repulsive force on the diffusion of $[Fe(CN)_6]^{3-/4-}$ through the molecular defects. Similar the R_{CT} for the GG_1PAMAM (1,379 Ω/cm) is 50 % lower than the unmodified electrode (R_{CT} 3,848 Ω/cm^2, Fig. 5, curve a).

Step 4: Deposition of AuNP on the GG1PAMAM monolayer

The GG1PAMAM modified electrode is dipped in the prepared AuNP solution for 2 h and reexamined in the phosphate buffer (pH 7.4) in the presence of 1 mM $[Fe(CN)_6]^{3-/4-}$. In this step, decreased peak current and increased ΔE_p have been observed indicating the decreased rate of electrocatalysis by the GG1PAMAM–AuNP composite, curve d. This could be related to the stabilization of AuNP by the graphene core of the GG1PAMAM.

Step 5: Immobilization of Now the GG1PAMAM–AuNP is

In this step, the GG1PAMAM–AuNP is incubated in 1 μM HS-ssDNA (capture probe) in NaCl for 2.30 h to immobilize the HS-ssDNA. This further decreases the redox activity of the $[Fe(CN)_6]^{3-/4-}$ due to the partial blocking effect of HS-ssDNA and the negative charge repulsion between the Hs-ssDNA and $[Fe(CN)_6]^{3-/4-}$, respectively, curve e.

Step 6: Hybridization of target DNA

Following the immobilization of the HS-ssDNA, the target DNA is allowed to interact with the ssDNA for 1.30 h in the sodium saline citrate buffer. This buffer is also termed as tris buffer. This is used to maintain the high salt concentration and pH 8.0 for stringent hybridization condition. The hybridized surface is further washed with blank phosphate buffer and studied in the phosphate buffer again under similar experimental conditions, curve f. This decreases further the redox activity of the $[Fe(CN)_6]^{3-/4-}$. The target hybridization increases the negative charge density and molecular crowding at the electrode/film interface. However, in the case of noncomplementary target hybridization, no significant change has been noticed. Therefore, the hybridized (complementary target hybridized) and un-hybridized (noncomplementary target hybridized) surfaces are discriminated by the electrochemical method.

Acknowledgments

Dr. V. Dharuman and K. Jayakumar acknowledge the Council of Scientific and Industrial Research, New Delhi, India for the financial support through project (CSIR No.03(1160)/10/EMR-II).

References

1. Tainaka K, Sakaguchi R, Hayashi H et al (2010) Design strategies of fluorescent biosensors based on biological macromolecular receptors. Sensors 10:1355–1376
2. Giepmans BN, Adams SR, Ellisman MH, Tsien RY (2006) The fluorescent toolbox for assessing protein location and function. Science 312:217–224
3. Johnsson N, Johnsson K (2007) Chemical tools for biomolecular imaging. ACS Chem Biol 2:31–38
4. Rao J, Dragulescu-Andrasi A, Yao H (2007) Fluorescence imaging *in vivo*: recent advances. Curr Opin Biotechnol 18:17–25
5. Cagnin S, Caraballo M, Guiducci C et al (2009) Overview of electrochemical DNA biosensors: new approaches to detect the expression of life. Sensors 9:3122–3148
6. Ozsoz MS (2012) Electrochemical DNA biosensors. Stanford Publishing, Singapore
7. Geim AK, Novoselov KS (2007) The rise of graphene nature materials. Nat Mater 6:183–191
8. Zhu Y, Murali S, Cai WW et al (2010) Graphene and graphene oxide—synthesis, properties, and applications. Adv Mater 22:3906–3924
9. Rao CNR, Sood AK, Subrahmanyam KS, Govindaraj A (2009) Graphene: the new two-dimensional nanomaterial. Angew Chem Int Ed 48:7752–7777
10. Eda G, Chhowalla M (2010) Current trends in shrinking the channel length of organic transistors down to the nanoscale. Adv Mater 22:20–32
11. Loh KP, Bao QL, Eda G, Chhowalla M (2010) Graphene oxide as a chemically tunable platform for optical applications. Nat Chem 2:1015–1024
12. Wu J, Agrawal M, Becerril HA et al (2010) Organic light-emitting diodes on solution-processed graphene transparent electrodes. ACS Nano 4:43–48
13. Jang H, Kim YK, Kwon HM et al (2010) A graphene-based platform for the assay of duplex-DNA unwinding by helicase. Angew Chem Int Ed 49:5703–5707
14. Wang Y, Li ZH, Hu DH et al (2010) Aptamer/graphene oxide nanocomplex for in situ molecular probing in living cells. J Am Chem Soc 132:9274–9276
15. Wang XH, Wang CY, Qu KG et al (2010) Ultrasensitive and selective detection of a prognostic indicator in early-stage cancer using graphene oxide and carbon nanotubes. Adv Funct Mater 20:3967–3971
16. Shi Y, Huang WT, Luo HQ, Li NB (2011) A label-free DNA reduced graphene based fluorescent sensor for highly sensitive and selective detection of hemin. Chem Commun 47: 4676–4678
17. Tang L, Wang AL, Loh JZ (2010) Graphene-based SELDI probe with ultrahigh exrtraction and sensitivity for DNA oligmer. J Am Chem Soc 132:10976–10977
18. Zhou M, Zhai Y, Dong S (2009) Electrochemical sensing and biosensing platform based on chemically reduced graphene oxide. Anal Chem 81:5603–5613

19. Lu CH, Yang HH, Zhu CL, Chen X, Chen GN (2009) A graphene platform for sensing biomolecules. Angew Chem Int Ed 48:4785–4787
20. Hu Y, Wang K, Zhang Q, Li F (2012) Decorated graphene sheets for label free DNA impedance biosensing. Biomaterials 33: 1097–1106
21. Huang PJJ, Liu J (2012) Molecular beacon lighting up on graphene-oxide. Anal Chem 84:4192–4198
22. Emilie D, Zhiyong Y, Kian PL (2011) Optimizing label-free DNA electrical detection on graphene platform. Anal Chem 83: 2452–2460
23. Shen JF, Shi M, Yan B et al (2004) Surface modification using photocrosslinkable chitosan for improving hemocompatibility. Colloids Surf B 38:47–53
24. Mohanty N, Berry V (2008) Graphene-based single-bacterium resolution biodevice and DNA transistor: interfacing graphene derivatives with nanoscale and microscale biocomponents. Nano Lett 8:4469–44763
25. Liu F, Choi JY, Seo TS (2010) Graphene oxide arrays for detecting specific DNA hybridization by fluorescence resonance energy transfer. Biosens Bioelectron 25:2361–2366
26. Wang ZJ, Zhou XZ, Zhang J et al (2009) Time-domain ab initio study of nonradiative decay in a narrow graphene ribbon. J Phys Chem C 113:14067–14070
27. Liu Y, Yu DS, Zeng C, Miao ZC, Dai LM (2010) Biocompatible graphene oxide-based glucose biosensors. Langmuir 26:6158–6160
28. Liu ZF, Jiang LH, Galli F et al (2010) A graphene oxide streptavidin complex for biorecognition—towards affinity purification. Adv Funct Mater 20:2857–2865
29. Jitendra SV, Sai VR, Soumyo M (2011) Dendrimers in biosensors—concept and applications. J Mater Chem 21:14367–14386
30. Ashavani K, Saikat M (2003) Investigation into the interaction between surface-bound alkylamines and gold nanoparticles. Langmuir 19:6277–6282
31. Lv W, Guo M, Liang MH et al (2010) Graphene DNA hybrids—self assembly and electrochemical detection performance. J Mater Chem 20:6668–6673
32. Zhang D, Liu X, Wang XJ (2011) Green synthesis of graphene oxide sheets decorated by silver nanoprims and their anti-bacterial properties. Inorg Biochem 105:1181–1186
33. Yang X, Xu M, Qiu W et al (2011) Graphene uniform decorated with gold nano dots—in situ synthesis, enhanced dispersibility and applications. J Mater Chem 21:8096–8103
34. Hu Y, Hua S, Li F et al (2011) Green synthesized gold nano particle decorated graphene sheets for label-free electrochemical impedance DNA hybridization biosensing. Biosens Bioelectron 26:4355–4361
35. Phama TA, Choib BC, Lima KT, Jeonga YT (2011) A simple approach for immobilization of gold nano particle on graphene oxide sheets by covalent bonding. Appl Surf Sci 257:3350–3357
36. Wang Y, Li Z, Wang J, Li J, Lin Y (2011) Graphene and graphene oxide—biofunctionalization and applications in biotechnology. Trends Biotechnol 5:205–212
37. Lee B, Chen Y, Duerr F et al (2010) Modification of electronic properties of graphene with self-assembled mono layers. Nano Lett 10:2427–2432
38. Li W, Tan C, Lowe MA, Abruna HD, Ralph DC (2011) Electrochemistry of individual monolayer graphene sheets. ACS Nano 5:2264–2270
39. Shao Y, Wang J, Wu H, Liu J, Aksay IA, Lin Y (2010) Graphene based electrochemical biosensors. A review. Electroanalaysis 22:1027–1036
40. Sharma R, Sharma Baik JH, Perera CJ, Strano MS (2010) Anamolously large reactivtity of single graphene layers and edges toward electron transfer chemistries. Nano Lett 10:398–405
41. Koehler FM, Jacobson A, Enasslin K, Stamper C, Stark WJ (2010) Selective chemical modification of graphene structures. Distinction between single and bilayer graphene. Small 6:1125–1130
42. Yao Y, Li Z, Moon KS, Agar J, Wong C (2011) Controlled growth of multilayer few layer and single layer graphene on metal substrates. J Phys Chem C 115:5232–5238
43. Bohem HP, Clauss A, Fisher GO, Hofmann U (1962) Thin carbon leaves. Z Naturforsch 17b:150–153
44. Dresselhaus MS, Dresselhaus G (1981) Intercalation compounds of graphite. Adv Phys 30:139
45. Novoselov KS, Geim AK, Morozov SV et al (2004) Electric field effect in thin carbon films. Science 306:666–669
46. Shuai W, Priscilla KA, Ziqian W, Ai Ling LT (2010) High mobility printable, and solution-processed graphene electronics. Nano Lett 10:92–98
47. Scott G, Song H, Minsheng W et al (2008) A chemical route to graphene for device applications. Nature 3:563–568
48. Yenny H, Valeria N, Mustafa L, Fiona M (2008) Nanotechnology. High-yield production of graphene by liquid-phase exfoliation of graphite. Nature 3:563–568
49. Joshua R, Xiaojun W, Kathleen T (2010) Nucleation of epitaxial graphene on SiC (0001). ACS Nano 4:153–158

50. Dreyer Daniel R, Park S, Bielawski CW, Ruoff RS (2010) The chemistry of graphene oxide. Chem Soc Rev 39:228–240
51. Gao Y, Chen XQ, Xu H, Zou YL (2010) Highly-efficient fabrication of nanoscrolls from functionalized graphene oxide by Langmuir-Blodgett method. Carbon 48:4475–4482
52. Kovtyukhova NI, Ollivier PJ, Martin BR (1999) Layer-by-layer assembly of ultrathin composite films from micron-sized graphite oxide sheets and polycations. Chem Mater 11:771–778
53. Hummers WS, Offeman RE (1958) Preparation of graphite oxide. J Am Chem Soc 80:1339–1340
54. Imabayashi S, Hobara D, Kakiuchi T (1997) Selective replacement of adsorbed alkane thiols in phase-separated binary self-assembled mono layer by elactrochemical partial desorption. Langmuir 13:4502–4504
55. Jasuji K, Linn J, Melton S, Berry V (2010) Microwave-reduced uncapped metal nano particles on graphene tuning catalytic, electrical and Raman properties. Phys Chem Lett 1: 1853–1860
56. Vinodgopal K, Neppolin B, Lightcap LV et al (2010) LettSonolytic design of graphene–Au nano composites. Simultaneous and sequential reduction of graphene oxide and Au(111). J Phys Chem 1:1987–1993
57. Rafaela FC, Renato SF, Lauro TK (2005) Polycrystalline gold electrode—a comparative study of pretreament procedure used for cleaning and thiol self-assembly mono layer formation. Eelectroanalysis 17:1251–1259
58. Li NB, Kwak J (2007) A penicillamine biosensor based on tyrosinase immobilized on nano-Au/PAMAM dendrimer modified gold electrode. Electroanalysis 19:2428–2436
59. Frens G (1973) Controlled nucleation for the regulation of particle size in monodsiperse gold suspension. Nat Phys Sci 241:20–22
60. Yi X, Huang-Xian J, Hong-Yuan C (2000) Direct electrochemistry of horseradish peroxidase immobilized on a colloid/cysteamine-modified gold electrode. Anal Biochem 278:22–28

Part VI

Advances in FISH

Chapter 18

Application of PNA Openers for Fluorescence-Based Detection of Bacterial DNA

Irina Smolina

Abstract

Peptide nucleic acid (PNA) openers have the unique ability to invade double-stranded DNA with high efficiency and sequence specificity, making it possible to detect short (about 20 bp), single-copy bacterial DNA sequences. PNA openers bind to a target signature site on one strand of bacterial DNA, leaving the other strand open for hybridization with a circularizable oligonucleotide probe. The assembled complex serves as a template for rolling circle amplification. The obtained amplicon is decorated with short, single-stranded DNA probes carrying fluorophores and detected via fluorescence microscopy.

Key words Peptide nucleic acid (PNA), Padlock probe, Rolling circle amplification (RCA), Bacterial detection, Fluorescence in situ hybridization (FISH)

1 Introduction

Fluorescence in situ hybridization (FISH) is a widely used method for many assays in clinical diagnostics. This technique makes it possible to accurately identify viruses and bacteria in clinical specimens that contain microorganisms, either directly or after culture enrichment [1–3]. The key step of FISH involves the detection of a nucleic acid region. Therefore, DNA molecules have typically been used to probe for the sequences of interest. However, since the turn of the century, an increasing number of laboratories have started to move on to the more robust DNA methods based on the application of DNA mimics, most notably peptide and locked nucleic acids (peptide nucleic acid—PNA and locked nucleic acid—LNA). Due to its high copy number, ribosomal RNA (rRNA) sequences are commonly used as targets for the fluorescently labeled probes. PNA FISH is clearly the most advanced technology with respect to applications of the rRNA-targeted FISH; the robustness of the method is evident in the commercially available diagnostic tests (www.advandx.com). There is a major shortcoming of

Dmitry M. Kolpashchikov and Yulia V. Gerasimova (eds.), *Nucleic Acid Detection: Methods and Protocols*, Methods in Molecular Biology, vol. 1039, DOI 10.1007/978-1-62703-535-4_18, © Springer Science+Business Media New York 2013

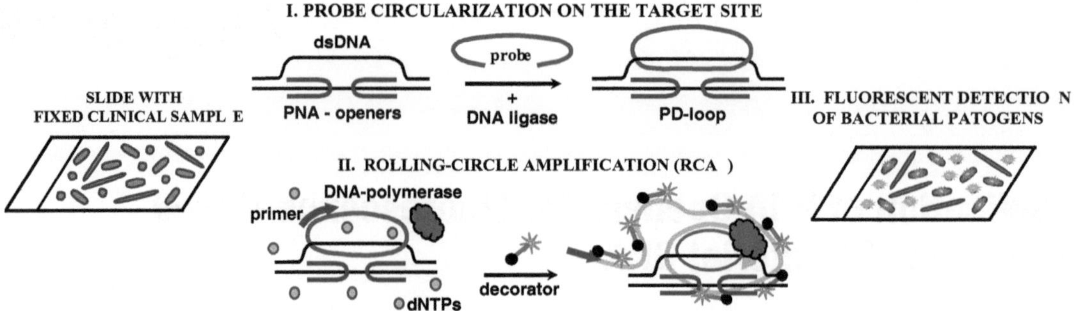

Fig. 1 PNA-based bacterial DNA detection scheme

the rRNA-FISH: the rRNA sequences of closely related strains, subspecies, or even different species are often identical and therefore cannot always be used as differentiating markers. As a result, rRNA-FISH cannot be used for drug-susceptibility testing [4].

rRNA has long been regarded as the only suitable molecule for bacterial FISH, but the development of potent signal amplification techniques has made it possible to move away from high copy number (10^4–10^5) rRNA targets. Catalyzed reporter deposition (CARD-FISH) and multiply labeled polynucleotide probes RING-FISH (recognition of individual genes) can be used to target a variety of other nucleic acids such as mRNA and tmRNA [5–9]. However, a major shortcoming of the CARD technique is the large size of the enzyme, which means that damaging permeabilization procedures are necessary, and the characteristic probe length in RING-FISH is rather long (usually more than 200 bases) for sequence-specific detection.

Here, we describe a new FISH-type approach that we have developed and validated during the recent years [10–12]. This method is used for the detection of short (about 20 nucleotides long) signature DNA sequences present in single copies per bacterial genome, such as nucleotide sequences of individual genes or gene fragments on chromosomal and/or plasmid DNA. This genomic sequence-based molecular technique can simultaneously detect bacteria and/or key genetic markers of drug resistance in order to identify and characterize various bacterial pathogens. The new PNA-based approach has a high specificity of detection (in contrast to RING-FISH), down to single-base substitutions, due to the short length of the target sequence. Specific regions of genomic duplex DNA are targeted, which provides sufficient versatility to design specific probes and makes it possible to discriminate bacterial strains with high selectivity. Figure 1 demonstrates the key steps of the detection method.

Pyrimidine PNAs are the main tools in our approach [13]. The strand-invading bis-PNAs (PNA openers) have a unique ability to pry open double-stranded DNA (dsDNA) via binding to one of the two DNA strands, leaving the other strand displaced. The local opening of a chosen DNA target site occurs when a pair of PNA openers binds one DNA strand within the DNA double helix (over a length of about 20–30 nt), leaving the other strand accessible for hybridization with synthetic oligonucleotide probes (Fig. 1, step I). Such a complex between dsDNA, a pair of PNAs, and the oligonucleotide probe is called a PD-loop [14, 15]. Statistically, on sites capable of forming PD-loop, are expected to be met once per several hundred base pairs of a random DNA sequence [12, 16].

After a PD-loop site is opened with a pair of PNA openers, a circular probe is assembled by T4 DNA ligase via circularization of an oligodioxyribonucleotide (ODN) with two termini complementary to the two halves of the displaced strand. Such circular probe assembly is exceedingly sequence specific since only chosen sites are opened by PNA openers and the remaining DNA maintains its duplex form and is inaccessible for mismatch hybridization [17, 18]. The circularized ODN forms about two turns around the displaced DNA strand and thus the circular probe remains correctly positioned along the double helix and will not slip away during various manipulations. Another advantage of the probe circularization consists in the fact that its assembly is possible only as a result of multiple coinciding events (the local opening of dsDNA by two PNA openers, the probe hybridization, and ligation). To uniquely define a specific fragment in genomic DNA, probes with a length of at least 15–20 nucleotides are required (e.g., two primers in PCR). Hybrids of such length are too stable to be sensitive to a base mispairing and require accurate design and precise temperature control to achieve good discrimination between SNPs. Two-component probes improve selectivity of nucleic acid recognition and highly increase the specificity of the ring formation, which provides a robust distinction between single-nucleotide variants under standard reaction conditions [19].

The assembled circular probe serves as a template for the rolling circle amplification (RCA) reaction, which is initiated by an appropriate primer. Isothermal RCA of the circular probe yields a long, single-stranded DNA concatomeric amplicon, in which the original PD-site is repeated more than a thousand times. The RCA amplification step provides the required sensitivity of detection on the level of a single copy of the target DNA site per genome [19, 20]. Furthermore, the large multiplexing capacity intrinsic to RCA enables parallel detection of different DNA marker sites. For fluorescence-based detection, the single-stranded amplicon produced by RCA is hybridized with the fluorophore-tagged decorator probe that yields a multiply fluorescently labeled product (Fig. 1, step II). These decorators are linear ODNs with fluorophores at their termini. Application of different fluorophores with resolved spectra, together with multiplex RCA, allows simultaneous labeling of various PD-sites as biomarkers for parallel detection and ensures reliable multiplex diagnosis of several specific DNA signature sites. The fluorescent signal is readily detected by standard techniques using fluorescence microscopy (Fig. 1, step III).

In summary, this method is based on a combination of the advanced PNA-based technology and the signal-enhancing RCA reaction to achieve a highly specific and sensitive assay. A specific PNA–DNA construct serves as an exceedingly selective and very effective biomarker, while RCA enhances the detection sensitivity and provides with a highly multiplexed assay system. We have shown that the PNA-based approach can be applied for the detection of bacteria fixed on slides. Figure 2 shows representative results of the detection of three types of bacteria: *E. coli*, *S. mutans*, and *B. subtilis*. The fixation procedure not only stabilizes cell morphology but also permeabilizes the cell membrane to allow all subsequent reagents and probes to diffuse to their intracellular targets [10, 11]. We have recently tested this new, more versatile, isothermal approach for rapid and sensitive diagnostics of clinically relevant organisms [12].

2 Materials

2.1 Fixed Slide Preparation

1. Cell fixative solution: Methanol:glacial acetic acid = 3:1 (v/v). To prepare 100 mL of the solution, mix 75 mL of methanol with 25 mL of glacial acetic acid.
2. Dehydration solution 1: 70 % ethanol.
3. Dehydration solution 2: 90 % ethanol.
4. Dehydration solution 3: 100 % ethanol.

2.2 PNA Invasion

1. PNA-binding (low-salt) buffer: 10 mM Na-phosphate, pH 6.8–7.2 (*see* **Note 1**). Dissolve 1.64 g of Na_2HPO_4–$7H_2O$ and 0.47 g of NaH_2PO_4 in 500 mL of H_2O. Adjust to the desired pH and bring the volume to 1 L with H_2O.
2. PNA opener solution: 2 μM each PNA opener in the PNA-binding buffer (*see* **Note 2**). Mix 2 μL of 40 μM opener 1 and 2 with 36 μL of PNA-binding buffer to prepare 40 μL of the PNA opener solution per slide. A pair of appropriate [7/8 + 7/8]-mer bis-PNAs with achiral and uncharged “classical” *N*-(2-aminoethyl)glycine backbone and pseudoisocytosine instead of cytosine in one-half are required (*see* **Note 3**).

2.3 ODN Ligation

1. ODN ligation mixture (*see* **Note 4**): T4 ligase buffer (×1), 5 U of T4 DNA ligase, 2 μM circularizable probe.
2. Phosphate-buffered saline (PBS): 137 mM NaCl, 2.7 mM KCl, 10 mM Na_2HPO_4/KH_2PO_4, pH of 7.4. To prepare 1 L of PBS, add 8 g of NaCl, 0.2 g of KCl, 1.44 g of Na_2HPO_4,

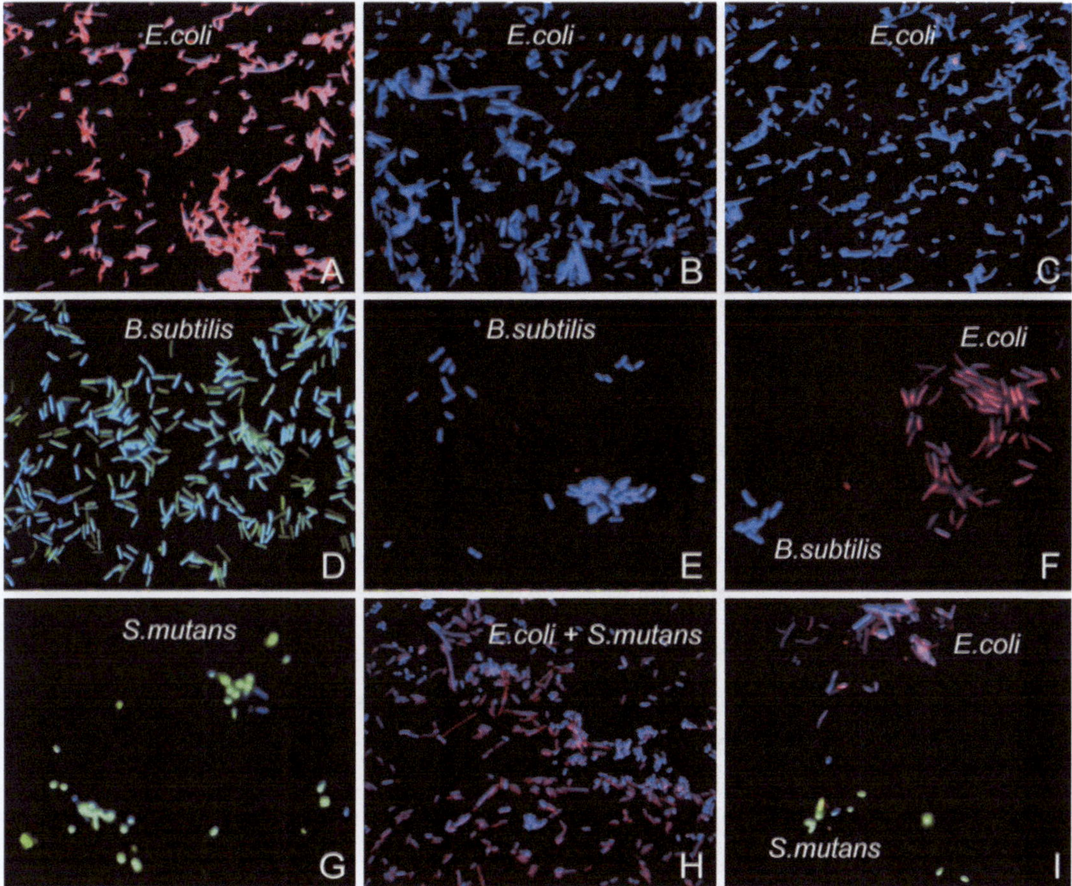

Fig. 2 Representative images of bacterial cells observed by fluorescent microscope in experiments performed according to the scheme in Fig. 1. Each image is a superposition of two separate images, with DAPI and Cy3 or DAPI and FITC signals pseudocolored in *blue* and *red* or *blue* and *green*, respectively. (**a**) *E. coli* cells to which the probes corresponding to an *E. coli* target site were applied. Virtually all cells displayed very bright spots. (**b**) No such spots were observed in numerous control experiments, in which any components of our procedure were omitted. Shown is the control in the absence of PNA openers. (**c**) *E. coli* cells to which the combination of probes specific to *B. subtilis* was applied: no signal. (**d**) *B. subtilis* cells to which the probes corresponding to a *B. subtilis* target site were applied. (**e**) The same procedure as in panel (**a**) with the combination of all probes specific to *E. coli* applied to *B. subtilis* cells: no signal. (**f**) The combination of probes specific to *E. coli* applied to a mixture of *E. coli* and *B. subtilis* cells. Only *E. coli* cells are labeled. (**g**) *S. mutans* cells to which the probes corresponding to an *S. mutans* target site were applied. Virtually all cells are colored in *green*. (**h**) The combination of probes specific to *E. coli* applied to a mixture of *E. coli* and *S. mutans* cells. Only *E. coli* cells are labeled. (**i**) Images of bacterial cells observed by fluorescent microscopy in experiments performed with a mixture of *E. coli* and *S. mutans* cells with the combination of probes specific to an *E. coli* target site and to an *S. mutans* target site. Figure obtained from ref. 10

and 0.24 g of KH_2PO_4 to 800 mL of H_2O. Adjust the pH to 7.4 with HCl. Add water to 1 L. Sterilize by autoclaving. Alternatively, purchase ready to use from a supplier.

2.4 Rolling Circle Amplification

1. SSC buffer (×20) (stock solution): 3 M NaCl, 300 mM sodium citrate. Dissolve 175.3 g of NaCl and 88.2 g of trisodium citrate dihydrate in 800 mL of distilled H_2O, adjust pH to 7.0 with HCl, and adjust the volume to 1 L with addition of distilled H_2O. Sterilize by autoclaving.
2. Buffer A: 100 mM Tris–HCl, 150 mM NaCl, 0.05 % Tween20.
3. Phi29 DNA polymerase buffer (×1): 50 mM Tris–HCl, pH 7.5, 10 mM $MgCl_2$, 10 mM $(NH_4)_2SO_4$, 4 mM dithiothreitol.
4. RCA reaction mixture: 2 μM primer, 10 U of Phi29 DNA polymerase, 200 μM dNTPs, 1× Phi29 DNA polymerase buffer (×1), 200 μg/mL BSA, 2 μM fluorescently labeled decorator probe.

2.5 Counterstaining, Mounting, and Visualization

1. T buffer (×4): SSC (×4), 0.05 % Tween 20.
2. Counterstaining mounting medium: 50 mg/mL of 4,6-diamidino-2-phenylindole (DAPI) in PBS.

2.6 Instrumentation

1. 15-mL sterile culture tubes.
2. Poly-lysine-coated microscope slides.
3. 24×50 mm coverslips.
4. Rubber cement.
5. Clear nail polish.
6. Slide jar with cover.
7. Shaker.
8. Humidified chamber.
9. Fluorescent microscope with the 100× objective.

3 Methods

3.1 Growth and Harvest of Bacterial Cells

1. Grow cells overnight in an appropriate medium in 15-mL sterile culture tubes (*see* **Note 5**).
2. Harvest cells by centrifuging the culture tubes at 3,700×*g* for 10 min.

3.2 Preparing Fixed Slides

1. Add freshly prepared ice-cold cell fixative solution to cells and incubate for 15 min on shaker at room temperature.
2. Wash the cells twice in 1 mL of cell fixative solution.
3. Resuspend cells in 0.2 mL of cell fixative solution and place a drop on a clean poly-lysine microscope slide.
4. Dehydrate the slides through a series of dehydration solutions 1–3 by incubating the slides in 40 mL of each solution for 2 min (*see* **Note 6**).

5. Drain and dry slides thoroughly (*see* **Note** 7).
6. Proceed to PNA invasion (*see* **Note 8**).

3.3 PNA Invasion

1. Wash the dehydrated slides twice in PNA-binding (low-salt) buffer to decrease ionic strength (*see* **Note 9**).
2. Drain and dry slides thoroughly (*see* **Note** 7).
3. Apply 40 μL of the PNA opener solution to each slide.
4. Cover each slide with a 24 × 50 mm coverslip.
5. Seal the coverslips with rubber cement using the dispenser tips.
6. Incubate the slides at 45 °C overnight in a humidified chamber.

3.4 ODN Ligation

1. Remove the coverslip by gently peeling back the rubber cement.
2. Wash the slides twice in PBS to remove the unbound PNA.
3. Drain and dry the slides thoroughly (*see* **Note** 7).
4. Apply 50 μL of the ODN ligation mixture to each slide.
5. Cover each slide with a 24 × 50 mm coverslip.
6. Incubate the slides at room temperature for 2 h in the humidified chamber.

3.5 Rolling Circle Amplification

1. Remove the coverslip and wash the slides twice in SSC buffer (×2) and once in buffer A (*see* **Note 10**).
2. Drain and dry the slides thoroughly (*see* **Note** 7).
3. Apply 50 μL of RCA mixture to each slide (*see* **Note 11**).
4. Cover each slide with a 24 × 50 mm coverslip.
5. Seal the coverslips with rubber cement using the dispenser tips.
6. Incubate the slides at 37 °C for 4 h in the humidified chamber, and wrap the chamber in aluminum foil (*see* **Note 12**).

3.6 Counterstaining, Mounting, and Visualization

1. Remove the coverslip from the slide by gently peeling off the rubber cement.
2. Wash the slides twice in buffer A and once in T buffer (×4).
3. Drain and dry slides thoroughly (*see* **Note** 7).
4. Apply a drop of mounting medium onto the slide.
5. Cover the slide with a 24 × 50 mm coverslip and wait for 5 min.
6. Seal the coverslips in place using clear nail polish.
7. Visualize the slides using the 100× objective of the fluorescent microscope with an appropriate filter (*see* **Note 13**).

4 Notes

1. Addition of 1 mM EDTA is recommended to bind bivalent cations.
2. The PNA openers can bind to a duplex DNA region carrying the 20–25 bp site that is able to form the extended PD-loop. This site must contain two 7/8-bp-long oligopurine PNA-binding sites located on the same DNA strand and separated by up to 10 bp of mixed purine–pyrimidine sequence.
3. The incorporation of two or three positively charged terminal lysines and use of noncharged linker of three egl units (egl = 8-amino-3,6-dioxaoctanoic acid) are recommended for better stability and specificity of PNA–DNA complex [21]. PNAs of the desirable sequence can be purchased from a number of vendors.
4. Circularizable ODN probes must carry sequences complementary to target DNA sites at the 3′- and 5′-ends. They must also carry the sequence complementary to the RCA primer and an arbitrary sequence in the middle, which has to be designed to hybridize specific decorator probes to the RCA product and to discriminate bacterial strains from one another.
5. Selection of medium depends on bacterial strain. For example, *E. coli* and *B. subtilis* are cultured in Luria-Bertani medium while *S. mutans* and *S. aureus* are cultured in brain heart infusion broth. Sufficiency of the cell growth can be determined by measuring OD_{600} of the sample.
6. Slide dehydration can be performed in 50 mL vertical slide jars.
7. Allow slides to air-dry at room temperature on a slide rack.
8. Slides can be stored at 4 °C until ready to use.
9. Wash slides in a 50 mL vertical slide jar; use about 40 mL of buffer.
10. Wash the slides in a 50 mL vertical slide jar, 5 min per wash, in shaker. Use 40 mL of SSC and buffer A.
11. From **step 3** on, the procedure should be performed in the dark.
12. RCA can proceed overnight to enhance signal.
13. This part of the procedure should be performed in the dark.

References

1. Mothershed EA, Whitney AM (2006) Nucleic acid-based methods for the detection of bacterial pathogens: present and future considerations for the clinical laboratory. Clin Chim Acta 363:206–220
2. Procop GW (2002) In situ hybridization for the detection of infectious agents. Clin Microbiol Newsl 24:121–125
3. Wagner M, Horn M, Daims H (2003) Fluorescence in situ hybridization for the iden-

tification and characterization of prokaryotes. Curr Opin Microbiol 6:302–309

4. Zwirglmaier K (2005) Fluorescence in situ hybridization—the next generation. FEMS Microbiol Lett 246:151–158
5. Amann R, Glockner FO, Neef A (1997) Modern methods in subsurface microbiology: in situ identification of microorganisms with nucleic acid probes. FEMS Microbiol Rev 20:191–200
6. Bakermans C, Madsen EL (2002) Detection in coal tar waste-contaminated groundwater of mRNA transcripts related to naphthalene dioxygenase by fluorescent in situ hybridization with tyramide signal amplification. J Microbiol Methods 50:75–84
7. Pernthaler A, Amann R (2004) Simultaneous fluorescence in situ hybridization of mRNA and rRNA in environmental bacteria. Appl Environ Microbiol 70:5426–5433
8. Wagner M, Schmid M, Juretschko S, Trebesius KH, Bubert A, Goebel W et al (1998) In situ detection of a virulence factor mRNA and 16S rRNA in Listeria monocytogenes. FEMS Microbiol Lett 160:159–168
9. Zwirglmaier K, Ludwig W, Schleifer KH (2004) Recognition of individual genes in a single bacterial cell by fluorescence in situ hybridization: RING-FISH. Mol Microbiol 51:89–96
10. Smolina I, Kuhn H, Lee C, Frank-Kamenetskii MD (2008) Fluorescence-based detection of short DNA sequences under non-denaturing conditions. Bioorg Med Chem 16:84–93
11. Smolina I, Lee C, Frank-Kamenetskii MD (2007) Detection of low-copy-number genomic DNA sequences in individual bacterial cells by using peptide nucleic acid-assisted rolling circle amplification and fluorescence in situ hybridization. Appl Environ Microbiol 73:2324–2328
12. Smolina IV, Miller NS, Frank-Kamenetskii MD (2010) PNA-based microbial pathogen identification and resistance marker detection: an accurate, isothermal rapid assay based on genome-specific features. Artif DNA PNA XNA 1:1–7
13. Nielsen PE, Egholm M, Berg RH, Buchardt O (1991) Sequence-selective recognition of DNA by strand displacement with a thymine-substituted polyamide. Science 254:1497–1500
14. Uhlmann E, Peyman A, Breipohl G, Will DW (1998) PNA: synthetic polyamide nucleic acids with unusual binding properties. Angew Chem Int Ed 37:2797–2823
15. Bukanov NO, Demidov VV, Nielsen PE, Frank-Kamenetskii MD (1998) PD-loop: a complex of duplex DNA with an oligonucleotide. Proc Natl Acad Sci U S A 95:5516–5520
16. Demidov VV, Frank-Kamenetskii MD (2004) Two sides of the coin: affinity and specificity of nucleic acid interactions. Trends Biochem Sci 29:62–71
17. Demidov VV, Kuhn H, Lavrentieva-Smolina IV, Frank-Kamenetskii MD (2001) Peptide nucleic acid-assisted topological labelling of duplex DNA. Methods 23:123–131
18. Kuhn H, Demidov VV, Frank-Kamenetskii MD (2000) An earring for the double helix: assembly of topological links comprising duplex DNA and a circular oligodeoxynucleotide. J Biomol Struct Dyn 17(Supp 1):221–225
19. Nilsson M (2006) Lock and roll: single-molecule genotyping in situ using padlock probes and rolling-circle amplification. Histochem Cell Biol 126:159–164
20. Zhang D, Wu J, Ye F, Feng T, Lee I, Yin B (2006) Amplification of circularizable probes for the detection of target nucleic acids and proteins. Clin Chim Acta 363:61–70
21. Egholm M, Christensen L, Dueholm KL, Buchardt O, Coull J, Nielsen PE (1995) Efficient pH-independent sequence-specific DNA binding by pseudoisocytosine-containing bis-PNA. Nucleic Acids Res 23:217–222

Chapter 19

DNA Probes for FISH Analysis of C-Negative Regions in Human Chromosomes

Evgeniy S. Morozkin, Tatyana V. Karamysheva, Pavel P. Laktionov, Valentin V. Vlassov, and Nikolay B. Rubtsov

Abstract

Fluorescent in situ hybridization (FISH) is a powerful technology for studying the chromosome organization and aberrations as well as for searching the homology between chromosomal regions in mammals. Currently, FISH is used as a simple, rapid, and reliable technique for analyzing chromosomal rearrangements and assigning chromosomal breakpoints in modern diagnosing of chromosomal pathology. In addition to cloned DNA fragments, the DNA probes produced by sequence-independent polymerase chain reaction are widely used in FISH assays. As a rule, the DNA probes generated from a genomic or chromosomal DNA by whole genome amplification are enriched for repetitive elements and, consequently, efficient FISH analysis requires that repetitive DNA hybridization is suppressed. The linker-adapter polymerase chain reaction (LA-PCR) using the genomic DNA hydrolyzed with *Hae*III and *Rsa*I restriction endonucleases allows the repetitive DNA fraction in DNA probe to be decreased and gene-rich DNA to be predominantly amplified. The protocol described here was proposed for production of the DNA probes for enhanced analysis of the C-negative regions in human chromosomes.

Key words C-negative regions, Linker-adapter polymerase chain reaction (LA-PCR), FISH, Supernumerary marker chromosomes

1 Introduction

The modern advance in molecular cytogenetics depends on the available DNA probes. Currently, the PCR techniques are widely used for whole genome amplification. The whole genome amplification methods include different techniques, such as primer extension preamplification PCR (PEP-PCR) [1], degenerate oligonucleotide primed PCR (DOP-PCR) [2], interspersed repetitive sequences-PCR (IRS-PCR; in particular, Alu-PCR) [3], as well as the techniques based on ligating adapters to fragments of original DNA with subsequent amplification of adapter-flanked DNA sequences (linker-adapter PCR, LA-PCR) [4]. These techniques are now used for generation of DNA probes derived from sorted

Dmitry M. Kolpashchikov and Yulia V. Gerasimova (eds.), *Nucleic Acid Detection: Methods and Protocols*,
Methods in Molecular Biology, vol. 1039, DOI 10.1007/978-1-62703-535-4_19, © Springer Science+Business Media New York 2013

or microdissected chromosomes [5, 6] and from DNA of a group of cells or even an individual cell [7, 8].

However, both the production of specific DNA probes for fluorescent in situ hybridization (FISH) and the FISH analysis of individual chromosomal regions are complicated by a high content of repetitive sequences in the eukaryotic genomes. Even in the case of uniform whole genome DNA amplification, which gives the DNA library adequately representing the initial DNA, the DNA probe thus obtained contains a considerable amount of repetitive sequences. Therefore, to obtain the FISH signal of unique DNA sequences, the hybridization of repeats should be suppressed via a prehybridization with a 50–100-fold excess of Cot-1 DNA [9].

We propose a protocol for production of DNA probes which can be efficiently used for cytogenetic analysis of C-negative chromosomal regions without suppressing DNA repeats. Based on the fact that a higher G+C content is characteristic of the unique sequences in the human genome, whereas the repeated sequences are, as a rule, enriched for A and T nucleotides [10], and the data on an in silico hydrolysis of human genomic DNA with various restriction endonucleases, we proposed to generate DNA probes by LA-PCR using the *Hae*III and *Rsa*I restriction endonucleases for enriching these probes in unique sequences [11]. The described protocol for producing DNA probes can be efficiently used for cytogenetic analysis of C-negative chromosomal regions without any suppression of DNA repeats, constituting heterochromatin. One of the possible applications for DNA probes specific to the C-negative regions of human chromosomes is detection of euchromatic DNA in the small supernumerary marker chromosomes for initial estimation of their clinical significance [11].

2 Materials

Ultrapure water (prepared by purifying deionized water to attain a sensitivity of 18 MΩ cm at 25 °C) and analytical grade reagents should be used for preparation of all solutions. All reagents should be prepared and stored at a room temperature (unless indicated otherwise). Note that some of the chemicals below are environmental toxins (e.g., formaldehyde and formamide). All manipulations with these chemicals should be confined to a chemical fume hood. Please ensure that these substances are collected and treated after use as a hazardous waste.

2.1 LA-PCR

1. TE-buffer: 10 mM Tris–HCl, 0.1 mM EDTA, pH 7.5.
2. Human genomic DNA (*see* **Note 1**).
3. Endonuclease *Hae*III (*see* **Note 2**).
4. Endonuclease *Rsa*I.

5. Endonuclease *Eco*32I.
6. DNA adapter: Mix in equimolar ratio the oligonucleotides 5′-ACGCTCGAGACGGTCGAT-3′ and 5′-pATCGACC-GTCTCGAGCGTTTTT-3′ to a final adaptor concentration of 15 μM in TE-buffer, heat at 95 °C for 5 min, and incubate at 37 °C for 30 min; store at −20 °C.
7. T4 DNA ligase.
8. 10× T4 DNA ligase buffer: 400 mM Tris–HCl, 100 mM $MgCl_2$, 100 mM DTT, 5 mM ATP; pH 7.8.
9. 50 % PEG 6000.
10. 10× dNTP mix: solution of 2 mM dATP, dTTP, dCTP, dGTP.
11. *Taq* DNA polymerase.
12. 10× *Taq* DNA polymerase buffer.
13. 50 mM $MgCl_2$.
14. Ethanol (70 and 100 %).
15. Benchtop thermostat (4–99 °C).
16. Centrifuge with refrigeration.
17. Thermocycler with heating lid.
18. Mini agarose gel electrophoresis apparatus.
19. UV transilluminator.

2.2 Probe Labeling

1. DNA sample obtained in Subheading 2.1.
2. *Taq* DNA polymerase.
3. 10× *Taq* DNA polymerase buffer.
4. 50 mM $MgCl_2$.
5. 10× dNTP solution for labeling: 2 mM dATP, 2 mM dCTP, 2 mM dGTP, 1 mM dTTP, 1 mM TAMRA-dUTP.
6. Thermocycler with heating lid.

2.3 Postlabeling Treatment of Labeled DNA Probe

1. Labeled DNA probe.
2. Salmon sperm DNA (10 mg/mL).
3. Ethanol (70 and 100 %).
4. 3 M sodium acetate, pH 5.2.
5. Hybridization buffer: 20 % dextran sulfate, 2× SSC, 0.05 % NP-40, 50 % formamide, pH 7.0. Store in aliquots at −20 °C.
6. Centrifuge with refrigeration.

2.4 Preparing Metaphase Chromosome Spread

1. Cell suspension.
2. Fixative solution: methanol:glacial acetic acid (3:1). Should be freshly prepared.

3. Microscope slides (*see* **Note 3**).
4. Microscope coverslips (24×24 mm).
5. Slide storage boxes.
6. Ethanol (100 %).
7. Heat block (30–100 °C).

2.5 Slide Pretreatment and Denaturation

1. 2× SSC.
2. 100 μg/mL RNase A in 2× SSC.
3. Ethanol (70, 80, and 100 %).
4. Pepsin solution: 0.05 % pepsin in 10 mM HCl.
5. Phosphate-buffered saline (PBS).
6. 50 mM $MgCl_2$ in PBS.
7. Solution A: 50 mM $MgCl_2$, 1 % formaldehyde in PBS.
8. Air bath (37 °C).

2.6 Fluorescent In Situ Hybridization

1. Labeled DNA probe.
2. Rubber cement.
3. Heat block (4–99 °C).
4. Moist chamber (*see* **Note 4**).
5. Air bath (42 °C).

2.7 Slide Washing and Mounting

1. 1× SSC.
2. Solution B: 4× SSC, 0.2 % NP-40.
3. PBS.
4. Ethanol (70, 80, and 100 %).
5. DAPI/antifade solution (*see* **Note 5**).
6. Water bath (45 and 60 °C).
7. Thermostated shaking platform (45 °C, 90–120 rpm).

2.8 Viewing and Interpretation

1. Fluorescence microscope with appropriate filter set specific for the fluorochromes to be viewed, e.g., Axioskop 2 PLUS (Zeiss).
2. High-resolution camera, e.g., CCD camera (CV M300, JAI Corporation).
3. Software for FISH analysis, e.g., ISIS4 (METASystems).

3 Methods

3.1 Preparing DNA Probe

1. Hydrolyze human genomic DNA with the restriction endonucleases *Hae*III and *Rsa*I for 15 min at 37 °C. Reaction mixture: 1 μg of human genomic DNA, 1× FastDigest buffer, 1 U

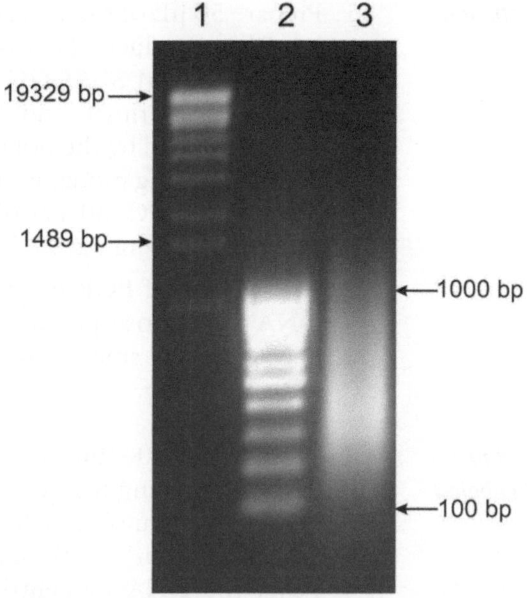

Fig. 1 Electrophoretic analysis (1 % agarose gel containing 0.001 % ethidium bromide) of the LA-PCR product obtained from genomic DNA isolated from peripheral leukocytes of human blood and hydrolyzed with the restriction endonucleases *Hae*III and *Rsa*I. Lanes: *1* and *2*, length markers of dsDNA; *3*, LA-PCR product

of *Hae*III restriction endonuclease, and 1 U of *Rsa*I restriction endonuclease in a total volume of 50 μL. Precipitate the digested DNA with 70 % ethanol, wash with 70 % ethanol, and air-dry. Finally, dissolve DNA in 20 μL of water.

2. Ligate the adapter to the hydrolyzed DNA for 16 h at 22 °C. Reaction mixture: 100 ng of DNA sample, 1.5 μM DNA adapter, 1× buffer for T4 DNA ligase, 5 % PEG 6000, 10 U of T4 DNA ligase, and 10 U of *Eco*32I restriction endonuclease in a total volume of 20 μL. Inactivate T4 DNA ligase by heating at 65 °C for 10 min.
3. Perform LA-PCR in 50 μL of the reaction mixture containing 1× dNTP mix, 2 mM $MgCl_2$, 2 μL of template DNA (prepared at **step 2**), 1.2 μM 5′-ACGCTCGAGACGGTCGAT-3′ oligonucleotide as a primer, and 1 U of *Taq* DNA polymerase in 1× Taq DNA polymerase buffer. Reaction conditions: initial stage of 5 min at 95 °C; 40 cycles of 30 s at 95 °C, 30 s at 62 °C, and 45 s at 72 °C; and final extension of 5 min at 72 °C.
4. Using 5 μL of PCR product, check the quality of amplified DNA by agarose gel electrophoresis. A correct PCR product gives a smear of DNA in the length range of 100–1,000 bp (Fig. 1).

3.2 Probe Labeling

1. Prepare 50 μL of the PCR reaction mixture containing 2 μL of LA-PCR product, 1× dNTP solution for labeling, 2 mM $MgCl_2$, 1.2 μM 5′-ACGCTCGAGACGGTCGAT-3′ oligonucleotide as a primer, and 1 U of *Taq* DNA polymerase in 1× buffer supplied by the polymerase manufacturer. Perform PCR in the following mode: initial stage of 5 min at 95 °C; 18 cycles of 30 s at 95 °C, 30 s at 62 °C, and 1 min at 72 °C; and final extension of 5 min at 72 °C.
2. Using 5 μL of PCR product, check the quality of amplified DNA by agarose gel electrophoresis. A correct PCR product gives a smear of DNA in the length range of 100–1,000 bp.

3.3 Postlabeling Treatment of Labeled DNA Probe

1. Precipitate 200–400 ng of the TAMRA-dUTP-labeled DNA probe containing 5 μg of salmon sperm DNA with 2.5 volume of 100 % ethanol and 0.1 volume of 3 M sodium acetate. Incubate for either 20 min at −80 °C or overnight at −20 °C. Pellet the DNA by centrifugation at 16,000 × *g* at 4 °C for 30 min and discard the supernatant.
2. Wash the pellet with 2.5 volume of ice-cold 70 % ethanol. Centrifuge at 16,000 × *g* at 4 °C for 15 min and discard the supernatant. Dry the pellet at a room temperature and dissolve in 20 μL of hybridization buffer (*see* **Note 6**).

3.4 Preparing Metaphase Chromosome Spread

FISH can be performed on metaphase spreads produced from any growing cells. Cell cultivation and treatment before harvesting mitotic cells depends on the used cell type. In our study, the metaphase chromosomes were prepared from human peripheral leukocytes according to the published protocol [12].

1. Drop 20–30 μL of cell suspension onto a clean glass slide from about 30 cm of height and air-dry (*see* **Notes 7** and **8**).
2. As fixative solution gradually evaporates, the surface of the slide becomes "grainy" (cells are visible). At this moment, place the slide on the heat block covered with a moist tissue paper (50 °C).
3. After the slide surface becomes "grainy," quickly drop 10–25 μL of fixative solution in several places on the slide. Place the slide on the heat block covered with a moist tissue paper (50 °C) and let the fixative completely dry out.
4. Check the number and quality of metaphase spreads by phase contrast microscopy (*see* **Note 9**).
5. Slides can be stored at a room temperature. Alternatively, slides can be stored "fresh" at −20 °C for months if sealed in storage boxes (*see* **Note 10**).

3.5 Slide Pretreatment and Denaturation

3.5.1 Pretreatment with RNase A

1. Flood the slide with 100 μL of 100 μg/mL RNase A in 2× SSC, cover with a coverslip, and incubate at 37 °C for 1 h in a moist chamber.
2. Wash the slide three times in 2× SSC for 5 min each at a room temperature.
3. Dehydrate the slide through 70, 80, and 100 % ethanol series for 3 min. Air-dry the slide for at least 10 min.

3.5.2 Pretreatment with Pepsin

1. Incubate the slide in pepsin solution at 37 °C for 15 min in a moist chamber.
2. Wash the slide for 5 min in PBS at a room temperature.
3. Wash the slide for 5 min in PBS with 50 mM $MgCl_2$ at a room temperature.
4. Fix chromosomes by incubating the slide in Solution A for 10 min at a room temperature.
5. Wash the slide in PBS for 5 min.
6. Dehydrate through the ethanol series (70, 80, and 100 %) for 3 min in each.

3.6 Fluorescent In Situ Hybridization

1. Add 20 μL of labeled DNA probe in hybridization mixture directly onto the slide, put a coverslip (avoiding bubbles) on the drop, and seal with rubber cement.
2. Incubate the slide on a heat block for 5 min at 75 °C.
3. Incubate the slide overnight at 42 °C in a moist chamber. Gently remove the rubber cement with forceps avoiding any movements of the coverslip.
4. Remove carefully the coverslip.

3.7 Slide Washing and Mounting

1. Transfer the slides to a preheated coupling jar containing 1× SSC for 5 min at 60 °C.
2. Transfer the slides to a coupling jar containing Solution B for 10 min at 45 °C on a shaking platform.
3. Put the slides in PBS solution for 5 min at a room temperature.
4. Rinse the slides with water for a few seconds.
5. Dehydrate the slides through ethanol series (70, 80, and 100 %) at 4 °C for 3 min in each and air-dry in a dark place.
6. Add 15 μL of DAPI/antifade solution and cover with a coverslip.
7. Store slides in the dark at 4 °C.

3.8 Viewing and Interpretation

1. Perform fluorescence microscopy with a microscope equipped with a CCD camera and a filter set for DAPI and the fluorophores used for labeling of the PCR products obtained in your experiment.

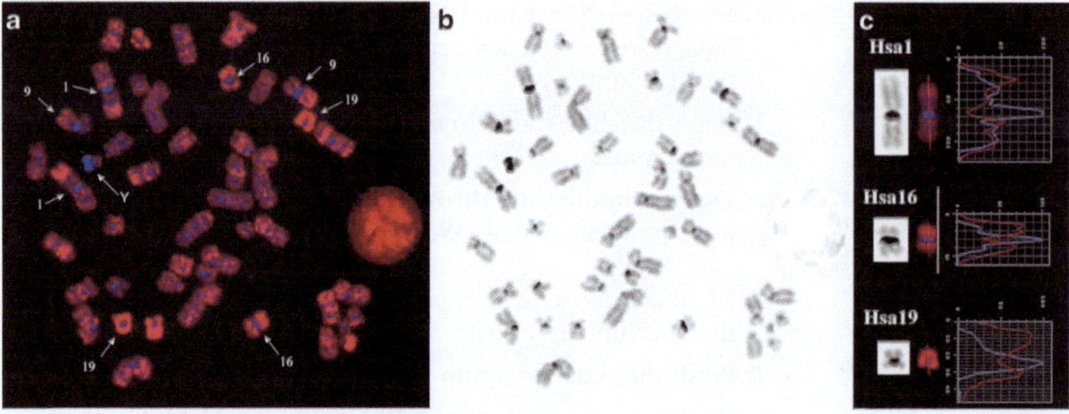

Fig. 2 FISH of the DNA probes prepared from genomic DNA of human peripheral leukocytes using a *Hae*III/*Rsa*I LA-PCR. (**a**) The DNA probes labeled with TAMRA-dUTP; (**b**) inverted DAPI banding of chromosomes; and (**c**) the profiles of signal intensities on human chromosomes 1, 16, and 19

2. Acquire and store the images of metaphase spreads in an appropriate format for further processing with software for CGH analysis.
3. Identify the chromosomes of metaphase spreads based on inverted DAPI banding patterns with the help of appropriate software according to the software instructions.
4. Check the profiles of fluorophore intensity along the chromosomes of individual metaphases.
5. Generate averaged profiles for the chromosome bands in a standard karyogram.
6. Analyze the ratio of fluorescent signal intensities of the fluorophores used for labeling of the analyzed DNA and DAPI (Fig. 2). The signal of DNA probes in C-positive regions should only insignificantly exceed the background level. An intensive FISH signal should be detectable in the C-negative chromosomal regions.

4 Notes

1. Isolate human genomic DNA from the source of interest using an appropriate isolation protocol. After isolation, the genomic DNA fragments should be longer than 10 kb. For preparing DNA probes, we used the genomic DNA isolated from human peripheral leukocytes.
2. We recommend using FastDigest™ endonucleases and FastDigest™ buffer for simultaneous hydrolysis of genomic DNA with *Hae*III and *Rsa*I in one tube. Step-by-step hydrolysis is also acceptable.

3. In general, a good metaphase spreading was achieved on all types of the used glass. Superfrost (Menzel) slides are commonly used in our laboratory. They do not require any extra cleaning steps or washes prior to use.
4. A plastic slide storage box containing a moist tissue paper can be used as a moist chamber.
5. We recommend using VECTASHIELD Mounting Medium with DAPI (Vector laboratories).
6. To improve dissolution of the labeled DNA probe, place the tube in a Thermomixer at 45 °C and vortex for 1 h until the pellet is dissolved. Otherwise, labeled DNA probe pellet can be stored at −20 °C for a long period.
7. For good spreading of the chromosomes, slides should be stored in cold water prior to dropping of cell suspension.
8. Cells should be dropped on the part of slide, which will be then covered with a 22×22 mm coverslip during hybridization. Mark the area of the dropped metaphase spreads on the back of the slide using a diamond pen.
9. The aim is to obtain pale gray chromosomes free from cytoplasmic material. The whole metaphases should cover a suitable area without or only a few overlapping chromosomes.
10. When defrosting the slides, leave the box unopened on the bench until a room temperature is reached to avoid water condensation, which may influence the quality of target material.

Acknowledgments

The study was supported by the Siberian Branch of the Russian Academy of Sciences (grants nos.VI.53.2.1, 6.1, and 75), and state contract no. 8289.

References

1. Zhang L, Cui X, Schmitt K, Navidi W, Arnheim N (1992) Whole genome amplification from a single cell: implications for genetic analysis. Proc Natl Acad Sci U S A 89:5847–5851
2. Telenius H, Carter NP, Bebb CE, Nordenskjold M, Ponder BA, Tunnacliffe A (1992) Degenerate oligonucleotide-primed PCR: general amplification of target DNA by single degenerate primer. Genomics 13:718–725
3. Nelson DL, Ledbetter SA, Corbo L, Victoria MF, Ramirez-Solis R, Webster TD et al (1989) Alu polymerase chain reaction: a method for rapid isolation of human-specific sequences from complex DNA sources. Proc Natl Acad Sci U S A 86:6686–6690
4. Ludecke HJ, Senger G, Claussen U, Horsthemke B (1989) Cloning defined regions of the human genome by microdissection of banded chromosomes and enzymatic amplification. Nature 338:348–350
5. Engelen JJ, Tuerlings JH, Albrechts JC, Schrander-Stumpel CT, Hamers AJ, De Die-Smulders CE (2000) Prenatally detected marker chromosome identified as an i(22)(p10) using (micro)FISH. Genet Couns 11:13–17
6. Gribble SM, Fiegler H, Burford DC, Prigmore E, Yang F, Carr P et al (2004) Applications of combined DNA microarray and chromosome sorting technologies. Chromosome Res 12: 35–43

7. Klein CA, Schmidt-Kittler O, Schardt JA, Pantel K, Speicher MR, Riethmuller G (1999) Comparative genomic hybridization, loss of heterozygosity and DNA sequence analysis of single cells. Proc Natl Acad Sci U S A 96:4494–4499
8. Wells D, Sherlock JK, Handyside AH, Delhanty JD (1999) Detailed chromosomal and molecular genetic analysis of single cells by whole genome amplification and comparative genomic hybridization. Nucleic Acids Res 27:1214–1218
9. Cremer T, Popp S, Emmerich P, Lichter P, Cremer C (1990) Rapid metaphase and interphase detection of radiation-induced chromosome aberrations in human lymphocytes by chromosomal suppression in situ hybridization. Cytometry 11:110–118
10. Abrusan G, Krambeck HJ (2006) The distribution of L1 and Alu retroelements in relation to GC content on human sex chromosomes is consistent with the ectopic recombination model. J Mol Evol 63:484–492
11. Morozkin ES, Loseva EM, Karamysheva TV, Babenko VN, Laktionov PP, Vlassov VV et al (2011) A method for generating selective DNA probes for the analysis of C-negative regions in human chromosomes. Cytogenet Genome Res 135:1–11
12. Henegariu O, Heerema N, Wright L, Bray-Ward P, Ward DC, Vance GH (2001) Improvements in cytogenetic slide preparation: controlled chromosome spreading, chemical aging and gradual denaturing. Cytometry 43:101–109

Part VII

Detection of MicroRNA

Chapter 20

Quantitative Analysis of MicroRNA in Blood Serum with Protein-Facilitated Affinity Capillary Electrophoresis

Maxim V. Berezovski and Nasrin Khan

Abstract

MicroRNAs play an important role in gene regulation and disease etiology and are blood-based biomarkers of diseases. Here, we describe a protein-facilitated affinity capillary electrophoresis (ProFACE) method for ultra-sensitive direct miRNA detection as low as 300,000 molecules in 1 mL of blood serum, using single-stranded DNA binding protein (SSB) and double-stranded RNA binding protein (p19) as separation enhancers. This method utilizes either the selective binding of SSB to a fluorescent single-stranded DNA/RNA probe or the binding of p19 to miRNA–RNA probe duplex.

Key words MicroRNA, SSB, p19 protein, Blood serum, Capillary electrophoresis, Laser-induced fluorescence detection

1 Introduction

Many bioanalytical methods have been established to identify and quantify miRNAs include Northern blotting [1], in situ hybridization [2], small RNA library sequencing [3], bead arrays [4], microarray hybridization [5], and reverse transcription (RT)-PCR [6]. While miRNA microarrays allow for massive parallel and accurate relative measurement of all known miRNAs, they lack sensitivity and very expensive to facilitate. PCR-based methods have variation with increasing cycle number [7] and require amplification, making them only semi-quantitative. Here we demonstrated a sensitive and quantitative miRNA detection method by protein-facilitated affinity capillary electrophoresis. Capillary electrophoresis with laser-induced fluorescent detection (CE-LIF) [8] is a promising technique for miRNA detection, where miRNAs are easily separated from most of the other biomolecules because of their strong negative charge. Proteins that were used for this method are single-stranded DNA binding protein (SSB) and RNA binding protein p19 from Carnation Italian ringspot virus. SSB and p19 enhance separation between the excess of the fluorescently labeled

Dmitry M. Kolpashchikov and Yulia V. Gerasimova (eds.), *Nucleic Acid Detection: Methods and Protocols*, Methods in Molecular Biology, vol. 1039, DOI 10.1007/978-1-62703-535-4_20, © Springer Science+Business Media New York 2013

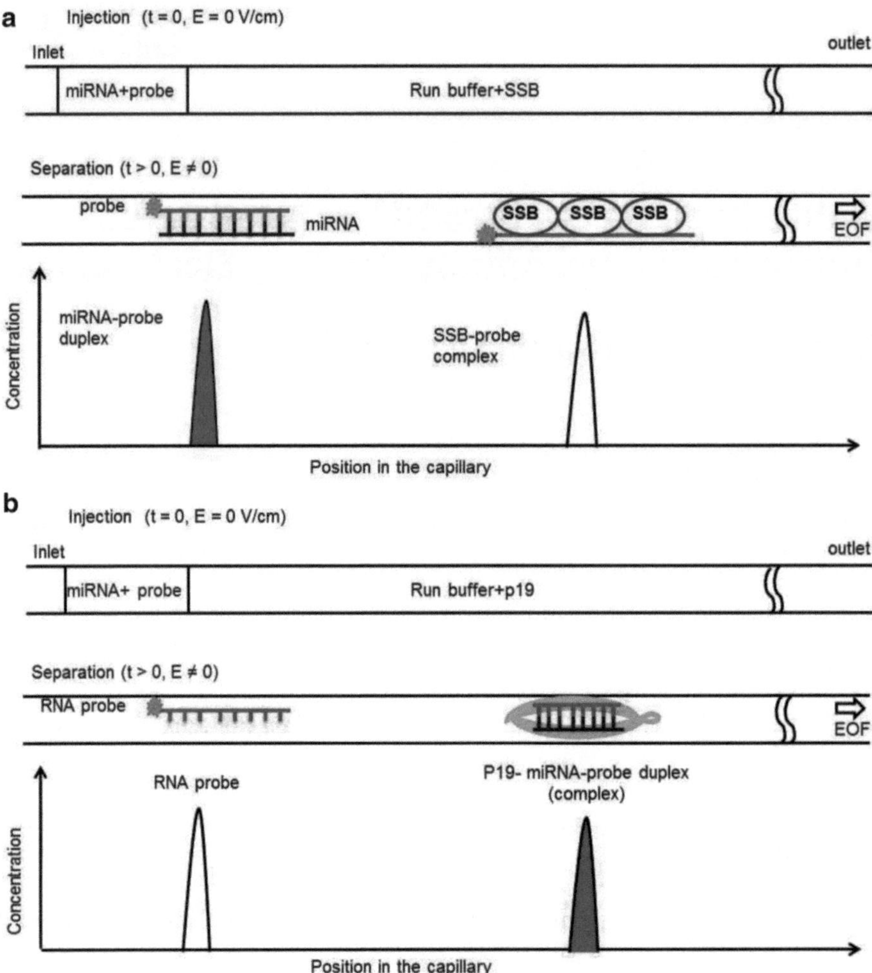

Fig. 1 Schematic diagram of the protein-facilitated affinity capillary electrophoresis (ProFACE) assay for miRNA detection. A mixture of a miRNA and fluorescently labeled probe is injected as a small plug into a capillary and subjected to electrophoresis. (**a**) When SSB is added in the run buffer, it binds only to the DNA or RNA probe and enhances the separation of the miRNA–probe duplex from the excess probe. (**b**) When p19 is present in the run buffer, it binds only to the miRNA–RNA probe duplex and shifts it far away from the free probe increasing resolution and sensitivity of the assay (reproduced and modified from ref. 12 with permission from ACS publications)

single-stranded DNA/RNA probe and the miRNA–probe duplex (Fig. 1). Earlier SSB has been used to distinguish single-stranded (ss) from double-stranded (ds) DNA using CE-LIF [9]. It binds to ssDNA (Kd < 50 nM) and ssRNA of eight bases or more in length but does not bind to dsDNA, dsRNA, or dsDNA–RNA hybrids [10]. The affinity of ssDNA to SSB is about ten times higher than that of ssRNA [11]. Due to the binding of the more neutral and bulky SSB, the probe migrates at a faster rate that significantly increases the separation time between the excess probe and

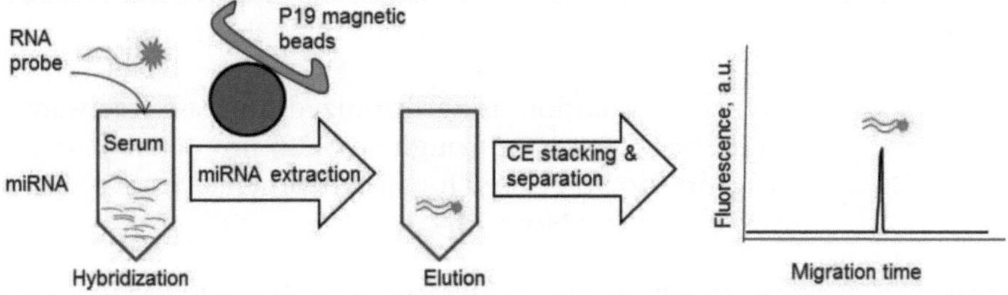

Fig. 2 Pre-concentration of miRNA from serum sample by p19 magnetic beads followed by detection of miRNA by ProFACE-LIF (reproduced and modified from ref. 12 with permission from American Chemical Society)

miRNA–probe duplex (Fig. 1a). On the other hand protein p19 binds only to double-stranded RNA in a size-specific manner. It does not bind to ssRNA, rRNA, mRNA, ssDNA, or dsDNA. p19 binds tightly to 21–23 nt long dsRNA. Binding is enhanced with the 5′ phosphate groups of miRNA due to the interaction between the tryptophan residues on the end capping helices of the p19 dimer which stabilizes the structure. Unlike SSB, p19 binds to the miRNA–probe duplex directly separating it from the excess probe in CE (Fig. 1b). Using these unique binding properties of p19 to miRNA, we developed a method that allowed us to detect ultra-low amount of miRNA from serum sample with p19-coated magnetic beads followed by CE-LIF detection [12] (Fig. 2). Without PCR amplification, as low as 300,000 molecules of miRNA in 1 mL of serum can be measured in a 20 min ProFACE run. The sensitivity of the method is comparable with existing PCR techniques. It can be parallelized to quantitatively detect multiple miRNA-based biomarkers in different biological samples.

Briefly, prior to CE separation miRNAs are incubated with excess of either a fluorescently labeled ssDNA probe or ssRNA probe. Depending on the probe, the annealing temperature is increased to 60 or 55 °C, and then slowly decreased to 20 °C to allow the proper hybridization of the probes with its complementary miRNA. A short plug of hybridized product is then hydrodynamically injected into a capillary and subjected to electrophoresis. Prior to injection, the capillary is prefilled with 25 mM borax buffer. In case of CE separation with SSB or p19, different amount of protein is added to the run buffer.

For serum experiments different concentration of miRNAs and probes are spiked in 1 mL of FBS and hybridized. Obtained hybridized product is then incubated with p19 magnetic beads in p19 binding buffer by shaking for 2 h at room temperature. After extensive washing with wash buffer miRNA duplex is eluted in small amount of elution buffer. Obtained miRNA duplex is then injected into a capillary with sample stacking and subjected to electrophoresis.

2 Materials

Prepare all solution using deionized nuclease-free water and filter all buffer solution through a 0.22 μm filter. Analytical grade reagents should be used for all experiments. Store prepared reagent as specified (*see* **Note 1**).

2.1 Capillary Electrophoresis (CE)

1. Capillary electrophoresis system with a laser source from Beckman Coulter (MDQ, PA800, PA800 plus).
2. Bare fused silica capillary (*see* **Note 2**).
3. Glass vials with caps.
4. Sample vials with strings and caps.

2.2 Other Related Equipments

1. PCR Thermocycler.
2. Magnetic separation rack.
3. Benchtop incubator.
4. Laboratory roller mixer.
5. 0.2 mL PCR tubes and 1.5 mL tubes.

2.3 CE Solutions

1. CE running buffer stock solution: 200 mM Sodium tetraborate (borax), pH 9.2. Add about 250 mL of water to a 1 L glass beaker (*see* **Note 3**). Weigh 76.274 g of sodium tetraborate and transfer to the beaker. Add water to bring the volume to 900 mL. Mix and bring final volume to 1 L by adding water.
2. CE running buffer working solution: 25 mM sodium tetraborate (borax), pH 9.2. Add 25 mL of 200 mM borax (from previous step) to 175 mL of water to obtain 200 mL of working solution.
3. CE wash solution: 100 mM NaOH. Weigh 4 g NaOH pallets and add slowly to 500 mL of water, stirring continuously. When pallets dissolve completely bring volume to 1 L by adding water.
4. CE wash solution: 100 mM HCl. Measure 8.33 mL of concentrated HCl (12 M) with graduated pipette. Put HCl into 1 L volumetric flask then add water until it reaches 1 L.

2.4 Storage and Incubation Solutions

1. Stock solution of 200 mM Tris-Acetate, pH 8.2: Add about 100 mL of water to a 500 mL glass beaker. Weigh 12.114 g of Tris base and transfer to the beaker (*see* **Note 4**). Add water to bring the volume to 400 mL. Mix and adjust pH with glacial acetic acid (*see* **Note 5**). Bring final volume to 500 mL by adding water.
2. Stock solution of 200 mM NaCl: Add about 250 mL of water to 1 L glass beaker. Weigh 11.68 g of NaCl and transfer to the beaker. Mix and add water to bring the final volume to 1 L.
3. Stock solution of 50 mM EDTA: Add about 25 mL of water to 100 mL glass beaker. Weigh 0.93 g of disodium salt of

EDTA and transfer to the beaker (*see* **Note 6**). Mix and add water to bring the final volume to 50 mL. Adjust pH to 8 with 10 M NaOH.

4. Incubation buffer: 50 mM Tris-acetate, pH 7.9, 50 mM NaCl, 1 mM EDTA. Add 12.5 mL of 200 mM Tris-acetate, 5 mL of 200 mM NaCl and 1 mL of 50 mM EDTA to 31.5 mL of water to have the final volume of 50 mL.
5. Storage buffer: 10 mM Tris, 0.1 mM EDTA. Add 2.5 mL of 200 mM Tris (*see* **Note 7**) and 0.1 mL of 50 mM EDTA to 47.4 mL of water to have the final volume of 50 mL.
6. 25× Wash buffer: 500 mM Tris–HCl, 2,500 mM NaCl, 25 mM EDTA.
7. 1× BSA-wash buffer: dilute the 25× wash buffer and 100× BSA stocks (both comes with the kit) with sterile water to a final concentration of 1× wash buffer and 1× BSA.

2.5 MicroRNA (miRNA) and Hybridization Probes

1. Resuspend synthetic miRNA (*see* **Note 8**) in storage buffer to obtain final concentration of 100 μM and store it at −20 °C (*see* **Note 9**).
2. Resuspend synthetic DNA and RNA probes complementary to miRNA (*see* **Note 10**) as described above (*see* **Note 9**).

2.6 Proteins for Affinity CE Separation

1. SSB—Single-stranded DNA binding protein (Sigma): Prepare the stock solution of SSB by resuspending in incubation buffer. Prepare 100 nM SSB from stock solution by diluting with incubation buffer.
2. p19 siRNA binding protein (BioLabs, New England, cat. # M0310L).

2.7 Blood Serum Components

1. p19 miRNA detection kit (BioLabs, New England, cat. # E3312S) (*see* **Note 11**).
2. 100 % fetal bovine serum.
3. tRNA from yeast.
4. 1× p19 elution buffer: 20 mM Tris–HCl, 100 mM NaCl, 1 mM EDTA, 0.5 % SDS.

3 Methods

Keep all miRNAs, probes, and proteins on ice during the procedures.

3.1 Hybridization of miRNA with DNA and RNA Probes

1. Prepare three 1 mL samples of 1 μM stock solution by adding 10 μL of 100 μM of each of the following: miRNA and probes (DNA and RNA) to 990 μL of incubation buffer to obtain three samples of 1 mL stock solutions.

2. Prepare three 1 mL samples of 200 nM working solution by adding 200 μL of 1 μM of each of the following: miRNA and probes (DNA and RNA) to 800 μL of incubation buffer to obtain three samples of 1 mL working solutions.
3. Mix together 25 μL of 200 nM miRNA and 50 μL of 200 nM DNA probe. Add 25 μL of incubation buffer to final volume of 100 μL. Hybridize the mixture in PCR thermocycler at 60 °C (*see* **Note 12**).
4. Mix together 25 μL of 200 nM miRNA and 50 μL of 200 nM RNA probe. Add 25 μL of incubation buffer to final volume of 100 μL. Hybridize the mixture in PCR thermocycler at 60 °C (*see* **Note 13**).

3.2 CE Separation Procedure for DNA Probe Without SSB

1. Prepare the capillary and the cartridge following the manufacturer instruction (*see* **Note 14**).
2. Take six glass CE vials and add the following into each vial separately: 1.5 mL of HCl, 1.5 mL of NaOH, 1.5 mL of H_2O and add 1.5 mL of 25 mM borax buffer (to three separate vials each). Arrange them on inlet buffer tray in the same order as they are prepared (*see* **Note 15**).
3. Take two more glass vials and fill one of them with 25 mM borax buffer. Add 0.3 mL of water to the other vial where the waste will be collected. Arrange them on outlet buffer tray (*see* **Note 16**).
4. Prepare a sample vial with spring on. Add 20 μL of hybridized miRNA–DNA sample (from Subheading 3.1, **step 3**) into a PCR tube. Place the tube into the sample vial. Place the sample vial on sample tray (*see* **Note 17**).

3.3 Creating a Method for CE Separation in Karat Software and Running the Sample

1. From the instrument setup window, select Initial Condition tab. Select "current" from auxiliary data channels. Insert 15 °C for cartridge and 4 °C for sample storage temperature. Analog output scaling factor keeps as 1.
2. From the instrument setup window, select the LIF Detector initial Conditions tab. Under electropherogram channel 1 select "acquisition enabled" and insert 1,000 RFU for dynamic range. Select normal for filter settings with peak width 16–25. Select direct for signal. Excitation and emission wavelengths are 488 nm and 520 nm, respectively. Keep data rate at 4 Hz for both channels.
3. From the instrument setup window, select the Time Program tab. Click in the "Event" column to display the down arrow. Click on the down arrow to display the popup menu. Scroll down and select Rinse… command. Set the parameters by selecting "pressure" for pressure type, with value of 20.0 psi for 2 min. Select the tray position from inlet and outlet box as

Table 1
Instrument setup for CE separation

	Time (min)	Event	Value	Duration	Inlet vial	Outlet vial	Summary
1		Rinse—pressure	20.0 psi	2.00 min	BI:A1	BO:A2	Forward
2		Rinse—pressure	20.0 psi	2.00 min	BI:B1	BO:A2	Forward
3		Rinse—pressure	20.0 psi	2.00 min	BI:C1	BO:A2	Forward
4		Rinse—pressure	20.0 psi	2.00 min	BI:D1	BO:A2	Forward
5		Inject—pressure	0.3 psi	2 s	SI:A1	BO:A1	Override, forward
6		Wait		0.10 min	BI:E1	BO:A1	
7	0.00	Separate—voltage	12.5 kV	15 min	BI:F1	BO:A1	0.17 min ramp, normal polarity
8	1.00	Autozero					

shown in Table 1. Keep the pressure direction to "forward". Press ok (*see* **Note 18**).

4. Following Table 1, create three more rows of rinsing event.
5. Then from the popup menu of Event column select Inject... command to create the fifth row. Select "pressure" for injection type with the values of 0.3 psi for 2 s. Keep pressure direction to forward and select the tray position as in Table 1.
6. For sixth row of Event select Wait... command. For duration insert 0.10 min and select the tray position as in Table 1.
7. Now select Separate... command from popup menu. Select "voltage" as a separation type with the value of 20.0 kV for 15.00 min at 0.17 ramp time. Select "At time" with 0.00 min. Select the tray position as in Table 1, and keep the polarity "Normal".
8. For the last row of Event select Autozero... command from popup menu and select at time inserting 1.0 min (*see* **Note 19**).
9. Save the method. To make a single run, click the blue arrow (single run button) located on the toolbar. The single run acquisition window will be displayed. Enter a sample ID for the run. Enter a path name where the data acquired for this run will be stored. Enter a file name to be used to save the data and then click "start" to start the run (Fig. 3).

3.4 CE Separation Procedure for DNA Probe with SSB

1. Use the same capillary if it is still ok or prepare the capillary following Subheading 3.2 [1].
2. Mix together, 75 μL of 200 mM borax and 300 μL of 100 nM SSB. Add 225 μL of water to final volume of 600 μL (*see* **Note 20**).

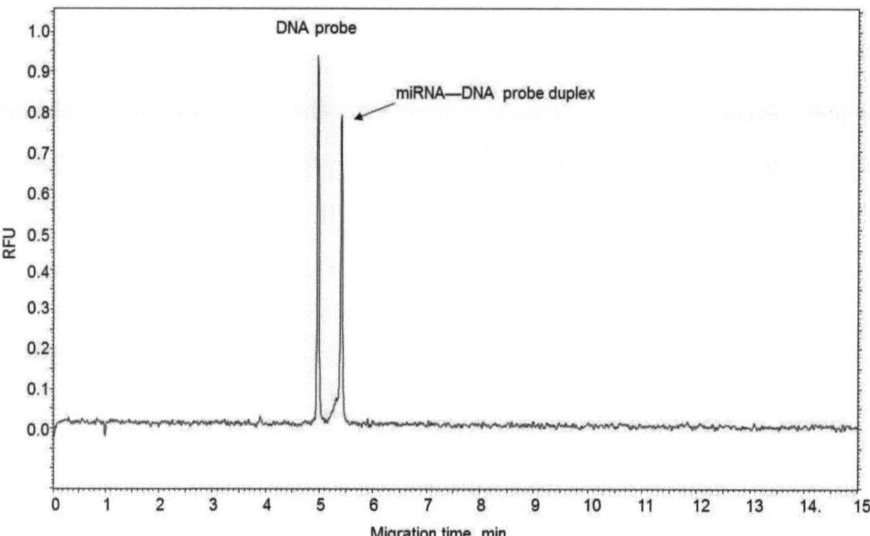

Fig. 3 Separation of miRNA–DNA probe duplex from the excess DNA probe without SSB. The sample contains 40 nM miRNA and 90 nM DNA probe

3. Take three glass CE vials and add the following into each vial separately: 1.5 mL of HCl, 1.5 mL of NaOH, and 1.5 mL of H_2O. Prepare three plastic vials with strings in. Add 200 μL of borax–SSB mixture (from **step 2**) to two PCR tubes. Place them into two of the prepared plastic vials (rinsing buffer and running buffer, respectively). Add 200 μL of 25 mM borax into another PCR tubes and place it into the third plastic vial (mocking buffer). Arrange them in inlet buffer tray as described earlier (*see* **Note 15**).
4. Take another plastic vial with string in. Add 200 μL of borax–SSB mixture (from **step 2**) to a PCR tube and place it into the prepared plastic vial. Add 0.3 mL of water another glass vial where the waste will be collected. Arrange them on outlet buffer tray (*see* **Note 16**).
5. Prepare a plastic sample vial with spring in. Add 20 μL of hybridized miRNA–DNA probe sample (from Subheading 3.1, **step 3**) into a PCR tube. Place the tube into the sample vial. Place the sample vial on sample tray (*see* **Note 17**).
6. Use the same method that was created previously (Subheading 3.3) to run the sample. Only chose a different file name to save the data and then click "start" to start the run (Fig. 4).

3.5 CE Separation Procedure for RNA Probe Without and with SSB

1. Follow the procedure as described in Subheadings 3.2, 3.3 and 3.4. Use miRNA–RNA probe hybridized sample (from Subheading 3.1, **step 4**) instead (Figs. 5 and 6, respectively).

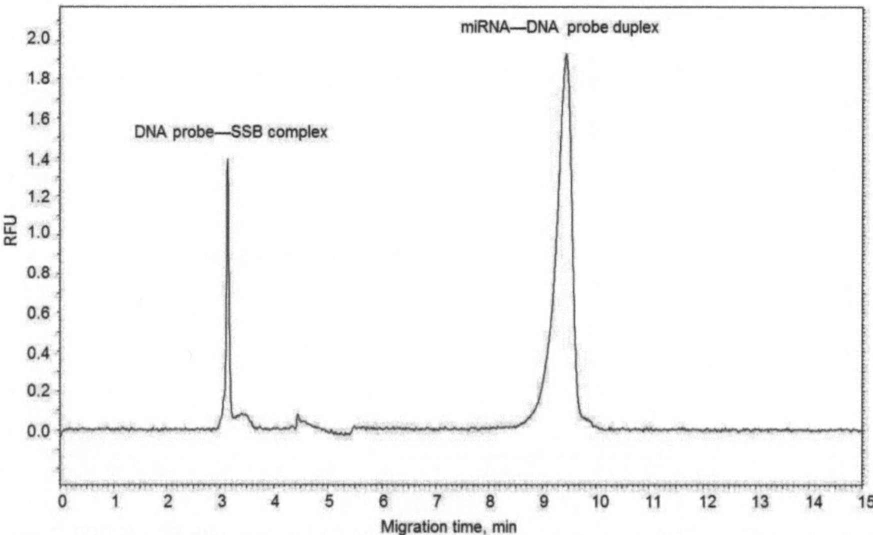

Fig. 4 Separation of miRNA–DNA probe duplex from the excess DNA probe with 50 nM SSB. The sample contains 40 nM miRNA and 90 nM DNA probe

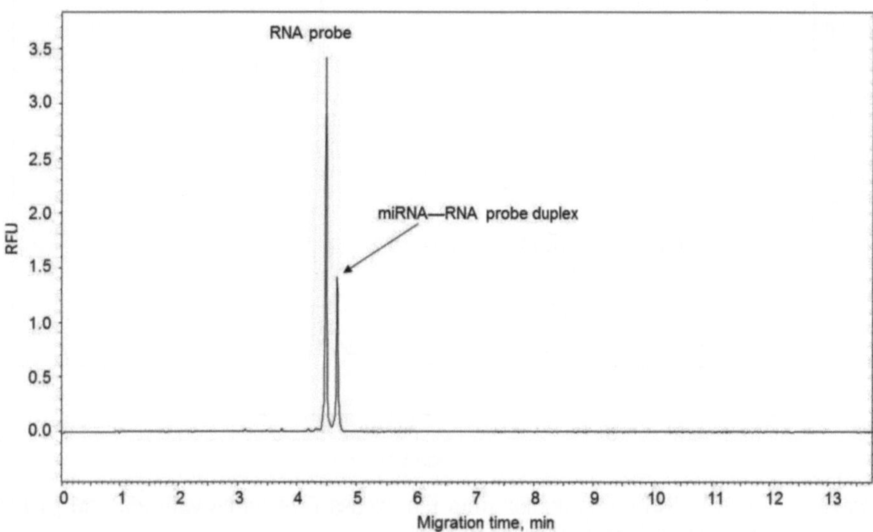

Fig. 5 Separation of miRNA–RNA probe duplex from the excess RNA probe without SSB. The sample contains 40 nM miRNA and 100 nM RNA probe

3.6 CE Separation Procedure for RNA Probe with p19

1. Mix together 75 μL of 200 mM borax and 5 μL of p19 protein. Add 520 μL of water to final volume of 600 μL (*see* **Note 21**).
2. Follow the procedure as described in Subheading 3.3 and 3.4 using miRNA–RNA probe hybridized sample instead (Fig. 7).

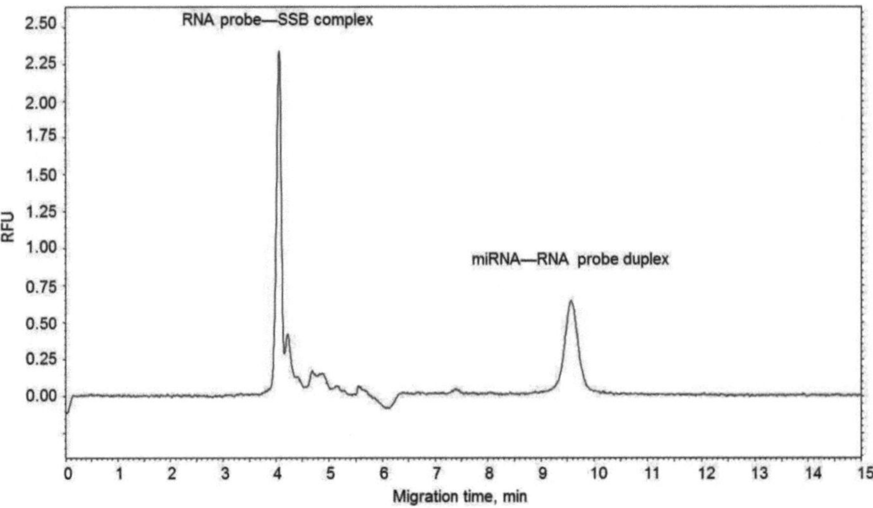

Fig. 6 Separation of miRNA–RNA probe duplex from the excess RNA probe with 50 nM SSB. The sample contains 40 nM miRNA and 100 nM RNA probe

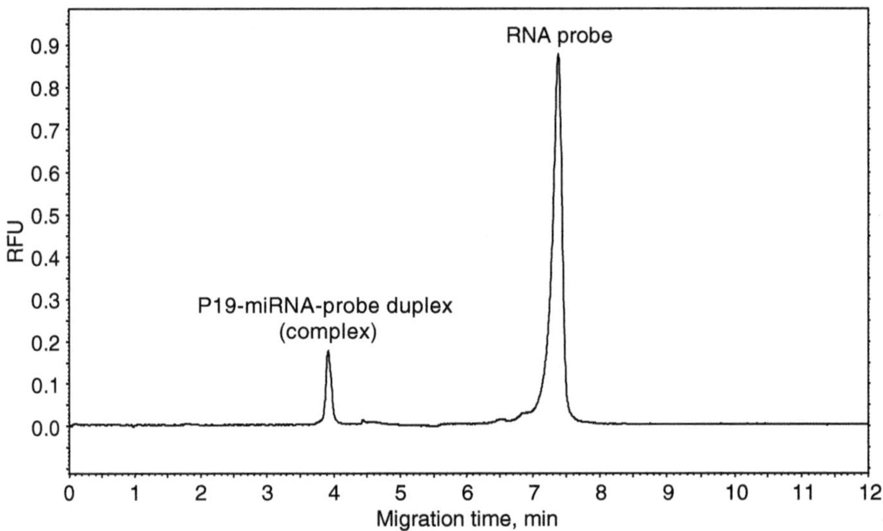

Fig. 7 Separation of miRNA–RNA probe duplex from the excess RNA probe with 250 nM p19 protein. The sample contains 40 nM miRNA and 100 nM RNA probe

3.7 MicroRNA Pre-concentration from Serum Using p19 Beads

1. For hybridization of miRNA with RNA probes in serum, mix together 1 mL of 100 % FBS, 0.55 μL of 200 nM miRNA (from Subheading 3.1, **step 2**), 5.5 μL of 200 nM miRNA probe (from Subheading 3.1, **step 2**), 1 μL of 100 μM tRNA (diluted from stock), 1 μL of murine RNase inhibitor (40 U, miRNA kit), and 91.95 μL of p19 binding buffer (*see* **Note 11**). Transfer the mixture into 11 PCR tubes, 100 μL each. Hybridize the mixture in PCR thermocycler at 75 °C (*see* **Note 22**). The final concentration would be 100 pM RNA and 1 nM RNA probe.

2. Control: Mix together 1 mL of 100 % FBS, 55 μL of 200 nM miRNA probe (from Subheading 3.1, **step 2**), 1 μL of 100 μM tRNA (diluted from stock), 1 μL of murine RNase inhibitor (40 U, miRNA kit), and 43 μL of p19 binding buffer (*see* **Note 11**). Transfer the mixture into 11 PCR tubes, 100 μL each. Hybridize the mixture in PCR thermocycler at 75 °C (*see* **Note 22**). The final concentration would be 10 nM RNA probe (*see* **Note 23**).
3. Add 20 μL of p19 beads into a 1.5 mL tube. Resuspend p19 beads with a brief vortex.
4. Place the tube with p19 beads in a magnetic separation rack. Remove the supernatant from beads.
5. Transfer all 1,100 μL of hybridized serum mixture of miRNA sample from PCR tubes to 1.5 mL tube with p19 magnetic beads. Incubate the mixture at room temperature by shaking on a benchtop roller mixer for 2 h.
6. Follow the same procedure as described above for control as well.
7. To remove the unbound RNA heat up the 1× BSA-wash buffer to 37 °C (*see* **Note 24**). Remove supernatant from dsRNA-p19 beads binding reaction using the Magnetic Separation Rack.
8. Wash the beads pellet in 500 μl of 1× BSA-wash buffer (*see* **Note 25**).
9. Repeat the wash seven times following the procedure as described above (*see* **Note 26**).
10. Follow the **steps 1–5** simultaneously for control as well.
11. Heat up 40 μl of 1× p19 elution buffer from p19 miRNA detection kit to 37 °C (*see* **Note 27**).
12. After the last wash, remove as much of the supernatant as possible without touching the beads pellet.
13. To elute isolated ds miRNA add 40 μl of pre-warmed 1× p19 elution buffer to the beads pellet and incubate for 10 min at room temperature with shaking followed by 10 more minutes incubation at 37 °C.
14. Carefully remove the supernatant containing miRNA-RNA probe hybrid from the tube using the magnetic rack and place into a clean PCR tube. Keep it frozen at −20 °C until ready for analysis.
15. Follow **steps 1–4** simultaneously for the control sample as well.

3.8 Detection of miRNA by Capillary Electrophoresis

1. Prepare the instrument and accessories following the procedures described as in Subheading 3.2. Follow through **step 1** (*see* **Note 28**) to **step 3**.
2. Add 20 μL of miRNA sample obtained from previous section to a PCR tube and place it into a plastic vial as described in Subheading 3.2, **step 4**.

Table 2
Instrument setup for CE separation with sample stacking

	Time (min)	Event	Value	Duration	Inlet vial	Outlet vial	Summary
1		Rinse—pressure	20.0 psi	2.00 min	BI:A1	BO:A2	Forward
2		Rinse—pressure	20.0 psi	2.00 min	BI:B1	BO:A2	Forward
3		Rinse—pressure	20.0 psi	2.00 min	BI:C1	BO:A2	Forward
4		Rinse—pressure	20.0 psi	2.00 min	BI:D1	BO:A2	Forward
5		Inject—pressure	1.5 psi	10 s	SI:A1	BO:A1	Override, forward
6		Inject—pressure	0.3 psi	2 s	BI:C1	BO:A1	No override, forward
7		Inject—pressure	0.3 psi	2 s	BI:E1	BO:A1	No override, forward
8		Wait		0.10 min	BI:E1	BO:A1	
9	0.00	Separate—voltage	12.5 kV	15 min	BI:F1	BO:A1	0.17 min ramp, normal polarity
10	1.00	Autozero					

3. Prepare the control following the same procedure as **step 2** (*see* **Note 29**).
4. For creating the method follow the steps as described in Subheading 3.3. For Time program, follow Table 2 instead of Table 1. Save the method. To make a single run, click the blue arrow (single run button) located on the toolbar. The single run acquisition window will be displayed. Enter a sample ID for the run. Enter a path name where the data acquired for this run will be stored. Enter a file name to be used to save the data and then click "start" to start the run (Fig. 8).

4 Notes

1. All CE wash solutions should be stored at room temperature. Stock buffer solution should be kept at 4 °C.
2. Capillary length should be 30 cm or 50 cm (20 or 40 cm from an injection to a detection point, respectively) with an o.d. of 365 μm and an i.d. of 75 μm.
3. In order to dissolve sodium tetraborate faster put the stirrer into the beaker with water first. Then slowly add in the sodium tetraborate crystal, warm up the solution to about 30 °C. Once the crystals are in, turn on the heat at a low temperature about 45–50 °C. Leave it stirring for 5–6 h until dissolved. But be sure to bring the solution to room temperature before adjusting pH.

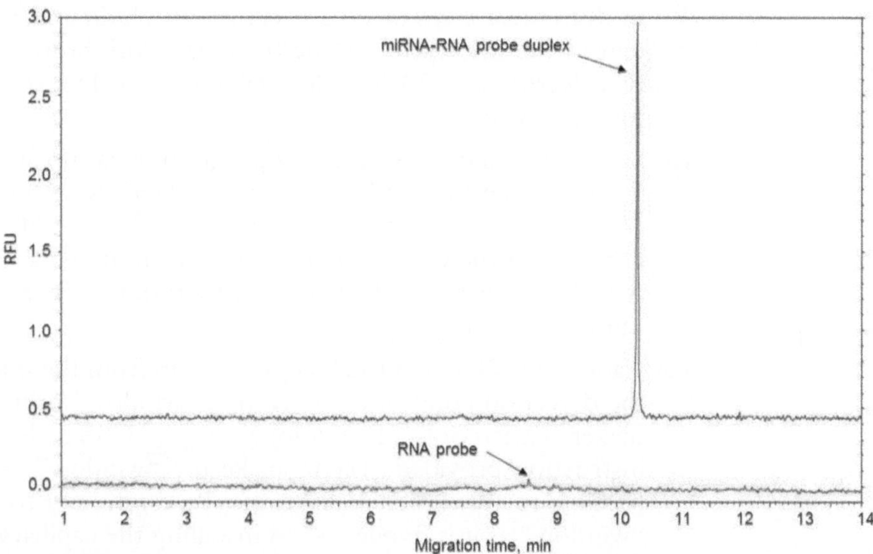

Fig. 8 Electropherogram of serum sample containing 50 fM miRNA with 1.25 nM RNA probe and 10 nM RNA probe alone as a control (*red*)

4. Follow the same procedure as above to dissolve Tris base.
5. Add acetic acid slowly to reach the desired pH, it might take about 2.5–3.0 mL.
6. Warming the water up to 40 °C along with stirring will make EDTA dissolve faster.
7. Prepare 200 mM Tris as in Subheading 2.4, **item 1** without adding acetic acid.
8. Synthetic miRNAs should be 21–23 nt long. All miRNAs and complementary probes should be ordered with 5′ phosphorylation modification.
9. For instance if you receive your oligo as lyophilized (freeze-dried) solids at a concentration of 25.2 nmol from manufacturer, resuspending it in 252 μL storage buffer will give a concentration of 100 μM. To prevent degradation of oligo by repeated freezing and thawing, oligo should be aliquoted into several tubes and freeze at −20 °C. The ones that are not required for immediate use freeze them at −80 °C for long-term use.
10. All DNA and RNA probes should be 3′ fluorescently labeled.
11. This kit includes: 1× p19 Beads, 1× p19 binding buffer, 1× p19 wash buffer, 1× p19 elution buffer, 10× BSA, RNase inhibitor. Store all the components at 4 °C.
12. Set up the following annealing program: Heat up at 60 °C for 2:00 min. Set the header temperature to 60 °C. Add 40 cycles

of decrement steps of 1 °C every 3 s and hold at 20 °C. Press enter when the cycle is done then wait until header temperature decreases to 20 °C before taking it out. Freeze at −20 °C until ready to use.

13. Set up the following annealing program: Heat up at 55 °C for 2:00 min. Set the header temperature to 55 °C. Add 35 cycles of decrement steps of 1 °C every 3 s and hold at 20 °C. Press enter when the cycle is done then wait until header temperature decreases to 20 °C before taking it out. Freeze at −20 °C until ready to use.
14. Capillary is 30 cm in total length, 20 cm from the injection to the detection point. Make detection "window" using window maker (MicroSolv Technology corp., USA). Follow the instruction provided. Try to make the "window" as small as possible to avoid the breakage, etc. Be careful not to bend the "window" as it is fragile. When installing the capillary into the cartridge, be very slow and gentle not to break. Once installed a cartridge with a new capillary into the machine, condition it by rinsing the capillary with 1 M NaOH solution for 15 min. Same capillary can be reused many times as long as it is not broken or clogged.
15. Place vials with 1.5 mL of HCl, 1.5 mL of NaOH, 1.5 mL of H_2O, and 1.5 mL of 25 mM borax buffer (three vials) into inlet tray from BI-A1 to BI-F1, respectively. First vial with borax buffer is a rinsing buffer (BI:D1), second vial with borax buffer is a mocking buffer (BI:E1), and the third one is a running buffer (BI:F1). Make sure caps are dry before placing them on.
16. Place the buffer vials on BO-A1 and waste vial in BO-A2. Make sure that caps are dry before placing them on.
17. Place the sample vial on SI-A1. Make sure that cap is dry before placing it on.
18. A row with inserted parameter will appear as shown in Table 1.
19. At the end, Time Program will look like Table 1.
20. In this experiment rinsing buffer and running buffer (inlet and outlet) are supplemented with 50 nM SSB.
21. In this experiment rinsing buffer and running buffer (inlet and outlet) are supplemented with p19 siRNA binding protein.
22. Set up the following annealing program: Heat up at 75 °C for 5 min and then decreased the temperature to 45 °C in decrement step of 1 °C every 3 s. Then hold at 45 °C for 7 h to allow the hybridization before decreasing the temperature to 20 °C.
23. In this experiment excess amount of single-stranded RNA probe is used as a control. Since single-stranded RNA does not bind to p19.

24. Keep the wash buffer in a benchtop incubator all the time to maintain the temperature at 37 °C during the wash process.
25. For each wash, before removing the supernatant shake the beads on a benchtop shaker for 5 min at room temperature.
26. To minimize any loss of the beads during the wash steps, allow the beads to settle to the bottom of the tube before using the magnetic rack. Beads are drawn to the side of the tube using the magnetic rack and the supernatant is carefully removed with a micropipette.
27. Put the elution buffer in the same benchtop incubator as previous step to warm up to 37 °C.
28. For this separation use the longer capillary (50 cm long) instead to achieve better resolution.
29. Run the control first then run the sample using the same CE method.

References

1. Lim LP, Lau NC, Weinstein EG, Abdelhakim A, Yekta S, Rhoades MW et al (2003) The microRNAs of *Caenorhabditis elegans*. Genes Dev 17:991–1008
2. Pena JT, Sohn-Lee C, Rouhanifard SH, Ludwig J, Hafner M, Mihailovic A et al (2009) miRNA *in situ* hybridization in formaldehyde and EDC–fixed tissues. Nat Methods 6:139–141
3. Lagos-Quintana M, Rauhut R, Lendeckel W, Tuschl T (2001) Identification of novel genes coding for small expressed RNAs. Science 294:853–858
4. Chen J, Lozach J, Garcia EW, Barnes B, Luo S, Mikoulitch I et al (2008) Highly sensitive and specific microRNA expression profiling using BeadArray technology. Nucleic Acids Res 36:e87
5. Krichevsky AM, King KS, Donahue CP, Khrapko K, Kosik KS (2003) MicroRNA array reveals extensive regulation of microRNAs during brain development. RNA 9:1274–1281
6. Chen C, Ridzon DA, Broomer AJ, Zhou Z, Lee DH, Nguyen JT et al (2005) Real-time quantification of microRNAs by stem-loop RT-PCR. Nucleic Acids Res 33:e179
7. Klein D (2002) Quantification using real-time PCR technology: applications and limitations. Trends Mol Med 8:257–260
8. Gassman E, Kuo JE, Zare RN (1985) Electrokinetic separation of chiral compounds. Science 230:813–814
9. Berezovski M, Krylov SN (2003) Using DNA-binding proteins as an analytical tool. J Am Chem Soc 125:13451–13454
10. Krauss G, Sindermann H, Schomburg U, Maass G (1981) Escherichia coli single-strand deoxyribonucleic acid binding protein: stability, specificity, and kinetics of complexes with oligonucleotides and deoxyribonucleic acid. Biochemistry 20:5346–5352
11. Overman LB, Bujalowski W, Lohman TM (1988) Equilibrium binding of Escherichia coli single-strand binding protein to single-stranded nucleic acids in the (SSB) 65 binding mode. Biochemistry 27:456–471
12. Khan N, Cheng J, Pezack JP, Berezovski MV (2011) Quantitative analysis of microRNA in blood serum with protein-facilitated affinity capillary electrophoresis. Anal Chem 83: 6196–6201

Chapter 21

High-Throughput Functional MicroRNA Profiling Using Recombinant AAV-Based MicroRNA Sensor Arrays

Wenhong Tian, Xiaoyan Dong, Xiaobing Wu, and Zhijian Wu

Abstract

There is a lack of methods for high-throughput functional microRNA (miRNA) profiling. In this chapter, we describe a recombinant adeno-associated virus-based miRNA sensor array (miRNA Asensor array), which is able to profile functional miRNAs in cultured cells. The preparation of an miRNA Asensor array and its usage are discussed.

Key words miRNA, miRNA activity, high-throughput, miRNA Asensor array

1 Introduction

The high-throughput profiling of microRNA (miRNA) expression and activity is critical for miRNA research. miRNA expression profiling can be achieved by microarray analysis [1], quantitative reverse transcription-polymerase chain reaction [2], and next-generation sequencing [3]. However, approaches for the high-throughput profiling of miRNA activity are lacking. miRNA activity, which is different from miRNA expression, correlates with mature miRNA function and is often affected by various factors, including the subcellular localization of miRNAs [4], the quantity of miRNA targets [5], and proteins that associate with the miRNA-induced silencing complex [6]. Therefore, functional miRNA profiling, which reflects miRNA activity, may confer some advantages over conventional miRNA profiling, if achievable.

Ideally, miRNA activity should be examined using the 3′-untranslated region (3′UTR) of its natural or predicted target genes as targets. However, individual miRNAs usually have more than one target gene, and the 3′UTR of genes often serve as targets for more than one miRNA [7]. Therefore, a perfect complementary sequence is often used in current assays of miRNA activity. This sequence is placed at the 3′UTR of a reporter expression cassette, forming a so-called miRNA sensor. miRNA activity is then

Dmitry M. Kolpashchikov and Yulia V. Gerasimova (eds.), *Nucleic Acid Detection: Methods and Protocols*, Methods in Molecular Biology, vol. 1039, DOI 10.1007/978-1-62703-535-4_21, © Springer Science+Business Media New York 2013

determined by comparing the reporter expression level produced by the miRNA sensor with that produced by the "control sensor," which carries an identical reporter cassette lacking the miRNA complementary target [8, 9]. As a prerequisite, miRNA sensors must be efficiently transfected into cells for reporter gene expression. In many cases, this is difficult, particularly when high-throughput transfection is required. We have overcome this problem by establishing an adeno-associated virus (AAV) reverse-infection array, which enables efficient and convenient high-throughput transduction for a variety of cell lines [10].

Based on our AAV reverse-infection array, we developed a dual-luciferase reporter system to examine miRNA activity in cultured cell lines in a high-throughput manner. This new system is referred to as an miRNA Asensor array [11]. Each miRNA Asensor contains two independent expression cassettes encoding a firefly luciferase (Fluc) and a *Gaussia* luciferase (Gluc). The former is used to calibrate the transduction efficiency, while the latter, which includes an miRNA perfect complementary target sequence in the 3′UTR, is used to monitor miRNA activity. The Asensor lacking the miRNA target sequence is designated as the control Asensor. The miRNA Asensor array is prepared by loading the Asensors into a 96-well plate. To calibrate the variation in loading of the different Asensors, the transduction coefficient (TC) of each miRNA Asensor must be determined. Using this array, we have acquired the functional profiles of 115 miRNAs for 12 cell lines. In this chapter, we describe the procedure for preparing an miRNA Asensor array and its application.

2 Materials

2.1 Vector Construction

1. pDRIVE-mPGK plasmid (InvivoGen, San Diego, CA, USA) (*see* **Note 1**).
2. Pyrobest DNA polymerase (Takara, Shiga, Japan).
3. Agarose.
4. TAE stock solution (50×): 2.5 M Tris-acetate, pH 8.0, 50 mM EDTA. Prepare 1× working solution by diluting TAE stock buffer (50×) with distilled water. Store at room temperature.
5. Power supply.
6. Mini horizontal gel electrophoresis cell.
7. Restriction enzymes *Kpn*I, *Bam*HI, *Bgl*II, *Hin*dIII, *Xho*I, and *Eco*RI (New England BioLabs, Inc., Ipswich, MA, USA).
8. Thermostatic water bath.
9. pGL4.14 (Promega Corp.) (*see* **Note 2**).
10. WD-9403E ultraviolet lamp.
11. DNA fragment purification kit.

12. T4 DNA ligase (New England BioLabs, Inc.)
13. T4 DNA ligase buffer (10×) (New England BioLabs, Inc.).
14. JM109 *Escherichia coli* competent cells (Takara).
15. Ampicillin stock solution: 100 mg/mL ampicillin. Dissolve 1 g of sodium ampicillin in 10 mL distilled water. Sterilize using a filter-sterilization unit. Store the solution in aliquots at −20 °C.
16. LB culture medium: 1 % (w/v) tryptone, 0.5 % (w/v) yeast extract, and 1 % (w/v) NaCl. Sterilize by autoclaving for 20 min at 15 psi on the liquid cycle or using a filter-sterilization unit. Store at 4 °C.
17. LB plate containing 100 μg/mL ampicillin. Dissolve 15 g of agar in 1.0 L of LB culture medium and sterilize by autoclaving for 20 min at 15 psi on the liquid cycle. Cool to 50 °C and add 1.0 mL of 100 mg/mL ampicillin. Pour into plates.
18. DNA ladder.
19. Benchtop incubator shaker.
20. Mini centrifuge.
21. Plasmid extraction kit.
22. Alkaline phosphatase, calf intestinal (CIP) (New England BioLabs, Inc.).
23. Annealing buffer (1×): 10 mM Tris–HCl, pH 7.5, 50 mM NaCl, 1 mM EDTA. Sterilize using a 0.2-μm filter and store at room temperature.
24. Qiagen plasmid midi kit (Qiagen, Inc., Valencia, CA).
25. TE buffer: 10 mM Tris–HCl, pH 8.0, 0.1 mM EDTA. Sterilize by autoclaving for 20 min at 15 psi on the liquid cycle or using a filter-sterilization unit. Store at room temperature.

2.2 AAV Cell Lines

1. BHK-21 [C-13] cells (ATCC #CCL-10).
2. Lipofectamine 2000 transfection reagent (Invitrogen, Carlsbad, CA, USA).
3. DMEM medium.
4. Fetal bovine serum.
5. G418 sulfate. Prepare a 100 mg/mL solution using ddH_2O. Sterilize using a 0.2-μm filter. Store at −20 °C.

2.3 Preparation of miRNA Asensor

1. A set of cosmids that contains the whole genome of HSV1 strain 17—SetC which is composed of cos6, cos28, cos14, cos56, and cos48 [12] (*see* **Note 3**).
2. pSub201 [13] (*see* **Note 4**).
3. Restriction enzyme (New England BioLabs, Inc.).
4. Lipofectamine 2000 transfection reagent (Invitrogen).
5. DMEM medium.

6. Fetal bovine serum.
7. DMEM/5 % FBS: 5 % FBS, 7.5 μg/mL human immunoglobulin in DMEM. Add 50 mL of FBS to 1.0 L of DMEM to obtain DMEM/5 % FBS. Then prepare DMEM/5 % FBS containing 7.5 μg/mL human immunoglobulin by adding 3 μL of 20 mg/mL pooled human immunoglobulin (Sigma) into 8 mL of DMEM/5 % FBS.
8. PBS^{2+}:137 mM NaCl, 2.7 mM KCl, 10 mM Na_2HPO_4, 1.8 mM KH_2PO_4, 1 mM $CaCl_2$, and 0.5 mM $MgCl_2$. Sterilize by autoclaving for 20 min at 15 psi on the liquid cycle or by filter-sterilization. Store at room temperature.
9. Methanol.
10. KaryoMax Giemsa Stain stock solution (Invitrogen).
11. Inverted microscope.
12. Chloroform.
13. NaCl.
14. PEG8000.
15. Benzonase nuclease (Merck) (*see* **Note 5**).
16. TE buffer (*see* Subheading 2.1, **item 25**).
17. 6-well plates.

2.4 Determination of AAV Titer by Dot-Blotting

1. 4 M LiCl.
2. 95 % ethanol.
3. 75 % ethanol.
4. Mini centrifuge.
5. Benzonase nuclease.
6. 0.1 M EDTA.
7. 10 mg/mL Proteinase K. Put 100 mg Proteinase K (Merck) into 8.0 mL distilled water. When completely dissolved, add more distilled water to the final volume of 10.0 mL. Store at −20 °C.
8. 0.5 M NaOH.
9. Linearized plasmid (*see* **Note 6**).
10. WD-9403E Ultraviolet lamp.
11. Nylon membrane, positively charged (Millipore, Billerica, MA, USA).
12. PCR DIG Probe Synthesis Kit (Roche, Basel, Switzerland).
13. Anti-Digoxigenin-AP, Fab fragment from sheep (Roche).
14. BCIP 4-toluidine salt.
15. 4-Nitro blue tetrazolium chloride solution (NBT).
16. PBS^{2+} (*see* Subheading 2.3, **item 8**).
17. TE buffer (*see* Subheading 2.1, **item 25**).

2.5 Preparation of miRNA Asensor Array

1. BioLux *Gaussia* Luciferase Flex Assay Kit (New England BioLabs, Inc.).
2. Luciferase assay system (Promega).
3. Liquidator 96 Manual Benchtop Pipetting System (Rainin Instruments LLC., Oakland, CA, USA).
4. Laminar flow tissue culture hood.
5. CO_2 Incubator.
6. 96-well plates.

2.6 Profiling miRNA Activity Using an miRNA Asensor Array

1. Centrifuge.
2. BioLux *Gaussia* Luciferase Flex Assay Kit (New England BioLabs, Inc.).
3. 0.25 % trypsin.

3 Methods

The production of an miRNA Asensor array is illustrated in Fig. 1.

3.1 Vector Construction

A flowchart for vector construction is shown in Fig. 2. The sequences of the primers used in this protocol are given in Table 1.

3.1.1 Construction of the Control Asensor Plasmid

1. Amplify the core region of the mPGK promoter by PCR using pDRIVE-mPGK as a template [14] (*see* **Note 7**). Use Pyrobest DNA polymerase for PCR in a 50-μL reaction system. Carry out the reaction at 94 °C for 5 min, followed by 30 cycles of 94 °C for 30 s, 62.2 °C for 30 s, and 72 °C for 40 s, with a final hold at 72 °C for 10 min. Load the products and DNA ladder in separate wells on an agarose gel and run at 150 V for 30 min. Excise the targeted DNA band and purify it using a DNA fragment purification kit.
2. Digest the purified PGK promoter fragment using *Kpn*I and *Bam*HI. In the meantime, digest the pGL4.14 plasmid with *Kpn*I and *Bgl*II. Incubate for 2 h at 37 °C. Load the digested products in separate wells on an agarose gel and run at 150 V for 30 min. Excise the targeted DNA bands (digested PGK promoter and pGL4.14) and purify them using a DNA fragment purification kit.
3. Ligate the digested PGK promoter fragment with the digested pGL4.14 fragment using T4 DNA ligase at a molecular ratio of 3:1 (*see* **Note 8**). Incubate at 16 °C for 16 h. Transform the ligation product into JM109 *E. coli* competent cells following the manufacturer's instructions. Spread 200 μL of transformed cells on LB plates containing 100 μg/mL ampicillin. Incubate the plates overnight at 37 °C.

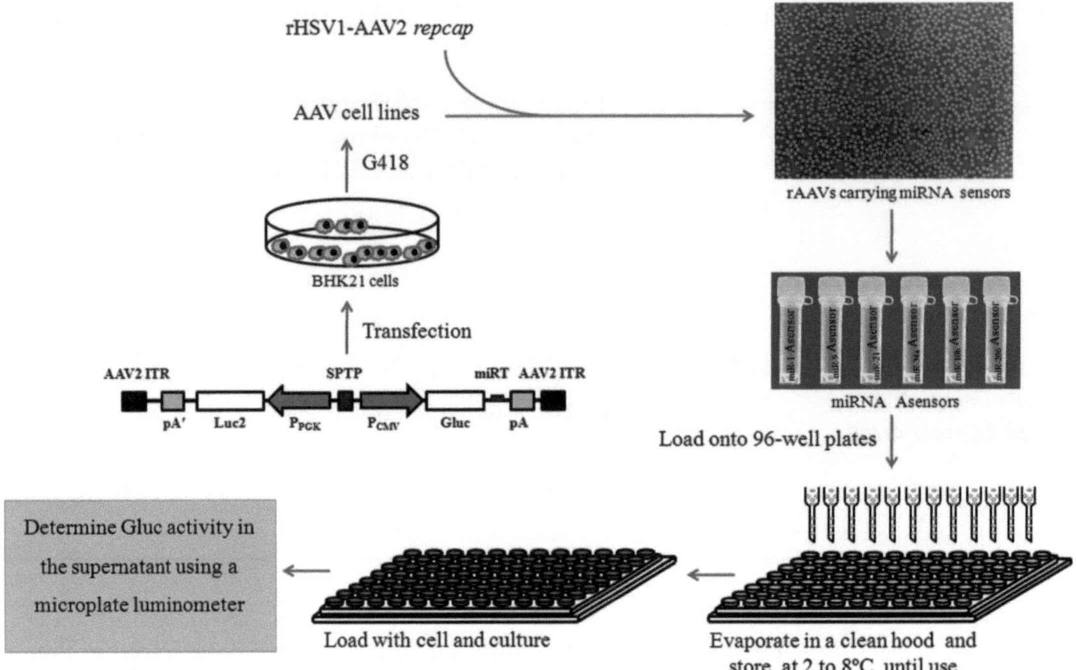

Fig. 1 Overview of the miRNA Asensor array approach. *SPTP* synthetic poly(A) signal/transcriptional pause site from the pGL4.14 vector, *miRT* perfect complementary miRNA target sequence. Adapted from ref. 15

4. Pick single colonies from the plates and inoculate each colony in 5 mL of LB medium containing 100 μg/mL ampicillin. Incubate at 37 °C for 12–16 h with vigorous shaking (220–300 rpm).
5. Harvest the bacterial cells and extract the plasmid using a plasmid extraction kit. Identify the plasmid by *Kpn*I and *Hin*dIII digestion and confirm it by DNA sequencing. Designate the correct clone pGL4.14-PGK. Amplify the Luc2 gene cassette by PCR using pGL4.14-PGK as a template. Use Pyrobest DNA polymerase for PCR in a 50-μL reaction system. Carry out the reaction at 94 °C for 5 min, followed by 30 cycles of 94 °C for 30 s, 58.7 °C for 30 s, and 72 °C for 160 s, with a final hold at 72 °C for 10 min. Load the products and DNA ladder in separate wells on an agarose gel and run at 150 V for 30 min. Excise the targeted DNA band (Luc2 gene cassette) and purify it using a DNA fragment purification kit.
6. Digest the Luc2 gene cassette and pAAV2neo-Gluc plasmid [15] with *Xho*I. Incubate the reactions at 37 °C for 2 h. Purify the digested products using a DNA fragment purification kit. Remove the 5′ phosphate of the digested pAAV2neo-Gluc fragment using CIP and purify it using a DNA fragment purification kit.
7. Ligate the digested Luc gene cassette with the digested pAAV2neo-Gluc using T4 DNA ligase at a molecular ratio of 3:1.

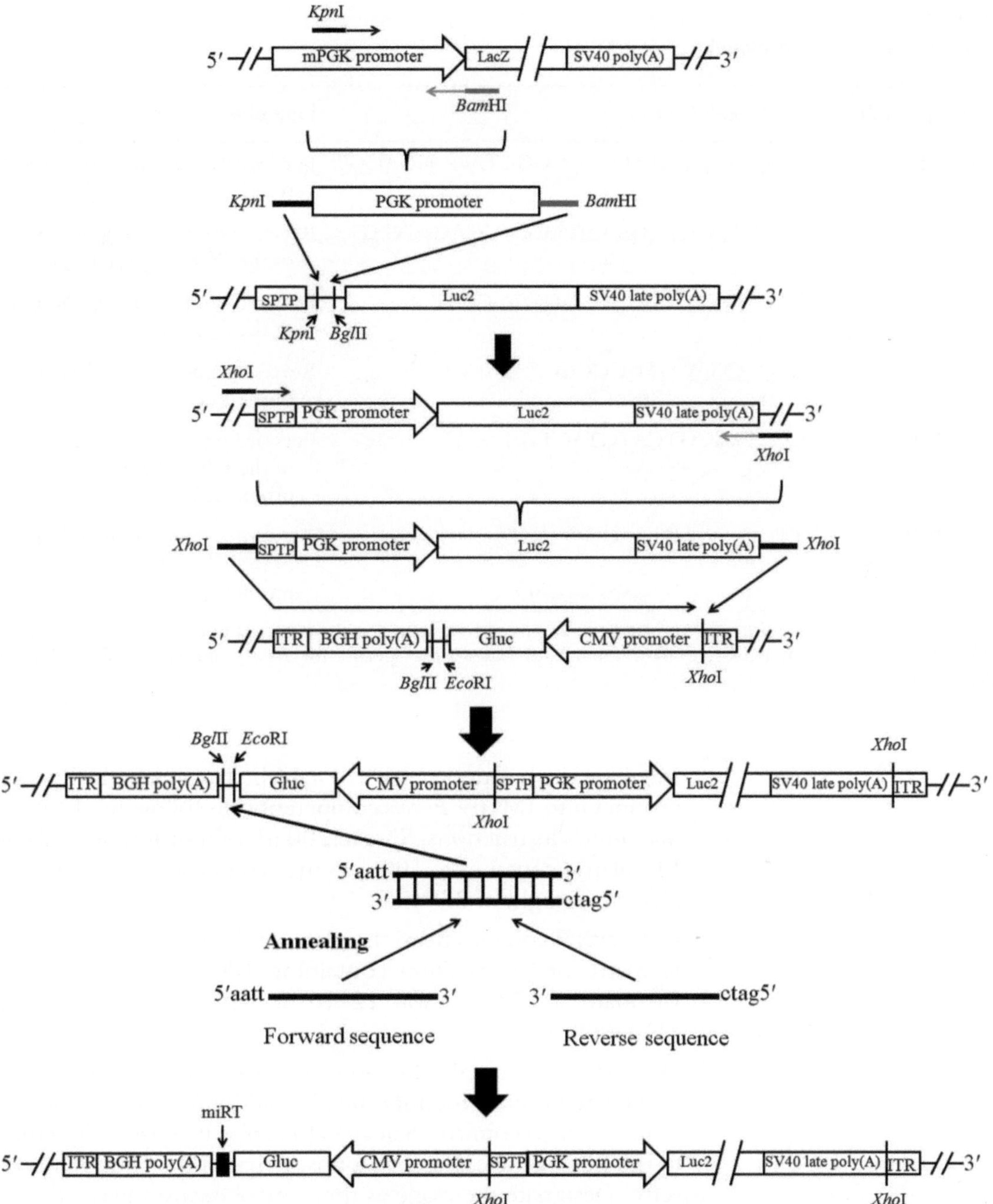

Fig. 2 The vector construction process. *mPGK promoter* mouse 3-phosphoglycerate kinase gene promoter region, *LacZ lacZ*-coding region, *SV40 poly(A)* SV40 polyA signal, *PGK promoter* core region of the mPGK promoter, *SPTP*, synthetic poly(A) signal/transcriptional pause site from the pGL4.14 vector, *Luc2* the firefly luciferase coding region, *SV40 late poly(A)* SV40 late-polyA signal, *ITR* AAV2 inverted terminal repeat, *CMV promoter* human cytomegalovirus promoter, *Gluc Gaussia* luciferase coding region, *BGH poly(A)* bovine growth hormone polyA signal. Adapted from ref. 15

Table 1
Primers used in this protocol

Primer name	5′→3′ Sequence	Function
PGK-f	CAGGGTACCATTCTACCGGGTAGGGGAGG	Forward primer for amplification of the PGK promoter
PGK-r	ATTGGATCCTCGAAAGGCCCGGAGATGAG	Reverse primer for amplification of the PGK promoter
Luc-f	GAACTCGAGTGGACAGGCCGCAAT	Forward primer for amplification of the Luc gene cassette
Luc-r	CAACTCGAGGAGCGCCCAATACGC	Reverse primer for amplification of the Luc gene cassette
CMV-P-f	CCCATAAGGTCATGTACTGGGCAT	Forward primer for amplification of the DIG-label CMV promoter probe
CMV-P-r	GTTCCCATAGTAACGCCAATAGGG	Forward primer for amplification of the DIG-label CMV promoter probe

[a]The restriction enzyme site used to insert the amplified fragment into the vector is underlined in each primer. The restriction sites in PGK-f, PGK-r, Luc-f, and Luc-r were specific for *Kpn*I, *Bam*HI, *Xho*I, and *Xho*I, respectively

Incubate the reaction at 16 °C for 16 h. Transform the ligation product into JM109 *E. coli* competent cells following the manufacturer's instructions. Spread 200 μL of transformed cells on LB plates containing 100 μg/mL ampicillin. Incubate the plates overnight at 37 °C.

8. Pick several colonies from the plates and inoculate each colony in 5 mL of LB medium containing 100 μg/mL ampicillin. Incubate at 37 °C for 12–16 h with vigorous shaking (220–300 rpm).
9. Harvest the bacterial cells and extract the plasmid using a plasmid extraction kit. Identify the plasmid by *Kpn*I digestion and sequence it to confirm. Select a clone in which the orientation of the Luc2 gene cassette is opposite that of the Gluc gene cassette. Designate the clone as the control Asensor plasmid.

3.1.2 Construction of miRNA Asensor Plasmids

1. Obtain mature miRNA sequences from http://www.mirbase.org. Deduce the sequences of perfect complementary targets for the miRNAs based on the principle of Watson–Crick base pairing (*see* **Note 9**). Add "5′aatt3′" to the 5′ end of the perfect complementary target to form the forward sequence, and add "5′gatc3′" to the 5′ end of the mature miRNA sequence to form the reverse sequence. Synthesize oligonucleotides for the forward and reverse sequences.

2. Dissolve the synthesized oligonucleotides in 1× annealing buffer. Anneal the forward and reverse oligonucleotides to form the adaptor. Prepare the annealing reaction as follows:

   Forward sequence solution (200 μM) 10 μL (100 μM final concentration).

   Reverse sequence solution (200 μM) 10 μL (100 μM final concentration).

   Carry out the reaction at 95 °C for 10 min, then gradually cool it to room temperature. Keep the reaction at room temperature for 2 h (*see* **Note 10**). Store the adapter at −20 °C until needed.

3. Digest the control Asensor plasmid with *Eco*RI and *Bgl*II. Incubate the reaction at 37 °C for 2 h. Purify the digested product using a DNA fragment purification kit. Ligate the treated control Asensor plasmid with the adaptor as follows:

Treated control Asensor plasmid (40 ng/μL)	1 μL
Adaptor (100 μM)	1 μL
T4 DNA ligase buffer (10×)	2 μL
T4 DNA ligase (400 U/μL)	1 μL
ddH_2O	15 μL

4. Incubate the ligation reaction at 16 °C for 16 h. Transform the ligation product into JM109 *E. coli* competent cells following the manufacturer's instructions. Spread 200 μL of transformed cells on LB plates containing 100 μg/mL ampicillin. Incubate the plates overnight at 37 °C.

5. Pick several colonies from the plates and inoculate each in 5 mL LB medium containing 100 μg/mL ampicillin. Incubate for 12–16 h at 37 °C with vigorous shaking (220–300 rpm).

6. Harvest the bacterial cells and extract the plasmid using a plasmid extraction kit. Identify the plasmid by sequencing.

3.2 AAV Cell Lines

1. Extract the control and miRNA Asensors plasmid DNA using a Qiagen Plasmid Midi Kit.

2. Transfect the control and miRNA Asensors plasmid DNA into BHK-21 cells using Lipofectamine 2000 according to the manufacturer's instructions. After 24 h, split the cells 1:10 and culture in DMEM containing 10 % FBS and 400 μg/mL G418 (*see* **Note 11**). Change the medium every 3 days. After 15 days, neomycin-resistant colonies will be present (*see* **Note 12**). Expand the colonies to obtain stable cell lines harboring AAV provirus for miRNA Asensors.

3.3 Preparation of the AAV Helper Virus rHSV1-rc/Δ UL2

Prepare the AAV helper virus rHSV1-rc/ΔUL2 as reported previously [16] based on a set of cosmids that contains the whole genome of HSV1 strain 17—SetC which is composed of cos6, cos28, cos14, cos56, and cos48 [12]. rHSV1-rc/ΔUL2 contains the AAV2 *rep* and *cap* gene cassettes and is used to package AAV vectors into AAV2 capsids.

1. Excise the 4.3 kb fragment of AAV2 *rep* and *cap* genes under control of their native promoters from pSub201 [13] and insert it into *Xba*I site of cos6 generating cos6-rcΔUL2.
2. Digest cos6-rcΔUL2, cos28, cos14, cos56, and cos48 with *Pac*I, respectively. Precipitate the digested product with ethanol and dissolve it in TE buffer.
3. Co-transfect the digested cos6-rcΔUL2, cos28, cos14, cos56, and cos48 into BHK-21 cells using Lipofectamine 2000 according to the manufacturer's instructions.
4. Replace the medium every 24 h until the cytopathic effect (CPE) appears. When full CPE is observed, collect the cells along with medium. Free and thaw the cells three times. Spin down the debris and collect the supernatant which contains rHSV1-rc/ΔUL2.
5. Propagate rHSV1-rc/ΔUL2 by infecting BHK-21 cells with the supernatant. When full CPE is observed, collect the cells along with medium. Free and thaw three times. Spin down the debris and collect the supernatant which contains rHSV1-rc/ΔUL2. Store the virus at −20 °C.

3.4 Determination of the rHSV1-rc/Δ UL2 Titer

Determine the rHSV1-rc/ΔUL2 titer using a plaque assay, as described previously [17].

1. Split appropriate number of BHK-21 cells into 6-well plates (*see* **Note 13**) using DMEM/5 % FBS so that ~90 % confluence will be reached in 24 h.
2. Serially dilute prepared rHSV1-rc/ΔUL2 by tenfold from 10^{-1} to 10^{-8} using DMEM/5 % FBS (*see* **Note 14**).
3. Aspirate medium from BHK-21 cell culture. Add 500 μL of each dilution of rHSV1-rc/ΔUL2 to the corresponding well of cells.
4. Swirl the plate every 30 min to allow virus to absorb. 2 h later, replace medium with DMEM/5 % FBS containing 7.5 μg/mL human immunoglobulin.
5. Incubate for ~2 days until plaques are visible.
6. Aspirate medium from plates. Rinse each well of monolayer cells twice with PBS^{2+} (*see* **Note 15**).
7. To fix the cells, add 500 μL of methanol to each well and leave it at room temperature for 5 min.

8. Dilute the KaryoMax Giemsa Stain stock solution 1:10 with distilled water. Aspirate the methanol and add 1 mL of the diluted Giesma stain to each well. Incubate for 20 min at room temperature.
9. Discard the stain solution. Rinse cells gently with cold tap water until the liquid is clear. Dry inverted plate on a paper towel.
10. Count plaques (circle or mark each plaque with a fine-point marker pen) viewing through an inverted light microscope at low (10×) magnification.

3.5 Preparation of AAV miRNA Asensors

1. Infect AAV cell lines of the control and miRNA Asensors with rHSV1-rc/ΔUL2 at a multiplicity of infection (MOI) of 0.1 (*see* **Note 16**). When the cytopathic effect of the virus is nearly complete (~48 h later), aspirate the medium and harvest the cells with PBS^{2+}.
2. Purify the AAV particles as reported previously [18]. For this purpose add 10 % (v/v) of chloroform to the collected cell suspension. Incubate at 37 °C with vigorous shaking (220–260 rpm) for 1 h until all cells are lysed (*see* **Note 17**). Add solid NaCl to lysates till the final concentration is 1 M and constantly shake for 30 min at room temperature. Centrifuge at 12,000 × *g* for 15 min at 4 °C. Harvest the supernatant.
3. Add an appropriate amount of solid PEG8000 to the supernatant to obtain a final concentration of 10 % (w/v). Cool the supernatant in ice water and let it stand still for 1 h. Recover the precipitated AAV particles by centrifugation at 10,000 × *g* for 15 min at 4 °C. Dissolve the pellets with PBS^{2+} buffer. Add Benzonase nuclease to a final concentration of 200 U/mL and incubate at room temperature for 30 min.
4. Add equal volume of chloroform to suspension and vigorously shake for 2 min. Separate the organic and aqueous phases by centrifugation at 12,000 × *g* for 5 min at 4 °C. Collect the aqueous phase containing the purified AAV particles.

3.6 Determination of the rAAV Titer by Dot-Blot Assay

1. Prepare a DIG-labeled CMV promoter probe using a PCR DIG probe synthesis kit. Set up the reaction as follows:

ddH_2O	41.25 μL
PCR DIG mix (10×)	5 μL
CMV-p-f (20 μM)	1 μL
CMV-p-r (20 μM)	1 μL
Enzyme mix	0.75 μL
Control Asensor plasmid (10 ng/μL)	1 μL

Carry out the reaction at 95 °C for 2 min, followed by 30 cycles of 95 °C for 30 s, 55 °C for 30 s, and 72 °C for 15 s, with a final hold at 72 °C for 7 min. Load 1 μL of the product onto an agarose gel and run it at 150 V for 30 min to verify the size. Add 5 μL of 4.0 M LiCl to the remaining product and mix. Next, add 150 μL of prechilled 95 % ethanol and incubate at −20 °C for 2 h. Centrifuge at 12,000 × *g* for 10 min. Wash the pellet twice with 75 % ethanol. Dry the pellet completely and dissolve it in 20 μL of TE. Store at −20 °C until use.

2. Digest the virus particles to release the DNA, as reported previously [19]. Specifically, add 10 μL purified AAV particle suspension in 50-μL reaction system where the final concentration of Benzonase nuclease is 200 U/mL and incubate at 37 °C for 30 min (*see* **Note 18**). Stop the digestion by adding 10 μL of 0.1 M EDTA to the reaction system and mixing. Release virion DNA by adding 60 μL PBS^{2+} containing 200 μg/mL proteinase K and incubate 1 h at 50 °C. Denature viral DNA by adding 120 μL of 0.5 M NaOH and incubate for 10 min at room temperature.
3. Prepare the standard and sample DNAs, as reported previously [19]. Specifically, prepare linearized plasmid (from the plasmid containing the probe sequence) at 1×10^{12} copies/mL for DNA concentration standards (*see* **Note 19**). Make twofold serial dilutions ranging from 1×10^{12} to 3.125×10^{10} copies/mL. Denature DNA standards by adding equal volume of 0.5 M NaOH to each dilution.
4. Load the standard and sample DNAs on a positively charged nylon membrane. Allow the membrane to air-dry completely. Fix the DNA to the membrane by exposing the side with DNA to a UV light source (254 nm) for 15.0 s.
5. Perform pre-hybridization and hybridization according to the manufacturer's Southern blot protocol. Test the DIG-labeled probe using anti-digoxigenin alkaline phosphatase-conjugated antibodies. Detect alkaline phosphatase activity in a color reaction using BCIP/NBT as the substrate (*see* **Note 20**).

3.7 Preparation of the miRNA Asensor Array

1. Select miRNA Asensors according to the experimental objectives.
2. Dilute the control and miRNA Asensors to 1.25×10^{10} virus genomes (vg)/mL in PBS^{2+}.
3. Load the control and miRNA Asensors onto a 96-well cell culture plate using the Liquidator 96 Manual Benchtop Pipetting System (Rainin Instruments LLC.) at 2.5×10^{8} vg/20 μL/well.
4. Evaporate the arrays in a laminar flow tissue culture hood then store them at 2–8 °C (*see* **Note 21**). Prepare multiple sets of miRNA Asensor array plates in a batch. Determine the

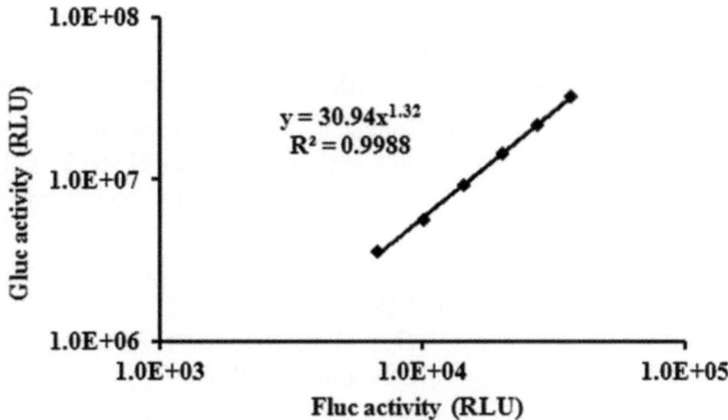

Fig. 3 Relationship between Fluc and Gluc activity in the control Asensor in BHK-21 cells. *RLU* relative light units. Adapted from ref. 15

transduction coefficients (TC) for each miRNA Asensor in an array (*see* **Note 22**). Different sets of miRNA Asensor array plates that are prepared in the same batch are considered to have similar TC values. To detect the TC value for each miRNA Asensor in the array, determine the relationship between Gluc and Fluc activity of the control Asensor when no miRNA repression occurs. Make serial twofold dilutions ranging from 5.00×10^8 to 1.56×10^7 vg/well of control Asensor in a 96-well plate. Load BHK-21 cells onto the plate at a density of 1×10^4 cells/well and culture them for 48 h (*see* **Note 23**). Measure the Gluc and Fluc activities using a Gluc and Fluc assay kit, following the manufacturers' instructions. Deduce the relationship between the levels of Gluc activity (G) and Fluc activity (F) (Fig. 3). Compute the TC, which is the ratio of G for the $\text{Asensor}_{\text{miRNA}}$ (G_{miRNA}) to that for the $\text{Asensor}_{\text{control}}$ (G_{control}) when no miRNA repression occurs, as follows:

$$TC = \frac{G_{miRNA}}{G_{control}} = \left(\frac{F_{miRNA}}{F_{control}}\right)^n,$$

where F_{control} and F_{miRNA} represent the F of the $\text{Asensor}_{\text{control}}$ and $\text{Asensor}_{\text{miRNA}}$, respectively. In the equation, n represents the relationship between F and G deduced from Fig. 3, as exemplified by 1.32 in Fig. 3.

3.8 Profiling miRNA Activity Using an miRNA Asensor Array

1. For adherent cells, digest the cells with 0.25 % trypsin to create a single-cell suspension. Determine the cell number and dilute the single-cell suspension to an appropriate density (*see* **Note 24**).

   For suspension cells, centrifuge the cells at $500 \times g$ for 5 min and resuspend them in culture medium. Determine the cell number and dilute the single-cell suspension to an appropriate density (*see* **Note 24**).

2. Load 200 μL of single-cell suspension into each well of the miRNA Asensor array. Culture the cells at 37 °C in an incubator with 5 % CO_2 for 48 h (*see* **Notes 25** and **26**).
3. Collect a 20 μL aliquot of supernatant from each well for a Gluc activity assay (*see* **Note 27**). Test for Gluc activity using a Gluc activity assay kit, following the manufacturer's instructions (*see* **Note 28**).
4. miRNA activity is represented as the relative inhibiting fold (RIF), which is the TC-calibrated ratio of *G* for the $Asensor_{control}$ ($G_{control}$) to that for the $Asensor_{miRNA}$ (G_{miRNA}). Compute the RIF value as follows:

$$\mathrm{RIF} = \left(\frac{G_{\mathrm{control}}}{G_{\mathrm{miRNA}}}\right) \times \mathrm{TC}.$$

5. Analyze the data. A sample miRNA activity profile for 12 cell lines is shown in Fig. 4.

4 Notes

1. The plasmid carries the mPGK promoter and is used as a template to amplify the fragment of mPGK promoter by PCR.
2. The pGL4.14 plasmid includes the synthetic firefly *luc2* gene which has been codon-optimized for more efficient expression in mammalian cells. It is a basic vector with no promoter and contains a multiple cloning region for insertion of a promoter of choice. mPGK promoter was inserted into the multiple cloning region of pGL4.14 plasmid to obtain the *luc2* gene cassette.
3. SetC cosmids were kindly provided by Dr. Davison (Glasgow University, UK). The digestion products of SetC cosmids were cotransfected into HSV1 sensitive cells (such as BHK-21) so that HSV1was rescued.
4. pSub201 plasmid was kindly provided by Dr. Samulski. It contains AAV2 *rep* and *cap* genes with their native promoters and polyA signals.
5. Benzonase nuclease can degrade all forms of DNA and RNA while having no proteolytic activity. It is used to degrade any form of DNA and RNA that is not packaged into virions.
6. Linearized plasmid is prepared by digestion of the plasmid containing the probe binding sequence, which is used as DNA standard in determination of the titer of recombinant AAV.
7. The core region of the mPGK promoter contains four Spl binding sites, a CCAAT box, a GC box, and a putative transcription start point, with a length of 553 nucleotides (nt).

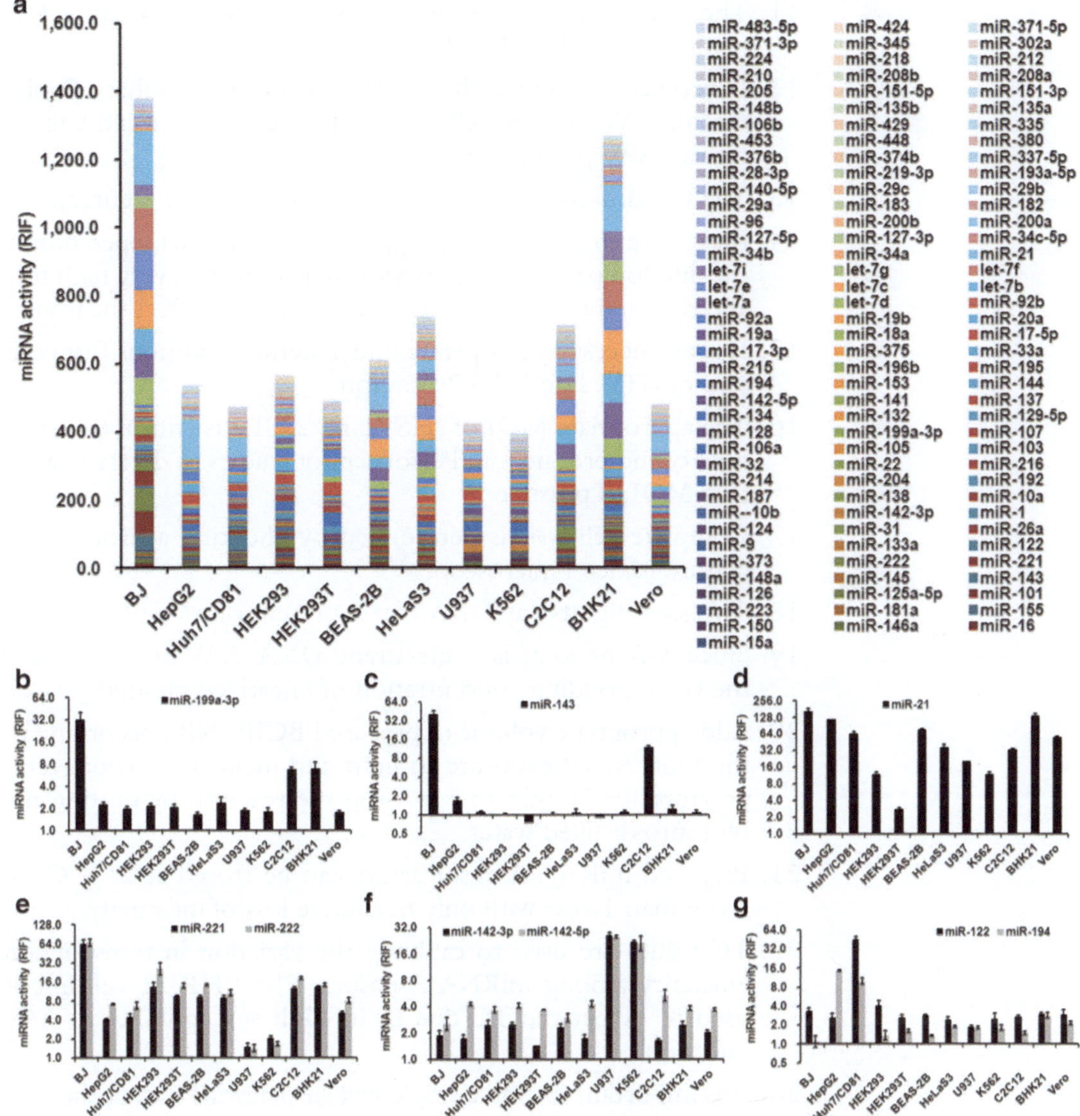

Fig. 4 Sample analysis of the miRNA activity profile for cell lines obtained from an miRNA Asensor array. The miRNA activity profiles of 12 cell lines are shown (**a**). Several miRNAs were selected and their activities in different cell lines were compared (**b**–**g**). *RIF* relative inhibiting fold. Adapted from ref. 15

8. To increase the cloning efficiency, the ratio of insert to vector fragment should range from 3:1 to 10:1.
9. In mammalian cells, most miRNA targets are not perfectly complementary to their mature sequences; one miRNA may have several targets. We use the perfect complementary target to investigate miRNA activity.
10. Let the samples cool gradually to room temperature and keep them at room temperature for at least 1 h.

11. The final concentration of G418 in the DMEM 10 % FBS medium should be 400–800 μg/mL.
12. If necessary, compare the rAAV yield from a number of colonies of AAV proviral cell lines and select the best cell line for future rAAV production.
13. The titration is performed in duplicate to ensure accuracy.
14. Make sure to use a clean, plugged pipet tip in each operation of the dilution process. Virus stocks are routinely of very high titer and repeated use of a pipet tip for dilution will affect the results.
15. It is not necessary to operate under sterile conditions from this step of rHSV1-rc/ΔUL2 titration.
16. The appropriate MOI of rHSV1-rc/ΔUL2 is critical for rAAV quality and production. Perform pilot studies to determine the best MOI, if possible.
17. Complete cell lysis is determined by checking a drop of the culture under a microscope.
18. This step digests any DNA that is present outside of virions.
19. Since AAV genome is single-strand DNA, AAV titer is twice of the corresponding concentration of linearized plasmid.
20. Add appropriate volume of prepared BCIP/NBT according to the blot. Avoid exposure to light and incubate at room temperature for 30 min to 1 h. Stop the reaction by putting the blot into distilled water.
21. Prepared miRNA Asensor arrays can be stored at 2–8 °C for more than 1 year with only negligible loss of infectivity.
22. TC values are used to calibrate the variation in transduction efficiency among miRNA Asensors. The BHK-21 cell line is used to detect the TC due to its high susceptibility to AAV infection.
23. It is important to create single-cell suspensions so that the cells can be counted accurately and dispersed evenly after loading onto the plates.
24. To increase the sensitivity of the assay, perform a pilot study to determine the appropriate quantity of cells loaded per well for each line.
25. For cell lines with a slow growth rate, a longer culture time (4–5 days) is required so that the RIF can reach its peak value.
26. After loading the cells, place the miRNA Asensor array plate in a wet box to reduce evaporation of the culture medium.
27. Centrifuge the plates, then remove 20 μL aliquots of cell-free medium from each well for the Gluc activity assay.
28. Carry out the Gluc activity assay at 25 °C.

Acknowledgment

This work was supported by a grant from the National High Technology Research and Development Program of China (No. 2012AA020810).

References

1. Miska EA, Alvarez-Saavedra E, Townsend M, Yoshii A, Sestan N, Rakic P et al (2004) Microarray analysis of microRNA expression in the developing mammalian brain. Genome Biol 5:R68
2. Katada T, Ishiquro H, Kuwabara Y, Kimura M, Mitui A, Mori Y et al (2009) microRNA expression profile in undifferentiated gastric cancer. Int J Oncol 34:537–542
3. Creighton CJ, Reid JG, Gunaratne PH (2009) Expression profiling of microRNAs by deep sequencing. Brief Bioinform 10:490–497
4. Chen B, Zhang B, Luo H, Yuan J, Skogerbø G, Chen R (2012) Distinct microRNA subcellular size and expression patterns in human cancer cells. Int J Cell Biol 2012, article ID 672462
5. Garcia DM, Baek D, Shin C, Bell GW, Grimson A, Bartel DP (2011) Weak seed-pairing stability and high target-site abundance decrease the proficiency of Lsy-6 and other microRNAs. Nat Struct Mol Biol 18:1139–1146
6. Fabian MR, Sonenberg N, Filipowicz W (2010) Regulation of mRNA translation and stability by microRNAs. Annu Rev Biochem 79:351–379
7. John B, Enright AJ, Aravin A, Tuschl T, Sander C, Marks DS (2004) Human microRNA targets. PLoS Biol 2:e363
8. Kan Z, Rouchka EC, Gish WR, States DJ (2001) Gene structure prediction and alternative splicing analysis using genomically aligned ESTs. Genome Res 11:889–900
9. Mayr C, Bartel DP (2009) Widespread shortening of 3′UTRs by alternative cleavage and polyadenylation activates oncogenes in cancer cells. Cell 138:673–684
10. Mansfield JH, Harfe BD, Nissen R, Obenauer J, Srineel J, Chaudhuri A et al (2004) MicroRNA responsive 'sensor' transgenes uncover Hox-like and other developmentally regulated patterns of vertebrate microRNA expression. Nat Genet 36:1079–1083
11. Skalsky RL, Samols MA, Plaisance KB, Boss IW, Riva A, Lopez MC et al (2007) Kaposi's sarcoma-associated herpesvirus encodes an ortholog of miR-155. J Virol 81:12836–12845
12. Cunningham C, Davison A (1993) A cosmid-based system for constructing mutants of herpes simplex virus type 1. Virology 197:116–124
13. Samulski RJ, Chang LS, Shenk T (1989) Helper-free stocks of recombinant adeno-associated viruses: normal integration does not require viral gene expression. J Virol 63:3822–3828
14. Dong X, Tian W, Wang G, Dong Z, Shen W, Zheng G et al (2010) Establishment of an AAV reverse infection-based array. PLoS One 5:e13479
15. Tian W, Dong X, Liu X, Wang G, Dong Z, Shen W et al (2012) High-throughput functional microRNAs profiling by recombinant AAV-based microRNA sensor arrays. PLoS One 7:e29551
16. Wu ZJ, Wu XB, Hou YD (1999) Construction of a recombinant herpes simplex virus which can provide packaging function for recombinant adeno-associated virus. Chin Sci Bull 44:715–719
17. Blaho JA, Morton ER, Yedowitz JC (2005) Herpes simplex virus: propagation, quantification, and storage. Curr Protoc Microbiol, Chapter 14, Unit 14E.1
18. Wu X, Dong X, Wu Z, Cao H, Niu D, Qu J et al (2001) A novel method for purification of recombinant adeno-associated virus vectors on a large scale. Chin Sci Bull 46:485–489
19. Haberman RA, Kroner-Lux G, McCown TJ, Samulski RJ (1999) Production of recombinant adeno-associated viral vectors and use in in vitro and in vivo administration. Curr Protoc Neurosci (Suppl 9), Unit 4.17

Chapter 22

The Use of Molecular Beacons to Detect and Quantify MicroRNA

Meredith B. Baker, Gang Bao, and Charles D. Searles

Abstract

Molecular beacons are oligonucleotide (DNA or RNA) probes that have become increasingly important tools for RNA sensitive detection both in vitro and in living cells. From their inception, molecular beacons have been used to determine the expression levels of RNA transcripts, but they also have the specificity to identify splice variants and single-nucleotide polymorphisms. Our group has performed extensive studies on molecular beacon design, molecular beacon hybridization assays, and cellular imaging of mRNA molecules. Compared to other methods for assessing RNA transcript expression, such as qRT-PCR, the beacon-based approach is potentially simpler, faster, more cost effective, and more specific.

Recently, our group demonstrated that molecular beacons can readily distinguish mature- and precursor microRNAs, and reliably quantify microRNA expression. MicroRNAs (miRNAs) are a class of short (19–25 nt), single-stranded, noncoding RNAs that regulate an array of cellular functions through the degradation and translational repression of mRNA targets. Importantly, tissue levels of specific miRNAs have been shown to correlate with pathological development of diseases. Thus, a rapid and efficient method of assessing miRNA expression is useful for diagnosing diseases and identifying novel therapeutic targets. Here, we describe the methods for designing and using molecular beacons to detect and quantify miRNA.

Key words microRNA, Molecular beacon, Fluorescence, RNA, Oligonucleotide probe

1 Introduction

Molecular beacons are single-stranded DNA or RNA oligonucleotide probes consisting of four elements: (1) a 5′-fluorophore for signal detection, (2) a 3′-quencher to block the fluorescence signal when the beacon is in a closed, unbound state and the 5′ and 3′ ends are in close proximity, (3) the hairpin loop segment that is complementary to the target RNA, and (4) a 4–7 bp sequence at each of the 5′ and 3′ ends, called a stem, that is self-hybridizing and stabilizes the overall hairpin shape of the closed beacon. In an inactive, unbound conformation the molecular beacon resembles a short-hairpin RNA brought together by binding between the two stem nucleotide sequences. In this conformation, the quencher diminishes

Dmitry M. Kolpashchikov and Yulia V. Gerasimova (eds.), *Nucleic Acid Detection: Methods and Protocols*, Methods in Molecular Biology, vol. 1039, DOI 10.1007/978-1-62703-535-4_22, © Springer Science+Business Media New York 2013

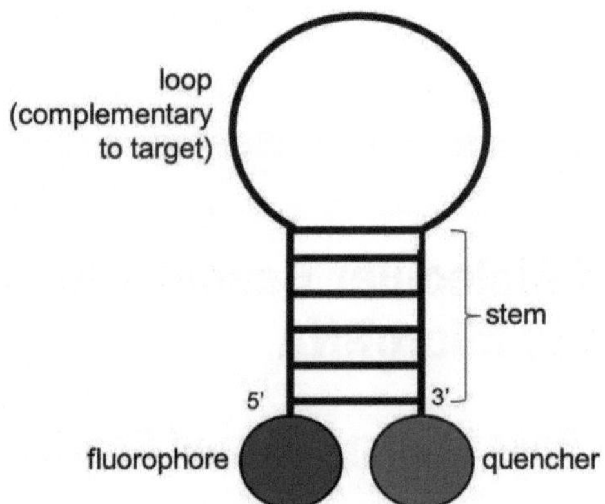

Fig. 1 Schematic illustration of a typical molecular beacon in closed/inactive configuration. The components of the beacon include: a 5′-fluorophore, a 3′-quencher, a hairpin loop segment that is complementary to the target RNA, a 4–7 bp sequence at each of the 5′ and 3′ ends, called a stem, that is self-hybridizing and stabilizes the overall hairpin shape of the closed beacon

the fluorescence signal (Fig. 1). In the presence of RNA sequence complementary to the beacon loop region, the hybridization of the target RNA to the molecular beacon opens the hairpin and physically separates the fluorophore from quencher, allowing a fluorescence signal to be emitted upon excitation [1–3] (Fig. 2).

In theory, the molecular beacon elicits a signal only upon direct hybridization to the complementary RNA sequence. This enables a molecular beacon to function as a sensitive and specific probe with a high signal-to-background ratio [4–11]. Furthermore, RNA assessment can be performed in one step without the need for reverse transcription (RT) or amplification steps, and excess probe does not have to be removed prior to measurement.

Recently, we demonstrated that molecular beacons can readily distinguish mature and pre-microRNAs (miRNAs), and reliably quantify miRNA expression [12]. miRNAs are short, single-stranded, noncoding RNAs that have been shown to posttranscriptionally regulate expression of genes involved in a diverse array of cellular functions. In human tissues, over 1,400 miRNAs have been identified [13]. Due to their role in the modulation of gene expression, miRNAs are regarded as pivotal regulators of normal development and physiology, as well as disease. Importantly, tissue levels of specific miRNAs have been shown to correlate with pathological development of diseases [14]. There is considerable active interest in determining whether miRNAs found in cells or bodily fluids can be used as biomarkers for disease states and disease progression [15, 16], and currently the abundances of specific miRNAs in disease states are being studied for their roles as biomarkers for diseases and indicators of patient prognosis [17–20].

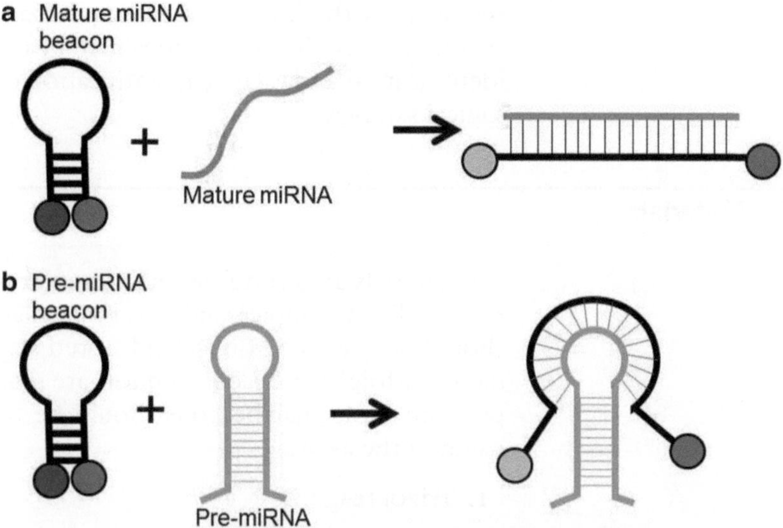

Fig. 2 Schematic illustration of molecular beacon hybridization assays. In the absence of complementary target, the molecular beacon designed for mature miRNA (mature beacon) or pre-miRNA (precursor beacon) forms a stem–loop structure that brings the quencher in close proximity to the fluorophore, thereby quenching the fluorescence emission. Hybridization of the beacon to its miRNA target opens the hairpin, which disrupts the stem of the beacon and physically separates the fluorophore from the quencher and allows fluorescence emission upon excitation. For some beacons, but not all, there may be one or more nucleotides that overlap between the beacon stem and the mature miRNA sequence. (**a**) Hybridization of the mature beacon to the mature miRNA target. (**b**) Hybridization of the precursor beacon to the pre-miRNA target. This beacon hybridizes to the loop sequence of the pre-miRNA hairpin structure

miRNAs in cells have three distinct forms: pri-miRNA, pre-miRNA, and mature miRNA. Most research studies have focused on only the expression of the mature forms of miRNAs. Although only the mature forms of the miRNA (approximately 22 nucleotides in length) are involved in posttranscriptional regulation of gene expression, quantifying the relative expression levels of mature miRNA and its precursors is important for understanding miRNA transcription, localization, and processing.

The most common method currently for detection and quantification of miRNAs is qRT-PCR with linear fluorescence probes (e.g., TaqMan). For assessing levels of distinct oligonucleotides, these linear probes are sufficient. However, the sequences of the mature forms of miRNAs are also present in the pri- and pre-forms of the miRNAs. Due to the biologically low abundance of miRNAs relative to other RNA sequences, any binding of probes to the pri- or pre-forms of the miRNA could skew PCR-based analyses of miRNA levels. Molecular beacons can distinguish between mature and precursor forms of miRNAs, as well as between miRNA family members with differences in sequence of one nucleotide [12]. Here, we describe how molecular beacons can be designed for identification and relative quantification of specific miRNAs in biological samples.

2 Materials

All materials are produced and stored at room temperature except for the RNA samples and the molecular beacons. RNA samples should be harvested on ice and stored at −80 °C to prevent degradation. Molecular beacon aliquots are stored at −20 °C. RNA samples and molecular beacons should be thawed completely before using in the assay.

1. Trizol reagent.
2. BCP (1-bromo-3-chloropropane).
3. Liquid nitrogen.
4. Centrifuge capable of 16,000 × *g* spin of 1.5 μL tubes at 4 °C.
5. Ceramic mortar and pestle.
6. PBS buffer: Dissolve 8 g of NaCl, 0.2 g of KCl, 1.4 g of Na_2HPO_4, and 0.24 g of KH_2PO_4 in 800 mL of distilled water. Adjust the pH to 7.4 by adding HCl and fill with distilled water to a volume of 1 L.
7. Molecular beacon aliquoted to 0.1 nmol/μL or 1 μg/μL.
8. 10 M NaCl: Dilute 584.4 g of NaCl in 1 L of nuclease-free water.
9. Corning® 384-Well Low Flange Black Flat Bottom Polystyrene TC-Treated Microplates.
10. Fluorescence plate reader.
11. PCR hood (optional).
12. Aluminum foil.
13. 37 °C incubator or closed oven that can maintain a 37 °C temperature.
14. Microliter pipettes that can transfer liquid samples of 1–50 μL.

3 Methods

All procedures are carried out at room temperature unless otherwise noted. The use of a PCR hood may reduce the risk of degradation of the RNA samples and molecular beacons. Alternatively, RNase Zap Wipes (Applied Biosystems) may be used to reduce the levels of RNase in the work area.

3.1 Design of Molecular Beacons

1. Determine the sequence of the miRNA of interest. The loop component of the molecular beacon should target the miRNA sequence, which will be either the mature sequence or the loop region of the precursor miRNA (Fig. 2). Therefore, the part of the beacon that hybridizes to the miRNA sequence should be complementary to this sequence, approximately 22 nucleotides.
2. Decide whether the molecular beacon will be DNA, RNA, or contain modified bases (*see* **Note 1**).
3. Add nucleotide bases to the 5′ and 3′ end of the complementary (beacon loop) sequence which will form the stem of the hairpin-shaped molecular beacon (*see* Fig. 1 and **Note 2**). If possible, use the existing complementary sequence of the miRNA as one end of the stem ("shared stem" beacon). The stem should be 4–7 nucleotides in length (*see* **Notes 3–5**).
4. Use a DNA/RNA folding program such as UNAfold (IDT, Inc.) to determine the secondary shape, free energy, and melting temperature of the proposed beacon design [21]. The shape should be hairpin with a strong stem. The change in Gibb's free energy for the closed beacon should be approximately −3.0 kcal/mol. The melting temperature should be between 50 and 60 °C. Although it is possible to make molecular beacons that do not fit these parameters, these are good starting points for optimal beacon design.
5. Add a fluorophore to the 5′ end of the beacon and a quencher to the 3′ end (*see* **Notes 6** and 7). These positions can be reversed, but the beacon is generated from the 3′ end. By placing the quencher at the 3′ end you will ensure that any incomplete beacons will have a quencher but no fluorophore as opposed to a fluorophore with no quencher.
6. Order the molecular beacon design from a scientific vendor, such as Sigma-Aldrich (www.sigmaaldrich.com), IDT (www.idtdna.com), Fisher (www.fishersci.com), Eurofins MWG Operon (www.operon.com), or Exiqon (www.exiqon.com).

3.2 Preparation of Molecular Beacons

1. Beacons arrive from the company lyophilized. Briefly spin down the tube containing the molecular beacon on a tabletop centrifuge.
2. Resolubilize the beacon in RNase-free, nuclease-free water to a final concentration of 0.1 nmol/μL or approximately 1.0 μg/μL.
3. Aliquot 5–20 μL of beacon into opaque microfuge tubes and store tubes at −20 °C.
4. For use during an experiment, thaw one beacon tube and dilute the stock 1:10 with RNase-free, nuclease-free water so that 1 μL will be 200 nM molecular beacon in 50 μL of total volume.

3.3 Isolation of RNA

1. Tissue should be frozen in sections of 0.5 cm^3 or less in liquid nitrogen.
2. Grind up tissue using a mortar and pestle on ice while the tissues are frozen.
3. Add 1.0 mL of Trizol reagent (*see* **Note 8**).
4. Rest the tissue in Trizol on ice for 5 min.
5. Add 100 μL BCP and vortex for 15 s or until the sample turns milky and opaque.
6. Rest the samples on ice for 10 min.
7. Spin down the samples for 15 min at 16,000 × *g* at 4 °C.
8. Pipette the upper layer carefully into a new tube (without including the white interface).
9. Add 0.7 mL of ice-cold isopropanol and invert tubes to mix.
10. Store the tubes at −20 °C for at least 20 min (*see* **Note 9**).
11. Spin down the tubes for 15 min at 16,000 × *g* at 4 °C.
12. Discard the isopropanol supernatant.
13. Add 1 mL ice-cold 70 % ethanol (made with distilled H_2O).
14. Spin down the tubes for 20 min at 16,000 × *g* at 4 °C.
15. Discard the ethanol and air-dry the pellet at room temperature for at least 5 min (*see* **Note 10**).
16. Resuspend the RNA in 30 μL of molecular grade water.

3.4 Setting Up the Plate

1. Measure the concentration of the total isolated RNA in all samples.
2. All reactions take place in a black 384-well flat-bottomed plate. If the plate reader can measure fluorescence in a 96-well plate, these plates may be used instead.
3. Calculate the volumes of each reagent using the following guidelines. The total volume is 50 μL. Use 1 μL of the diluted molecular beacon stock (200 nM). At least 1 μg of total RNA sample should be used for each reaction. In order to maintain the salt concentration of the reaction, add an appropriate volume of 10 M NaCl to compensate for the volume of RNA added (*see* **Note 11**). Last, complete the 50 μL reaction with 1× PBS buffer. To compare concentration levels across multiple RNA samples, use an equal amount of total RNA in each well (*see* **Note 11**).
4. Add the required volume of 1× PBS buffer and 10 M NaCl to each of the reaction wells, and then add the RNA sample to each well. Last, add the molecular beacon to each reaction well in the plate. Mix the contents of each well by pipetting up and down slowly (*see* **Notes 12–14**).

5. Include background controls on the plate by saving some wells for molecular beacon added to buffer components only (*see* **Notes 15** and **16**).
6. Cover the entire plate with aluminum foil and incubate the plate at 37 °C for at least 20 min. Plates may be incubated overnight with little change in signal.
7. Place the plate in a plate reader and determine the fluorescence intensity of each well at the desired wavelength. Most plate readers can perform this task in less than 5 min.

4 Notes

1. Modified bases include locked nucleic acid bases (LNA), 2′-*O*-methylated bases, phosphorothioated bases, or unnatural bases. Molecular beacons with these modified bases are less susceptible to degradation (e.g., 2′-*O*-methyl) and/or enhanced stability of hybridization (e.g., LNA) to target RNA. However, the LNA beacons tend to have higher background signal, likely due to difficulty in forming hairpin structure.
2. Do not begin a molecular beacon sequence with a guanine at the 5′ end nearest the fluorophore. Guanine can quench some fluorescence signal.
3. Cytosine–guanine interactions are stronger than adenine–thymine interactions and are generally used more often in stem sequences. If a 5-nt all GC stem is too strong (difficulty with hairpin opening), AT nucleotides may be substituted to weaken the stem.
4. Runs of G or C bases are stronger than interspersed bases and may make the stem stronger.
5. Remember that RNA–RNA interactions are stronger than DNA–RNA interactions, so when designing RNA molecular beacons, the stem should be shorter or weaker than a DNA beacon.
6. Not all fluorophores are equal. Some dyes are temperature sensitive, more expensive to produce, or easily bleached. Ensure that the fluorophore chosen fits the need and the conditions for molecular beacon use. We have used Cy3, Cy5, 6-FAM, Texas Red, and HEX dyes.
7. Ensure that the quencher is adequate for the fluorophore chosen. We have used Iowa Black quencher (IDT Technologies) and Black Hole Quencher 1 (Biosearch Technologies).
8. Go directly from the frozen mortar and pestle to the Trizol buffer to prevent degradation of RNA.
9. The samples mixed with isopropanol can be stored at −20 °C overnight. If longer storage than overnight is required, store

the tubes at −80 °C. For samples with low RNA concentrations, storage of the samples at −20 °C overnight will increase the RNA yield.

10. The tubes may also be spun down again to remove any excess supernatant with a fine pipette.
11. The less volume needed to obtain 1 μg of RNA, the better because the RNA sample does not contain any salt. For example, if the 50 μL reaction sample is composed of 25 μL of RNA, then the salt concentration of the reaction is half of what it should be. Add 0.375 μL of 10 M NaCl to compensate for the loss of salt due to the addition of the 25 μL RNA sample.
12. Load the plate with the lights off to reduce the loss of fluorescence signal.
13. Thaw all reagents before loading the plate.
14. Perform each reaction in triplicate to compensate for loading or pipetting errors or plate reader inconsistencies.
15. Most plate readers measure fluorescence in relative fluorescence units (RFU), so all measurements are relative.
16. A standard curve using microRNA mimic of known concentration may be added to the reaction plate for each experiment to estimate total concentrations of microRNAs.

Acknowledgments

This work was supported by the National Heart Lung and Blood Institute of the National Institutes of Health as a Program of Excellence in Nanotechnology (HHSN268201000043C to G.B.) and a VA Merit Award (I01 BX000704 to C.D.S.).

References

1. Tyagi S, Kramer FR (1996) Molecular beacons: probes that fluoresce upon hybridization. Nat Biotechnol 14:303–308
2. Tyagi S, Bratu DP, Kramer FR (1998) Multicolor molecular beacons for allele discrimination. Nat Biotechnol 16:49–53
3. Vet JA, Majithia AR, Marras SA, Tyagi S, Dube S, Poiesz BJ, Kramer FR (1999) Multiplex detection of four pathogenic retroviruses using molecular beacons. Proc Natl Acad Sci USA 96:6394–6399
4. Bao G, Rhee WJ, Tsourkas A (2009) Fluorescent probes for live-cell RNA detection. Annu Rev Biomed Eng 11:25–47
5. Guo J, Ju J, Turro NJ (2012) Fluorescent hybridization probes for nucleic acid detection. Anal Bioanal Chem 402:3115–3125
6. Nitin N, Bao G (2008) NLS peptide conjugated molecular beacons for visualizing nuclear RNA in living cells. Bioconjug Chem 19:2205–2211
7. Nitin N, Santangelo PJ, Kim G, Nie S, Bao G (2004) Peptide-linked molecular beacons for efficient delivery and rapid mRNA detection in living cells. Nucleic Acids Res 32:e58
8. Santangelo P, Nitin N, Bao G (2006) Nanostructured probes for RNA detection in living cells. Ann Biomed Eng 34:39–50
9. Santangelo PJ, Nix B, Tsourkas A, Bao G (2004) Dual FRET molecular beacons for mRNA detection in living cells. Nucleic Acids Res 32:e57
10. Tsourkas A, Behlke MA, Bao G (2002) Structure-function relationships of shared-stem

and conventional molecular beacons. Nucleic Acids Res 30:4208–4215

11. Tsourkas A, Behlke MA, Xu Y, Bao G (2003) Spectroscopic features of dual fluorescence/luminescence resonance energy-transfer molecular beacons. Anal Chem 75:3697–3703
12. Baker MB, Bao G, Searles CD (2012) In vitro quantification of specific microRNA using molecular beacons. Nucleic Acids Res 40:e13
13. Kozomara A, Griffiths-Jones S (2011) miRBase: integrating microRNA annotation and deep-sequencing data. Nucleic Acids Res 39:D152–D157
14. Chang TC, Mendell JT (2007) microRNAs in vertebrate physiology and human disease. Annu Rev Genomics Hum Genet 8:215–239
15. Gilad S, Meiri E, Yogev Y, Benjamin S, Lebanony D, Yerushalmi N, Benjamin H, Kushnir M, Cholakh H, Melamed N et al (2008) Serum microRNAs are promising novel biomarkers. PLoS One 3:e3148
16. Osman A (2012) MicroRNAs in health and disease—basic science and clinical applications. Clin Lab 58:393–402
17. Le HB, Zhu WY, Chen DD, He JY, Huang YY, Liu XG, Zhang YK (2012) Evaluation of dynamic change of serum miR-21 and miR-24 in pre- and post-operative lung carcinoma patients. Med Oncol 29(5):3190–3197
18. Long G, Wang F, Duan Q, Chen F, Yang S, Gong W, Wang Y, Chen C, Wang DW (2012) Human circulating microRNA-1 and microRNA-126 as potential novel indicators for acute myocardial infarction. Int J Biol Sci 8:811–818
19. Saikumar J, Hoffmann D, Kim TM, Gonzalez VR, Zhang Q, Goering PL, Brown RP, Bijol V, Park PJ, Waikar SS et al (2012) Expression, circulation, and excretion profile of microRNA-21, -155, and -18a following acute kidney injury. Toxicol Sci 129:256–267
20. Zhou X, Marian C, Makambi KH, Kosti O, Kallakury BV, Loffredo CA, Zheng YL (2012) MicroRNA-9 as potential biomarker for breast cancer local recurrence and tumor estrogen receptor status. PLoS One 7:e39011
21. Markham NR, Zuker M (2008) UNAFold: software for nucleic acid folding and hybridization. Methods Mol Biol 453:3–31

Part VIII

RNA Imaging in Live Cells

Chapter 23

Sequence-Specific Imaging of Influenza A mRNA in Living Infected Cells Using Fluorescent FIT–PNA

Susann Kummer, Andrea Knoll, Andreas Herrmann, and Oliver Seitz

Abstract

Significant efforts have been devoted to the development of techniques allowing the investigation of viral mRNA progression during the replication cycle. We herein describe the use of sequence-specific FIT–PNA (Forced Intercalation Peptide Nucleic Acids) probes which contain a single intercalator serving as an artificial fluorescent nucleobase. FIT–PNA probes are not degraded by enzymes, neither by nucleases nor by proteases, and provide for both high sensitivity and high target specificity under physiological conditions inside the infected living host cell.

Key words FIT–PNA, Influenza A, mRNA detection, Living cells, Fluorescence microscopy

1 Introduction

Methods to study the biosynthesis of nucleic acids, such as RNA, in living cells, their intracellular transport, subcellular localization, and degradation are of great interest for understanding cellular networks, their regulation and malfunction [1–3]. In particular, assessing the the RNA expression of pathogens of pathogens provides insight into the time course of infection, the formation of pathogens components, and the assembly of new pathogens.

Fluorescently labeled oligonucleotide probes enable the visualization of RNA targets in fixed [4, 5] and in living [3, 6, 7] cells. The commonly used probe technologies rely on either the multiple linkage of fluorescent proteins or the distance dependent interaction between a fluorophore and a fluorescence quencher [2, 8, 9]. Typically, the last are designed such that formation of the probe–target complex will change the fluorescence properties owing to an altered distance between the two chromophores. However, probe degradation or unintended, yet unavoidable, protein binding can also affect the fluorophore-quencher distance. We have recently introduced the peptide nucleic acid (PNA)-based FIT-probes, which

Dmitry M. Kolpashchikov and Yulia V. Gerasimova (eds.), *Nucleic Acid Detection: Methods and Protocols*, Methods in Molecular Biology, vol. 1039, DOI 10.1007/978-1-62703-535-4_23, © Springer Science+Business Media New York 2013

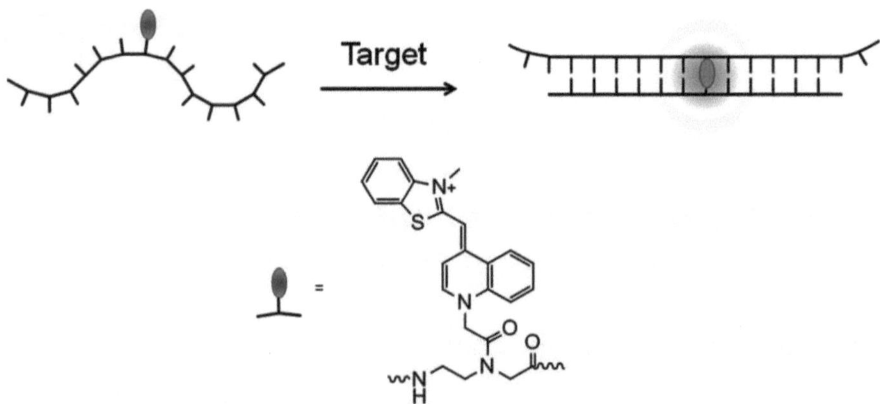

Fig. 1 Schematic depiction of the FIT–PNA principle and the chemical structure of TO. In the absence of the target, the fluorescence of FIT–PNA probes is low. Base stacking in the probe–target duplex upon hybridization results in the dramatic enhancements in the fluorescence of the cationic chromophore

circumvent off-target effects caused by nonspecific binding of proteins or degradation by nucleases [10–12].

In a PNA molecule the nucleobases (purine and pyrimidine) are attached to an uncharged pseudopeptide backbone, which provides stronger binding and more specific hybridization properties than DNA/DNA or DNA/RNA hybrids [13]. FIT–PNA probes contain an intercalating fluorophore which occupies the position of a canonical nucleobase [14, 15]. For example, the asymmetric cyanine dye thiazole orange (TO) was linked as base surrogate to PNA [16–19]. In the absence of target, the fluorescence of FIT–PNA probes is low because twisting around the central TO methine bridge depletes the excited state. Base stacking in the probe–target duplex will hinder the twisting motions. As a result, the cationic chromophore experiences dramatic enhancements in fluorescence upon hybridization with the target (Fig. 1). Up to 30-fold increases in fluorescence intensity have been achieved. Of note, neighboring base mismatches attenuate the fluorescence enhancement [17]. Thus, FIT–PNA probes discriminate base mismatches under conditions where both matched and mismatched probe–target complexes coexist.

Caused by their structural characteristics (e.g., pseudopeptide backbone) FIT–PNA probes are not subject to degradation, neither by nucleases nor by proteases. This eliminates false-positive signaling which occurs, e.g., when doubly labeled oligonucleotide probes are exposed to cytosolic enzymes. PNAs do not interact with cationic lipids or related transfection reagents. Therefore the delivery of PNAs into living MDCK cells requires helpers such as cell penetrating peptides, inbuilt positively charged residues, partially complementary DNA strands or streptolysin O (SLO), a streptococcal hemolytic exotoxin, which mediates reversible plasma membrane penetration [20].

In this chapter, we present a protocol for visualizing viral mRNA in living infected MDCK cells in a sequence-specific manner using FIT–PNA probes.

2 Materials

2.1 Synthesis of FIT–PNA (2 μmol)

1. 0.5 mL reactors for the synthesizer.
2. NovaSyn®TGR resin loaded with Fmoc-L-lysine (150 μmol/g, Novabiochem).
3. Fmoc cleavage solution: dimethylformamide (DMF)/piperidine (4:1) [v/v]. Mix 400 μL of DMF with 100 μL of piperidine.
4. HCTU solution: 0.6 M 2-(6-chloro-1-*H*-benzotriazole-1-*yl*)-1,1,3,3-tetramethylaminium hexafluorophosphate (HCTU) in *N*-methylpyrrolidone (NMP). Dissolve 12.4 mg of HCTU in 50 μL of NMP.
5. NMM solution: 4 M *N*-methylmorpholine (NMM) in DMF. Mix 8.8 μL of NMM with 11.2 μL of DMF.
6. PNA monomer solution: 0.2 M Fmoc/Bhoc-protected PNA monomer in NMP.
7. TO-PNA monomer solution: 0.2 M Fmoc-protected TO-PNA monomer in NMP.
8. Amino acid solution: 0.3 M Fmoc-L-lysine(Boc)-OH in NMP.
9. Capping solution: acetic anhydride/2,6-lutidine/DMF (5:6:89) [v/v/v].
10. Cleavage solution: 29 mM cysteine methyl ester hydrochloride in TFA/triisopropylsilane/H_2O (18:1:1) [v/v/v]. For 1 mL cleavage solution mix 900 μL of TFA, 50 μL of triisopropylsilane, 50 μL of H_2O, and 5 mg of cysteine methyl ester hydrochloride.
11. Ethanol, diethyl ether.
12. Water-equilibrated SepPak® C18 cartridge (Waters).
13. Gradient solution 1: 20 % acetonitrile, 0.1 % trifluoroacetic acid (TFA) in H_2O. Mix 400 μL of acetonitrile, 2 μL of TFA, 1,598 μL of water.
14. Gradient solution 2: 30 % acetonitrile, 0.1 % TFA in H_2O. Mix 600 μL of acetonitrile, 2 μL of TFA, 1,398 μL of water.
15. Gradient solution 3: 40 % acetonitrile, 0.1 % TFA in H_2O. Mix 800 μL of acetonitrile, 2 μL of TFA, 1,198 μL of water.
16. RP-C18 A polaris (PN A 2000-250 × 100) columns (Varian).
17. Eluent A: 1 % acetonitrile, 0.1 % TFA in H_2O. Mix 10 mL of acetonitrile and 1 mL of TFA. Add water up to 1 L. Degas.
18. Eluent B: 99 % acetonitrile, 0.1 % TFA in H_2O. Mix 10 mL of water ad 1 mL of TFA. Add acetonitrile up to 1 L. Degas.

2.2 Spectral Characterization of FIT–PNA

1. Buffered solution: 100 mM NaCl, 10 mM NaH_2PO_4, pH 7.0. Dissolve 1.17 g of NaCl and 0.276 g of NaH_2PO_4 in 200 mL of ultrapure water. Degas the solution and adjust pH to 7.0 with 1 M NaOH. Store at 4 °C (*see* **Note 1**).
2. Complementary RNA solution: 5 μM of complementary RNA.

2.3 Virus Growth and Purification

1. Embryonated chicken eggs.
2. Incubation chamber that allows constant temperature of 37 °C and rotation of the eggs.
3. Teflon-coated homogenizer.
4. Ultracentrifuge.

2.4 Cell Growth and Infection

1. DMEM with supplements: Dulbecco's minimum essential medium (DMEM) supplemented with 10 % fetal bovine serum (FBS), 1 mM L-glutamine, 1 % penicillin/streptomycin.
2. Stock streptolysin O (SLO) solution: 1 mg/mL (~25,000–50,000 U) of SLO in water. Add the amount of ultrapure water recommended by the supplier to the lyophilized powder of streptolysin O and mix to obtain a solution. Split the solution into the aliquots of 10–20 μL and store at −20 °C (*see* **Note 2**).
3. Final SLO solution: 5 U of SLO in 500 μL of 25 mM HEPES, pH 7.4, 10 mM DTT in Dulbecco's phosphate-buffered saline (DPBS).

2.5 Instrumentation

1. ResPep parallel synthesizer (*see* **Note 3**).
2. HPLC system.
3. Fluorescence Spectrometer with thermostat.
4. UV–Vis Spectrometer.
5. Laminar hood.
6. Humid incubator (for cell culture).
7. Confocal Microscope.

3 Methods

All steps involving viruses have to be carried out at room temperature if not stated otherwise regarding the S2 safety instructions (*see* **Note 4**).

3.1 Synthesis of FIT–PNA

A similar synthesis protocol has already been published [21]. In the published protocol, FIT–PNAs were synthesized by using automated solid-phase synthesis and purified by high pressure liquid chromatography (HPLC).

1. Allow the resin (13.5 mg) to swell in DMF (300 μL) for 30 min in 0.5-mL reactors, than transfer it to the synthesizer and wash with DMF (3 × 200 μL).
2. *Fmoc cleavage*: add 2 × 200 μL of Fmoc cleavage solution to the resin and allow it to stay for 2 min after each addition, and then wash the resin three times with 200 μL of DMF.
3. *Coupling of PNA-building blocks*: Charge a vessel with HCTU solution (12 μL), NMM solution (5 μL), and PNA monomer solution (40 μL). After pre-activation for 2 min transfer this mixture to the resin and allow it to react for 30 min. Then wash the resin with DMF (3 × 200 μL).
4. *Coupling of Fmoc-Aeg(TO)-COOH*: Charge a vessel with HCTU solution (12 μL), NMM solution (6 μL), and TO-PNA monomer solution (40 μL). After pre-activation for 2 min transfer this mixture to the resin and allow it to react for 30 min. Then wash the resin with DMF (3 × 200 μL).
5. *Coupling of lysine*: Charge a vessel with HCTU solution (18 μL), NMM solution (6 μL), and amino acid solution (40 μL). After pre-activation for 2 min transfer this mixture to the resin and allow it to react for 30 min. Then wash the resin with DMF (3 × 200 μL).
6. *Capping*: Add 200 μL of the capping solution to the resin and allow it to stay for 3 min. Wash the resin with DMF (2 × 200 μL).
7. *Synthesis of the whole sequence*: Repeat **step 2**, **3**, and **6** as cycles with each PNA monomer until the TO-monomer has to be introduced.
8. Perform **step 2**, 2× **step 4**, and final **step 6** for the introduction of the TO-monomer.
9. Perform **step 2**, 2× **step 3**, and final **step 6** with each PNA monomer to complete assembly of the PNA sequence.
10. To add a N-terminal lysine residue (increase solubility and target affinity) perform **step 2**, 2× **step 5**, and final **step 6** with Fmoc-L-lysine(Boc)-OH.
11. *Final washing*: 5 × 200 μL DMF, 3 × 200 μL ethanol.
12. *Cleavage*: Pass the cleavage solution through the dried resin in 30 min (5 × 200 μL). Wash the resin with TFA (2 × 200 μL). Concentrate the combined filtrates (about 1.4 mL) in vacuo at 40 °C to 30 μL.
13. *Purification*: Add ice-cold diethyl ether (1 mL) to 30 μL of the concentrated filtrates. Collect the precipitate by centrifugation (6 × *g*, 30 s) and dispose the supernatant. Repeat this procedure to wash the precipitate. Dissolve the remaining pellet in 200 μL of water and pass it through a water-equilibrated SepPak®C18 cartridge. Carry out the gradient elution of the synthesized FIT–PNA from the cartridge: consequently pass

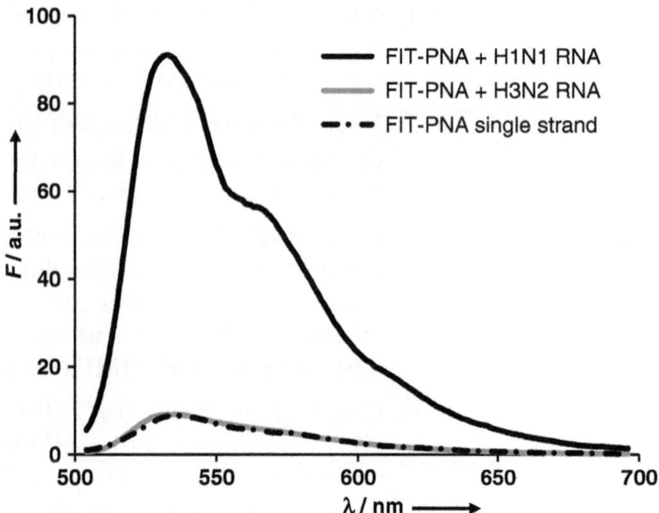

Fig. 2 Fluorescence spectra of H1N1 specific a) FIT-PNA 1a (H-Lys-cagtta-Aeg(TO)-tatgccgttg-Lys-NH2; λex = 485 nm) and b) MB 2 (λex = 559 nm) before and after addition of matched H1N1 RNA (5`-CAACGGCAUAAUAACUG-3`)3a or 3b) or mismatched H3N2 RNA (5′-UCAAAAGAUAAUAACAA-3′) 4 at 37°C. Conditions: 1μM probe and 10 μM target in 100 mM NaCl, 10 mM NaH2P04, pH 7.0. For sequence details of probes and oligonucleotides please review Kummer et al. (2011) [22]

2 mL of gradient solution 1, gradient solution 2, and gradient solution 3 through the cartridge. Collect the colored eluates of FIT–PNA. Purify the eluates by semi-preparative HPLC using a Varian Polaris C18 A 5μ (PN A 2000-250-100) column and a 3–30 % gradient of eluent B in eluent A for 30 min, flow rate 6 mL/min. Evaporate in vacuum at room temperature to dryness. Dissolve the purified FIT–PNA in 300 μL of water. Analyze by MALDI-TOF/MS.

14. *Yield determination*: Take a 5 μL aliquot of the purified FIT–PNA solution and dilute it in water up to 1 mL. Measure the optical density of the diluted solution at 260 nm by using a quartz cuvette with a 10-mm path length. Calculate the amount of FIT–PNA (*see* **Note 5**).

3.2 Spectral Characterization of FIT–PNA

To evaluate (and verify) the fluorescence enhancement upon target hybridization, measure the fluorescence emission of FIT–PNA in the absence and presence of a complementary artificial target sequence (Fig. 2).

1. Prepare 1 μM of FIT–PNA probe in the buffered solution.
2. Transfer the probe solution into a fluorescence quartz cuvette (3 × 3 mm).
3. Record TO fluorescence spectrum by excitation at 485 nm ($Slit_{Ex}$ = 10 nm, $Slit_{Em}$ = 5 nm).

4. Add 10 μM (final concentration) complementary RNA to the FIT–PNA probe solution, heat up to 95 °C for 1 min and cool down to 37 °C within 5 min. After 5 min incubation at 37 °C record the fluorescence spectrum with the same parameters used above (*see* **Note 6**).

3.3 Virus Growth and Purification

1. Cultivate embryonated chicken eggs in an incubation chamber maintaining 37 °C and allowing the automated rotation of the eggs.
2. At day 11 after fertilization inject 200 μL of allantoic fluid (dilution 1:500,000 of a former preparation) into the allantoic cavity of each egg (*see* **Note 9**).
3. Let the virus propagate for 48 h and collect the allantoic fluid of the inoculated eggs. Centrifuge the allantoic fluid at 3,000 × *g* for 30 min to clean it from cell debris.
4. Purify virus particles by pelleting them at 100,000 × *g* for 90 min and resuspend the pellet in DPBS. Use a Teflon-coated homogenizer to homogenize the virus solution.
5. Determine the titer of hemagglutinin (HA; spike protein of influenza A virus) employing the hemagglutination assay with human blood [23].
6. Store the aliquoted virus stocks at −80 °C.

3.4 Cell Culture

The SLO-based protocol for cellular uptake of FIT–PNAs is suitable to many mammalian cell lines. However, other cell lines may require adaptation of the protocol (*see* **Notes 7** and **8**).

1. Use MDCK cells in DMEM (without phenol red) supplemented with 10 % FBS, 1 mM L-glutamine, and 1 % penicillin/streptomycin in a humid incubator at 37 °C and 5 % CO_2. Seed cells subjected to virus infection in 35 mm glass bottom dishes and incubate as aforementioned to 80 % confluence.
2. Count cells before seeding to calculate the multiplicity of infection either manually by using a Neubauer counting chamber or automatically employing a cell coulter counter.
3. On day of staining, remove all medium from the cells and replace it with 1 mL of fresh DMEM without supplements.

3.5 Infection and Staining

In the following the influenza A infection protocol is merged with the FIT–PNA probe staining procedure.

1. Infect MDCK cells with the virus stock solution using a multiplicity of infection (M.O.I.) of 100, i.e., one cell is virtually infected by 100 virus particles (*see* **Note 10**). To this end, add the appropriate volume of the virus stock solution directly into the supernatant of your glass bottom dish (*see* **Note 11**).

2. Incubate the infected cells at 37 °C for 1 h to let the virus attach to the cell surface.
3. Replace the virus containing supernatant with 1 mL of fresh DMEM with supplements and continue incubation at 37 °C until the requested time point (*see* **Note 12**).
4. Activate 500 μL of final SLO solution (*see* Subheading 2.4, **item 3** and **Note 2**) to fulfill its activity.
5. Prepare 250 nM FIT-PNA in 500 μl activated final SLO solution.
6. For staining, add the SLO-containing FIT-PNA solution (*see* Subheading 3.5, **step 5**) to the cells for 30 min at 37 °C.
7. Pre-warm 2 mL of DMEM without supplements at 37 °C.
8. To inactivate and thus reseal the SLO treated cells, carefully add 2 mL of the pre-warmed DMEM and incubate for 30 min at 37 °C.

3.6 Fluorescence Microscopy Imaging

1. Before fluorescence microscopy imaging, change the medium of the cells (*see* Subheading 3.5, **step 8**) to 1 mL of DMEM without supplements in order to reduce background. Keep the cells at 37 °C during measurement.
2. Adjust the laser power and excitation wavelength to obtain the signal of the desirable intensity (*see* **Note 13**).
3. Measure untreated, unstained as well as uninfected cells as control any time (Fig. 3).

4 Notes

1. Degasing the solution is required to avoid formation of bubbles inside the cuvettes used for spectroscopy.
2. SLO is an enzyme, and it can lose its activity upon multiple freeze and thaw cycles. To fulfill its activity, SLO needs to be activated at 37 °C for 90 min before use.
3. Reaction volumes are specific for the Intavis system.
4. Give attention to hazardous biological waste such as influenza A-infected cells, contaminated pipet tips, or plastic ware. These may be harmful to health and environment. Please consider the instruction of your country for disposal of these materials (Germany: http://www.gesetze-im-internet.de/biostoff/index. html).
5. To calculate the amount of FIT–PNA, use the formula $A_{260} = C \times \varepsilon_{260} \times l$, where A_{260} is the absorbance of the solution at 260 nm, C is the concentration of FIT–PNA in the solution, ε_{260} is the extinction coefficient of the solution at 260 nm, l is the cuvette's path length in cm. To determine the extinction

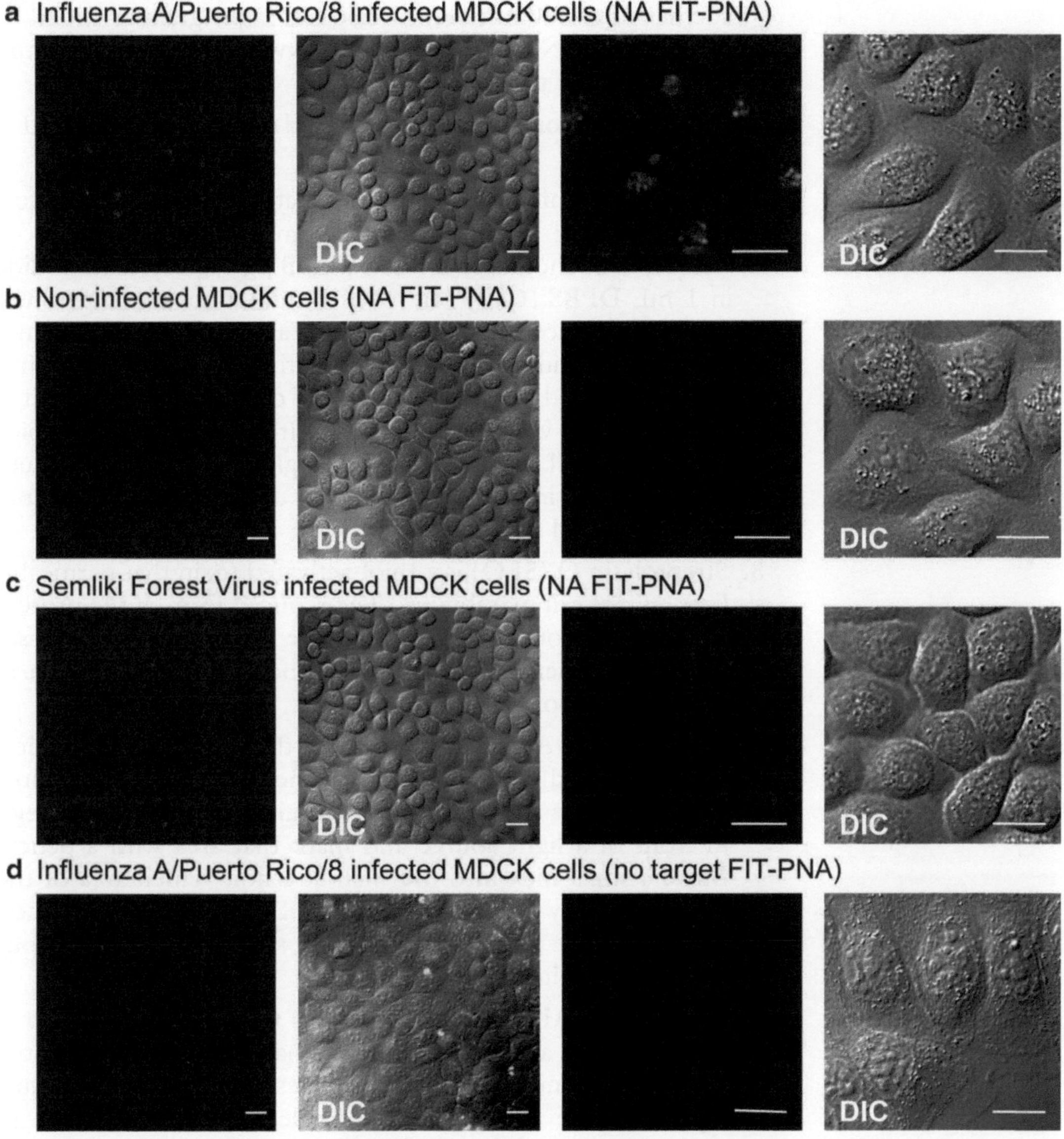

Fig. 3 (**a–d**) Confocal laser scanning microscopy images of living MDCK cells stained with the FIT-PNA probe specific for influenza A NA (H-Lys(PEG)-catgtta-Aeg(TO)-tatgccgttg-Lys-NH2) (**a–c**) or the no target FIT-PNA probe (H-Lys(PEG)-cgttt-Aeg(TO)-taatcgtctc-Lys-NH2) (**d**) 4.5 h after infection (M.O.I. multiplicity of infection 100). FIT-PNAs were excited with a 488 nm argon ion laser in living MDCK cells. Images were recorded with an inverted confocal laser scanning microscope at 37°C and $\lambda ex = 488$ nm. The two sets of panels on the right represent a threefold magnification of the left two. White bars: 10 μm (DIC differential interference contrast). For details please review Kummer et al. (2011) [22]

coefficient at 260 nm use the calculation program provided by Integrated DNA Technologies (www.eu.idtdna.com) and approximate as thymine.

6. Buffer-only fluorescence (background) should be subtracted at any time.
7. Due to virus infection or SLO-mediated transient cell perturbation the cell viability might be impaired. For testing cell viability cells should be treated with 5 μg of propidium iodide in 1 mL DPBS (Ca^{2+}/Mg^{2+}) for 5 min at room temperature and washed once with fresh DPBS (Ca^{2+}/Mg^{2+}) prior imaging. Propidium iodide is a DNA intercalating fluorophore which is commonly used for identifying dead or apoptotic cells as it is excluded from viable cells. For a positive control, i.e., propidium iodide cell staining, use cells (glass bottom cell culture dish, 35 mm) incubated with 10 μl of 1 N HCl for 20 min resulting in cell apoptosis.
8. Streptolysin O (SLO) is a thiol-activated toxin that permeabilizes animal cell membranes. The reduced form of the protein binds as a monomer to membrane cholesterol and subsequently polymerizes into large arc- and ring-shaped structures surrounding pores of >12 nm.
9. For influenza A virus propagation the allantoic inoculation was performed. In a chicken egg the allantoic cavity is situated below the air sac. Localize the air sac by placing the egg in front of a light source and mark that area with a pencil. Make a small nick into the shell at a non-veined area of the allantoic cavity and a hole using a Dremel motorized tool. Otherwise the pressure in the air sac will press out your injected inoculum.
10. Such a high M.O.I. is required for two reasons: first, virus replication is associated with formation of a rather large amount of noninfective viral particles. The HA titer test just represents the amount of HA molecules which is used to recalculate the virus titer. However, this does not provide information on the exact number of intact, i.e., infectious viral particles in your solution. Second, for studying the temporal progression of viral replication it is advantageous to infect all cells at once.
11. Influenza A/PR/8 does not need any trypsin treatment for an efficient infection. This may be required while working with other influenza strains and has to be checked.
12. Keep in mind that the infection process is not interrupted during the SLO treatment so that you have to consider the staining procedure into the infection duration. For example, if you want to measure your cells 5 h post infection you have to start the SLO treatment 4 h after adding the virus solution over the

cells because the entire staining procedure takes 1 h (SLO treatment and resealing, each 30 min).

13. Intercalating dyes like thiazole orange and its derivatives exhibit a high signal-to-background ratio but tend to have lower quantum yields than fluorescent proteins. Therefore use an appropriate laser power and optimal excitation wavelength.

References

1. Jayagopal A, Halfpenny KC, Perez JW, Wright DW (2010) Hairpin DNA-functionalized gold colloids for the imaging of mRNA in live cells. J Am Chem Soc 132:9789–9796
2. Bao G, Rhee WJ, Tsourkas A (2009) Fluorescent probes for live-cell RNA detection. Annu Rev Biomed Eng 11:25–47
3. Rhee WJ, Bao G (2009) Simultaneous detection of mRNA and protein stem cell markers in live cells. BMC Biotechnol 9:30
4. Raj A, van den Bogaard P, Rifkin SA, van Oudenaarden A, Tyagi S (2008) Imaging individual mRNA molecules using multiple singly labeled probes. Nat Methods 5:877–879
5. Manolakos E, Kefalas K, Neroutsou R, Lagou M, Kosyakova N, Ewers E et al (2010) Characterization of 23 small supernumerary marker chromosomes detected at pre-natal diagnosis: the value of fluorescence in situ hybridization. Mol Med Rep 3:1015–1022
6. Santangelo PJ (2010) Molecular beacons and related probes for intracellular RNA imaging. Wiley Interdiscip Rev Nanomed Nanobiotechnol 2:11–19
7. Kihara T, Yoshida N, Kitagawa T, Nakamura C, Nakamura N, Miyake J (2010) Development of a novel method to detect intrinsic mRNA in a living cell by using a molecular beacon-immobilized nanoneedle. Biosens Bioelectron 26:1449–1454
8. Wu B, Piatkevich KD, Lionnet T, Singer RH, Verkhusha VV (2011) Modern fluorescent proteins and imaging technologies to study gene expression, nuclear localization, and dynamics. Curr Opin Cell Biol 23:1–8
9. Dirks RW, Molenaar C, Tanke HJ (2001) Methods for visualizing RNA processing and transport pathways in living cells. Histochem Cell Biol 115:3–11
10. Agrawal S (1996) Antisense oligonucleotides: towards clinical trials. Trends Biotechnol 14:376–387
11. Good L, Nielsen PE (1998) Inhibition of translation and bacterial growth by peptide nucleic acid targeted to ribosomal RNA. Proc Natl Acad Sci USA 95:2073–2076
12. Good L, Nielsen PE (1997) Progress in developing PNA as a gene-targeted drug. Antisense Nucleic Acid Drug Dev 7:431–437
13. Egholm M, Buchardt O, Christensen L, Behrens C, Freier SM, Driver DA et al (1993) PNA hybridizes to complementary oligonucleotides obeying the Watson-Crick hydrogen-bonding rules. Nature 365:566–568
14. Paroo Z, Corey DR (2003) Imaging gene expression using oligonucleotides and peptide nucleic acids. J Cell Biochem 90:437–442
15. Seitz O, Bergmann F, Heindl D (1999) A convergent strategy for the modification of peptide nucleic acids: novel mismatch-specific PNA-hybridization probes. Angew Chem Int Ed Engl 38:2203–2206
16. Kohler O, Jarikote DV, Seitz O (2005) Forced intercalation probes (FIT Probes): thiazole orange as a fluorescent base in peptide nucleic acids for homogeneous single-nucleotide-polymorphism detection. Chembiochem 6: 69–77
17. Kohler O, Seitz O (2003) Thiazole orange as fluorescent universal base in peptide nucleic acids. Chem Commun 23:2938–2939
18. Jarikote DV, Köhler O, Socher E, Seitz O (2005) Divergent and linear solid-phase synthesis of PNA containing thiazole orange as artificial base. Eur J Org Chem 15:3187–3195
19. Bethge L, Jarikote DV, Seitz O (2008) New cyanine dyes as base surrogates in PNA: forced intercalation probes (FIT-probes) for homogeneous SNP detection. Bioorg Med Chem 16:114–125
20. Wang W, Cui ZQ, Han H, Zhang ZP, Wei HP, Zhou YF et al (2008) Imaging and characterizing influenza A virus mRNA transport in living cells. Nucleic Acids Res 36:4913–4928
21. Socher E, Jarikote DV, Knoll A, Roglin L, Burmeister J, Seitz O (2008) FIT probes: peptide nucleic acid probes with a fluorescent base surrogate enable real-time DNA quantification and single nucleotide polymorphism discovery. Anal Biochem 375:318–330
22. Kummer S, Knoll A, Socher E, Bethge L, Hermann A, Seitz O (2011) Fluorescence Imaging of Influenza H1N1 mRNA in Living Infected Cells Using Single-Chromophore FIT-PNA. Angew. Chemie. Int. Ed. 50: 1931–1934
23. Bamforth J (1949) Blood agglutination tests. Practitioner 163:36–42

Chapter 24

Application of Caged Fluorescent Nucleotides to Live-Cell RNA Imaging

Akimitsu Okamoto

Abstract

A caged fluorescent nucleic acid probe, which contains a nucleotide modified with one photolabile nitrobenzyl unit and two hybridization-sensitive thiazole orange units, has been synthesized for area-specific fluorescence imaging of RNA in a cell. The probe emits very weak fluorescence regardless of the presence of the complementary RNA, whereas it shows hybridization-sensitive fluorescence emission after photoirradiation for uncaging. Such probes are generated via several chemical synthetic steps and are applicable to area-specific RNA imaging in a cell. Only probes that exist in the defined irradiation area are activated via uncaging irradiation.

Key words Caged nucleic acid, Photolysis, Fluorescent probe, RNA imaging

1 Introduction

Sequence-specific binding of fluorescence-labeled DNA probes is often used for RNA imaging in a cell [1, 2]. In this case, the fluorescence intensity reflects the concentration of fluorescent probes in the cell, but does not always reflect the concentration of the target RNA, because fluorescent dyes inherently emit fluorescence even in the absence of their target RNA. It is important to overcome this disadvantage to achieve visualization and to monitor a specific RNA located in a limited area in the cell of interest. For such fluorescence imaging, fluorescent probes require (1) an increase in fluorescence by target-specific binding and a decrease in extraneous fluorescence by nonspecific binding, for exclusive visualization of the target RNA, (2) low background fluorescence, for suppression of the fluorescence from excess dye, and (3) high area-specific fluorescence, for spatiotemporal observation of the target RNA. These probes were designed to give us control over their fluorescence. An ideal fluorophore only emits a fluorescent signal when it recognizes a target molecule. Designing a fluorophore that

Dmitry M. Kolpashchikov and Yulia V. Gerasimova (eds.), *Nucleic Acid Detection: Methods and Protocols*, Methods in Molecular Biology, vol. 1039, DOI 10.1007/978-1-62703-535-4_24,

can be area-specifically activated by remote control, such as by photoirradiation, is also desirable.

The most popular area-specific photocontrols are fluorescence recovery after photobleaching (FRAP) [3] and fluorescence loss in photobleaching (FLIP) [4]. The inverse FRAP (iFRAP) method is occasionally used instead of these types of pinpoint fluorescence bleaching, as it provides area-specific fluorescence emission in which all the fluorescence of the specimen is bleached with the exception of a small region of interest [5]. Molecular caging systems are also effective for area-specific activation: the photolysis of a photolabile precursor of ATP is one of the well-known examples of photoactivation of a molecular function [6, 7].

In addition to such photoirradiation-based area-specific fluorescence control methods, the addition of target-specific fluorescence controls to the fluorescent probes should be a significant key feature in the design of the next generation of RNA imaging approaches; such probes would yield negligible emission in the single-stranded state, before recognition of a target RNA, and strong emission upon forming a duplex with a target RNA. Several fluorescent nucleic acid probes [8, 9] have been developed for RNA imaging in a cell. Furthermore, to achieve higher flexibility of probe sequence design and sensitive control of fluorescence emission, there is a need to design a fluorescence-controlled probe in which the functions of the area-specific probe initiated by photoactivation and the hybridization-sensitive fluorescence emission are packaged into one nucleotide.

In this paper, we report the preparation of a caged hybridization-sensitive fluorescent probe and its application to intracellular RNA monitoring [10]. Fluorescent probes containing a nucleotide with a photolabile group and hybridization-sensitive fluorescent dyes (Fig. 1) emit very weak fluorescence, regardless of the presence of the complementary RNA, whereas they show hybridization-sensitive fluorescence emission after photoirradiation for uncaging (Fig. 2). Such probes are applicable to live-cell RNA imaging via point photoactivation.

2 Materials

The route of the chemical synthesis is drawn in Fig. 3.

2.1 Synthesis of Compound 2 and Diastereomeric Resolution

1. (−)-Camphanic chloride.
2. 4,5-Dimethoxy-α-methyl-2-nitrobenzyl alcohol (**1**) [11].
3. Pyridine.
4. Ethyl acetate.
5. Saturated NaCl aqueous solution.
6. $MgSO_4$.

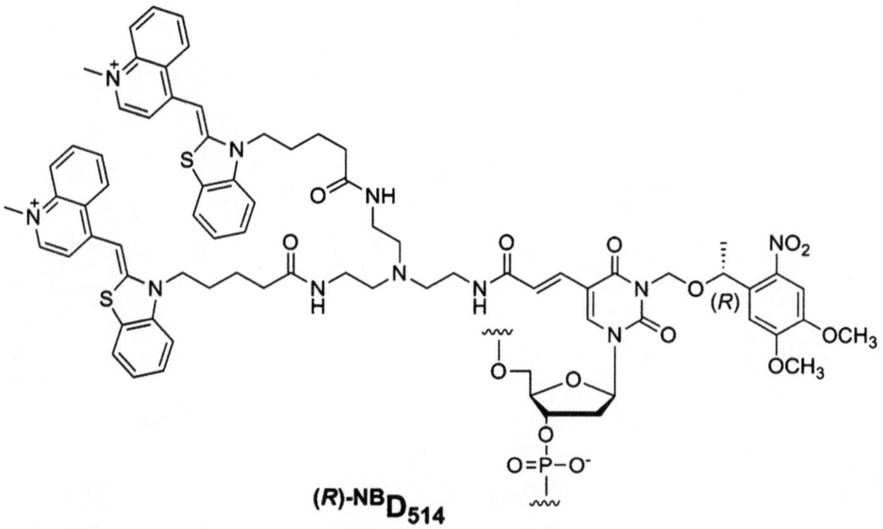

Fig. 1 Structure of $^{(R)\text{-NB}}D_{514}$

7. Silica gel.
8. Dichloromethane.
9. Hexane.

2.2 Synthesis of Compound (R)-1

1. KOH (Wako).
2. Water.
3. *N*,*N*-Dimethylformamide (DMF).
4. Ethyl acetate.
5. Saturated NaCl aqueous solution.
6. $MgSO_4$.
7. Silica gel.
8. Dichloromethane.
9. Hexane.

2.3 Synthesis of Compound (R)-3

1. Tetrahydrofuran (THF).
2. Sodium hydride (60 % dispersion in liquid paraffin).
3. Sodium iodide.
4. Chloromethyl methyl sulfide.
5. Ethyl acetate.
6. Saturated NaCl aqueous solution.
7. $MgSO_4$.
8. Silica gel.
9. Dichloromethane.
10. Hexane.

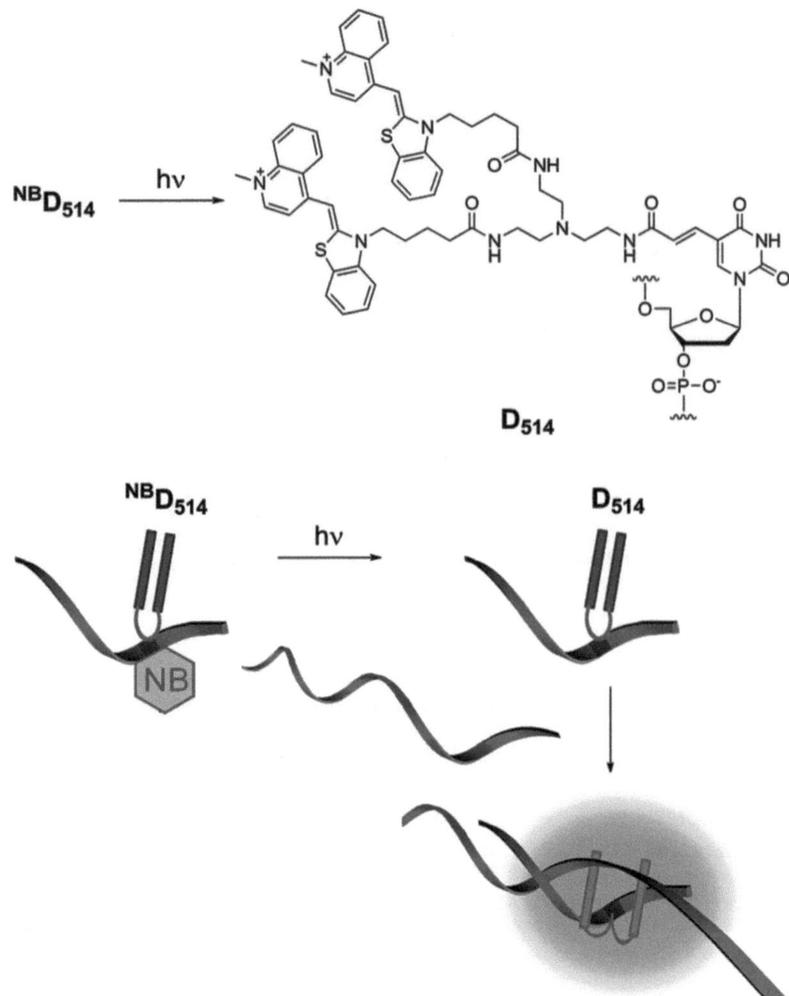

Fig. 2 Uncaging and emission control of $^{NB}D_{514}$-incorporated DNA probes

2.4 Synthesis of Compound (R)-5

1. Nucleoside **4** [12].
2. Sulfuryl chloride.
3. Dichloromethane.
4. Cesium carbonate.
5. DMF.
6. Ethyl acetate.
7. Saturated NaCl aqueous solution.
8. $MgSO_4$.
9. Silica gel.
10. Methanol.
11. Triethylamine.

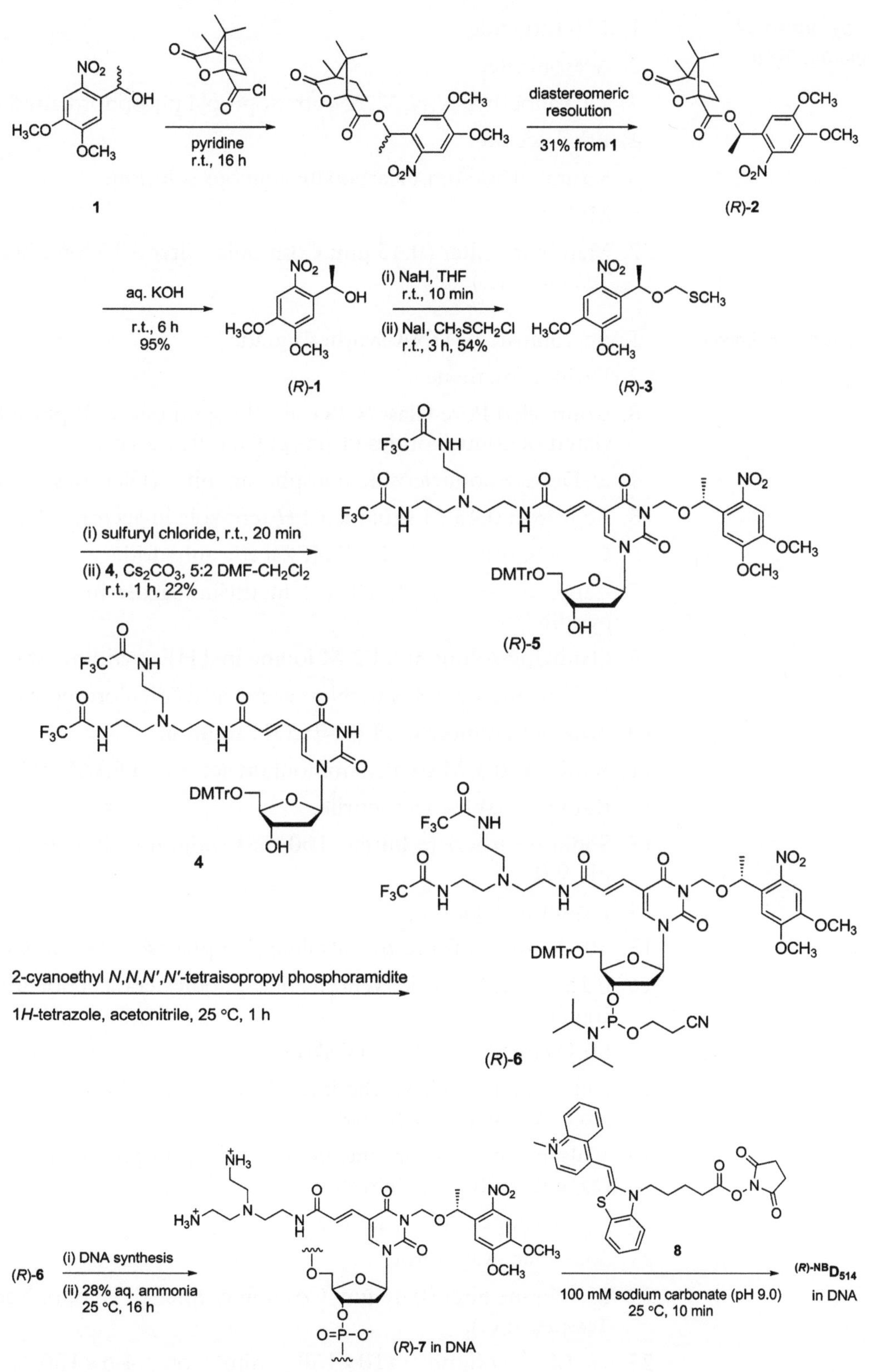

pyridine
r.t., 16 h

diastereomeric
resolution

31% from **1**

1

(*R*)-**2**

aq. KOH

r.t., 6 h
95%

(i) NaH, THF
r.t., 10 min

(ii) NaI, CH_3SCH_2Cl
r.t., 3 h, 54%

(*R*)-**1**

(*R*)-**3**

(i) sulfuryl chloride, r.t., 20 min

(ii) **4**, Cs_2CO_3, 5:2 DMF-CH_2Cl_2
r.t., 1 h, 22%

(*R*)-**5**

4

2-cyanoethyl *N*,*N*,*N*′,*N*′-tetraisopropyl phosphoramidite

1*H*-tetrazole, acetonitrile, 25 °C, 1 h

(*R*)-**6**

(*R*)-**6**

(i) DNA synthesis

(ii) 28% aq. ammonia
25 °C, 16 h

(*R*)-**7** in DNA

8

100 mM sodium carbonate (pH 9.0)
25 °C, 10 min

(*R*)-^{NB}D_{514}
in DNA

Fig. 3 Synthetic route for a caged fluorescent probe

2.5 Synthesis of Compound (R)-6

1. 1*H*-Tetrazole.
2. Acetonitrile.
3. 2-Cyanoethyl-*N*,*N*,*N*′,*N*′-tetraisopropyl phosphordiamidite.
4. Ethyl acetate.
5. Saturated sodium bicarbonate aqueous solution.
6. $MgSO_4$.
7. Membrane filter (0.45 μm; Cosmonice filter S 13 mm; Nacalai Tesque, Inc.)

2.6 Probe Synthesis

1. Acetonitrile of DNA-synthesis grade.
2. Dichloromethane.
3. Controlled Pore Glass (CPG) for the synthesis of 3′-phosphorylated oligonucleotides (1 μmol; Glen Research).
4. 2′-Deoxyribonucleoside phosphoramidites (Glen Research).
5. Activator: 0.45 M sublimed 1*H*-tetrazole in acetonitrile.
6. Capping solution A: 1:1 THF/acetic anhydride.
7. Capping solution B: 10 % 1-methylimidazole in 8:1 THF/pyridine.
8. Oxidizing solution: 0.02 M iodine in THF/pyridine/water.
9. Deblocking mix: 3 % trichloroacetic acid/dichloromethane.
10. Aqueous ammonia: 28 % NH_4OH solution.
11. Buffer A: 0.1 M triethylammonium acetate (TEAA), pH 7.0.
12. Buffer B: 100 % acetonitrile.
13. Sodium carbonate buffer: 100 mM sodium carbonate buffer, pH 9.0.
14. Liquid nitrogen.
15. 50 U/mL Calf-intestine alkaline phosphatase (Nippon Gene).
16. 0.15 U/mL Snake-venom phosphodiesterase (Boehringer Ingelheim).
17. 50 U/mL P1 nuclease (Wako).
18. Automated DNA synthesizer (ABI 392 DNA/RNA synthesizer, Applied Biosystems).
19. Vials and bottles for attachment of the phosphoramidites and reagents to the synthesizer.
20. Screw-capped tube (Assist).
21. SpeedVac evaporator.
22. Membrane filter (0.45 μm; Cosmonice filter W 13 mm; Nacalai Tesque, Inc.).
23. HPLC column (10 × 150 mm or 4.6 × 150 mm; CHEMCOBOND 5-ODS-H column; Chemco).

24. Centrifuge tubes (Falcon).
25. Dewar flask.
26. Freeze dryer.
27. Dye **8** [12].
28. Standard solution: 0.1 mM dA, 0.1 mM dC, 0.1 mM dG, and 0.1 mM dT.

2.7 Imaging and Uncaging

1. Cell culture medium: DMEM, 10 % heat-inactivated fetal bovine serum, 50 U/mL penicillin, and 50 mg/mL streptomycin
2. CO_2 gas.
3. Incubation system (INU, Tokai Hit).
4. Pneumatic injector (FemtoJet express, Eppendorf).
5. Glass needles (FemtoTip, Eppendorf).
6. 3-D manipulators (Narishige).
7. Motorized inverted microscope (Axio Observer Z1; Zeiss), equipped with a 63× objective (PlanApochromat NA 1.4, oil immersion) and a confocal unit (LSM510 META; Zeiss).
8. Operating software Zen2008.

3 Methods

The organic synthesis is relatively short and straightforward. However, careful attention to the details of the basic organic synthesis procedure is required. The preparation of the chemicals needs prior experience with routine chemical laboratory techniques, such as solvent evaporation, extraction, TLC, and column chromatography (*see* **Notes 1–3**). Carry out all procedures at room temperature, unless otherwise specified.

^{1}H, ^{13}C, and ^{31}P NMR spectra are measured on a Varian NMR system 500. Coupling constants (J values) are reported in Hz. The chemical shifts are shown in ppm downfield from tetramethylsilane. An external H_3PO_4 standard ($\delta = 0.00$) is used for ^{31}P NMR measurements. EI and ESI mass spectra are recorded on a JEOL JMS-T100LC and JMS-700V apparatuses, respectively.

3.1 Synthesis of Compound 2 and Diastereomeric Resolution (See Note 4)

1. Weigh 11.5 g (50.6 mmol) of 4,5-dimethoxy-α-methyl-2-nitrobenzyl alcohol (**1**) (*see* **Note 5**) and add to an oven-dried 500 mL round-bottom flask equipped with a large Teflon-coated magnetic stir bar and sealed with a rubber septum under an N_2 atmosphere.
2. Add 100 mL of pyridine into the flask.
3. Weigh 11.0 g (50.6 mmol) of (−)-camphanic chloride and add to the flask.

4. Stir the reaction mixture for 16 h.
5. Concentrate under reduced pressure in a vacuum pump.
6. Dissolve the residue in 200 mL of ethyl acetate and 200 mL of saturated NaCl.
7. Pour the mixture into a 1 L separatory funnel and wash the solution twice with 200 mL portions of saturated NaCl.
8. Dry over anhydrous $MgSO_4$ and filter off the drying reagent.
9. Concentrate under reduced pressure in a rotary evaporator.
10. Pack a chromatography column by mixing 250 g of dry silica gel with 1.5 L of 80 % dichloromethane–hexane and pour the silica gel into the column as a slurry.
11. Add the reaction products to the top of the packed chromatography column.
12. Elute the column using 80 % dichloromethane–hexane and collect 100 mL fractions.
13. Identify (*R*)-**2** via TLC by eluting thin-layer silica gel plates on glass backing with 80 % dichloromethane–hexane. Visualization can be achieved using a 254 nm UV lamp.
14. Evaporate the solvents of the combined fractions using a rotary evaporator and dry the white powder further using a vacuum pump for 2 h.
15. Recrystallize three times in dichloromethane and hexane; this gives (*R*)-**2** (6.32 g, 15.5 mmol, 31 % yield) as a white powder.
16. Characterize the product (*R*)-**2** using ^{1}H and ^{13}C NMR.

The white powder obtained can be stored neat at ambient temperature without decomposition over a period of 6 months.

Yield, 6.32 g (31 %, 99 % d.e.; 29 % recovery of an unseparated diastereomeric mixture of (*R*)-**2** and (*S*)-**2**): ^{1}H NMR ($CDCl_3$) δ 7.62 (s, 1H), 7.13 (s, 1H), 6.67 (q, J=6.4, 1H), 3.963 (s, 3H), 3.955 (s, 3H), 2.43 (ddd, J=13.5, 10.9, 4.2, 1H), 2.13–2.07 (m, 1H), 1.97–1.91 (m, 1H), 1.75–1.69 (m, 4H), 1.11 (s, 3H), 1.04 (s, 3H), 0.86 (s, 3H); ^{13}C NMR ($CDCl_3$) δ 178.0, 166.3, 153.5, 148.0, 139.5, 132.2, 107.9, 107.5, 90.6, 69.5, 56.3, 56.2, 54.6, 54.1, 30.7, 28.7, 21.9, 16.6, 16.5, 9.4; HRMS (ESI) calcd for $C_{20}H_{25}NO_8Na$ ([M+Na]$^+$) 430.1478, found 430.1473; $[\alpha]_D^{27}$ −180.2 °C (c 0.202, MeOH).

3.2 Synthesis of Compound (R)-1 (See Note 4)

1. Weigh 5.72 g (14.0 mmol) of (*R*)-**2** and add to an oven-dried 100 mL round-bottom flask (flask A).
2. Add 16 mL of DMF into the flask.
3. Weigh 3.62 g (65 mmol) of KOH and add to an oven-dried 10 mL round-bottom flask (flask B).

4. Add 4 mL of water into the flask.
5. Add the solution of flask B to flask A using a syringe.
6. Stir the reaction mixture for 6 h.
7. Add 200 mL of ethyl acetate and 200 mL of saturated NaCl.
8. Pour the mixture into a 1 L separatory funnel and wash the solution twice with 200 mL portions of saturated NaCl.
9. Dry over anhydrous $MgSO_4$ and filter off the drying reagent.
10. Concentrate under reduced pressure in a rotary evaporator.
11. Add 50 mL of dichloromethane and then 500 mL of hexane.
12. Collect the yellow precipitate and dry under reduced pressure.
13. Characterize the product (*R*)-**1** using ^{1}H and ^{13}C NMR.

The yellow powder obtained can be stored neat at ambient temperature without decomposition over a period of 6 months.

Yield, 3.03 g (95 %): ^{1}H NMR ($CDCl_3$) δ 7.55 (s, 1H), 7.31 (s, 1H), 5.55 (q, *J*=6.3, 1H), 4.00 (s, 3H), 3.94 (s, 3H), 2.58 (br, 1H), 1.55 (d, *J*=6.3, 3H); ^{13}C NMR ($CDCl_3$) δ 153.7, 147.6, 139.5, 136.9, 108.4, 107.6, 65.6, 56.34, 56.27, 24.2; HRMS (EI) calcd for $C_{10}H_{13}NO_5$ $[M]^+$ 227.0794, found 227.0798; $[\alpha]_D^{27}$ −151.4 °C (*c* 1.010, MeOH).

3.3 Synthesis of Compound (R)-3

1. Weigh 480 mg (12.0 mmol) of sodium hydride (60 %, dispersion in paraffin liquid) and add to an oven-dried 50 mL round-bottom flask (flask A) equipped with a large Teflon-coated magnetic stir bar and sealed with a rubber septum under an N_2 atmosphere.
2. Add 2 mL of hexane into the flask and remove the supernatant using a syringe. Repeat twice.
3. Add 5 mL of THF into the flask.
4. Weigh 909 mg (4.0 mmol) of (*R*)-**1** and add to an oven-dried 10 mL round-bottom flask (flask B) equipped with a large Teflon-coated magnetic stir bar and sealed with a rubber septum under an N_2 atmosphere.
5. Add 5 mL of THF into the flask.
6. Add the solution of flask B to flask A using a syringe.
7. Stir the reaction mixture for 10 min.
8. Add 600 mg (4.0 mmol) of sodium iodide into the flask.
9. Add 503 μL (6.0 mmol) of chloromethyl methyl sulfide into the flask.
10. Stir the reaction mixture for 3 h.
11. Concentrate under reduced pressure in a rotary evaporator.
12. Add 80 mL of ethyl acetate and 80 mL of saturated NaCl.

13. Pour the mixture into a 500 mL separatory funnel and wash the solution twice with 80 mL portions of saturated NaCl.
14. Dry over anhydrous $MgSO_4$ and filter off the drying reagent.
15. Concentrate under reduced pressure in a rotary evaporator.
16. Pack a chromatography column by mixing 25 g of dry silica gel with 1 L of 25 % hexane–dichloromethane and pour the silica gel into the column as a slurry.
17. Add the reaction products to the top of the packed chromatography column.
18. Elute the column using 25 % hexane–dichloromethane and collect 20 mL fractions.
19. 17. Identify (*R*)-**3** via TLC by eluting thin-layer silica gel plates on glass backing with 25 % hexane–dichloromethane. Visualization can be achieved using a 254 nm UV lamp.
20. Evaporate the solvents of the combined fractions using a rotary evaporator and dry the yellow oil further using a vacuum pump for 2 h.
21. Characterize the product (*R*)-**3** using ^1H and ^{13}C NMR.

The yellow oil obtained can be stored neat at ambient temperature without decomposition over a period of 6 months.

Yield, 619 mg (54 %): ^1H NMR ($CDCl_3$) δ 7.59 (s, 1H), 7.21 (s, 1H), 5.55 (q, J=6.4, 1H), 4.62 (d, J=11.4, 1H), 4.34 (d, J=11.4, 1H), 4.00 (s, 3H), 3.95 (s, 3H), 2.16 (s, 3H), 1.54 (d, J=6.4, 3H); ^{13}C NMR ($CDCl_3$) δ 153.6, 147.6, 140.3, 134.5, 108.3, 107.3, 73.0, 70.4, 56.2, 56.1, 23.1, 14.0; HRMS (ESI) calcd for $C_{12}H_{18}NO_5S$ ($[M+Na]^+$) 310.0725, found 310.0722; $[\alpha]_D^{28}$ −42.6 °C (c 0.530, MeOH).

3.4 Synthesis of Compound (R)-5

1. Weigh 439 mg (1.53 mmol) of (*R*)-**3** and add to an oven-dried 10 mL round-bottom flask (flask A) equipped with a large Teflon-coated magnetic stir bar and sealed with a rubber septum under an N_2 atmosphere.
2. Add 3 mL of dichloromethane into the flask.
3. Weigh 193 μL (2.38 mmol) of sulfuryl chloride and add to the flask.
4. Stir the reaction mixture for 20 min.
5. Concentrate under reduced pressure in a rotary evaporator.
6. Add 2 mL of dichloromethane into the flask.
7. Weigh 1.47 g (1.60 mmol) of **4** (*see* **Note 6**) [12] and 920 mg (4.77 mmol) of cesium carbonate and add to an oven-dried 30 mL round-bottom flask (flask B) equipped with a large Teflon-coated magnetic stir bar and sealed with a rubber septum under an N_2 atmosphere.
8. Add 5 mL of DMF into the flask.

9. Add the solution of flask A to flask B using a syringe.
10. Stir the reaction mixture for 1 h.
11. Add 200 mL of ethyl acetate and 200 mL of saturated NaCl.
12. Pour the mixture into a 1 L separatory funnel and wash the solution twice with 200 mL portions of saturated NaCl.
13. Dry over anhydrous $MgSO_4$ and filter off the drying reagent.
14. Concentrate under reduced pressure in a rotary evaporator.
15. Pack a chromatography column by mixing 15 g of dry silica gel with 1 L of 2 % methanol and 2 % triethylamine in dichloromethane and pour the silica gel into the column as a slurry.
16. Add the reaction products to the top of the packed chromatography column.
17. Elute the column using 2–10 % methanol and 2 % triethylamine in dichloromethane and collect 20 mL fractions.
18. Identify (*R*)-**5** via TLC by eluting thin-layer silica gel plates on glass backing with 2–10 % methanol and 2 % triethylamine in dichloromethane. Visualization can be achieved using a 254 nm UV lamp.
19. Evaporate the solvents of the combined fractions using a rotary evaporator and dry the white powder further using a vacuum pump for 2 h.
20. Characterize the product (*R*)-**5** using ^{1}H and ^{13}C NMR.

The white powder obtained can be stored neat at ambient temperature without decomposition over a period of 6 months.

Yield, 412 mg (22 %; Unconverted **4** 1.03 g (70 %)): ^{1}H NMR ($CDCl_3$) δ 7.87 (t, J=5.3, 2H), 7.77 (s, 1H), 7.49 (s, 1H), 7.43–7.41 (m, 2H), 7.32–7.27 (m, 9H), 7.01 (d, J=15.6, 1H), 6.86–6.83 (m, 4H), 6.47 (d, J=15.6, 1H), 6.12 (t, J=6.5, 1H), 5.40 (d, J=9.9, 1H), 5.36 (q, J=6.4, 1H), 5.31 (d, J=9.9, 1H), 5.24 (t, J=6.1, 1H),4.42–4.39 (m, 1H), 4.06–4.04 (m, 1H), 3.91 (s, 3H), 3.86 (s, 3H), 3.76 (s, 6H), 3.48 (dd, J=10.5, 3.9, 1H), 3.34–3.28 (m, 5H), 3.13–3.05 (m, 2H), 2.67–2.59 (m, 4H), 2.47–2.39 (m, 3H), 2.15–2.09 (m, 1H), 1.54 (d, J=6.4, 3H); ^{13}C NMR ($CDCl_3$) δ 166.8, 160.7, 158.7, 157.8 (q, J=37.7), 153.5, 149.5, 147.7, 144.4, 140.1, 138.3, 135.6, 135.4, 135.3, 132.3, 129.82, 129.79, 128.1, 127.9, 127.2, 121.1, 115.8 (q, J=287.5), 113.4, 113.3, 109.2, 108.6, 107.1, 86.6, 86.1, 86.0, 73.5, 71.6, 69.9, 63.3, 56.3, 56.1, 55.2, 55.0, 53.2, 46.0, 41.1, 38.1, 37.8, 23.7; HRMS (ESI) calcd for $C_{54}H_{59}N_7O_{15}F_6Na$ ($[M+Na]^+$) 1,182.3871, found 1,182.3884; $[\alpha]_D^{27}$ −113.5 °C (c 0.985, MeOH).

3.5 Synthesis of Compound (R)-6 (See Note 7)

1. Weigh 412 mg (0.36 mmol) of (*R*)-**5** and 50 mg (0.71 mmol) of 1*H*-tetrazole and add to an oven-dried 10 mL round-bottom flask equipped with a large Teflon-coated magnetic stir bar and sealed with a rubber septum under an N_2 atmosphere.

2. Add 5 mL of acetonitrile into the flask.
3. Weigh 338 μL (1.07 mmol) of 2-cyanoethyl-*N*,*N*,*N*′,*N*′-tetraisopropyl phosphordiamidite and add to the flask.
4. Stir the reaction mixture for 1 h.
5. Add 25 mL of ethyl acetate and 25 mL of saturated sodium bicarbonate.
6. Pour the mixture into a 100 mL separatory funnel and wash the solution twice with 25 mL portions of saturated NaCl.
7. Dry over anhydrous $MgSO_4$ and filter off the drying reagent.
8. Concentrate under reduced pressure in a rotary evaporator.
9. Add 5 mL of acetonitrile and concentrate under reduced pressure in a rotary evaporator. Repeat twice.
10. Add 2.4 mL of acetonitrile and pass through a 0.45 μm filter.
11. Use the resulting solution for automated DNA synthesis without further purification.

^{31}P NMR ($CDCl_3$) δ 149.639, 149.307; HRMS (ESI) calcd for $C_{63}H_{76}N_9O_{16}PF_6Na$ ($[M+Na]^+$) 1,382.4950, found 1,382.4944.

3.6 DNA Autosynthesis (See Note 8)

1. Use a 0.1 M solution of the phosphoramidite (*R*)-**6** in acetonitrile, prepared in Subheading 3.5 (*see* **Note 9**).
2. Set up a CPG support column.
3. Install the reagents and solvents in the synthesizer: amidites, activator, capping solutions A and B, oxidizing solution, deblocking solution, dry acetonitrile, and 28 % aqueous ammonia.
4. Start the automated solid-phase oligonucleotide synthesis.
5. Elongate the desired oligonucleotide and standard 2′-deoxyribonucleoside phosphoramidites in DMTr-OFF and auto-cleavage mode.
6. Collect the cleavage mixture and transfer it into a screw-capped tube.
7. Vortex the mixture and incubate at 25 °C for 16 h.
8. Evaporate the ammonia completely using a SpeedVac evaporator.
9. Filter the mixture through a 0.45 μm membrane filter prior to HPLC purification.
10. Purify the synthetic DNA via reverse-phase HPLC using the following recommended conditions:

    Gradient: 5–45 % Buffer B in Buffer A for 32 min

    Flow rate: 3.0 mL/min

    Detection: 254 nm

11. Collect the fraction containing the probe into a centrifuge tube and evaporate most of the acetonitrile using a SpeedVac evaporator.
12. Freeze the fraction in the tube by immersion in a Dewar bath using liquid nitrogen and concentrate using a freeze dryer.
13. Check the molecular weight of the oligonucleotides using MALDI–TOF mass spectrometry.

3.7 Incorporation of Dyes into DNA Probe

14. Prepare 0.1 mL solution of deprotected DNA in sodium carbonate buffer and add to a 15 mL plastic bottle.
15. Prepare 40 mM solution of the succinimidyl ester of thiazole orange pentanoic acid (**8**) (*see* **Note 10**) [12] in DMF and add to the plastic bottle.
16. Incubate at 25 °C for 10 min.
17. Dilute with 5 mL of cold ethanol for ethanol precipitation.
18. Keep at −78 °C for 1 h and centrifuge at 4 °C for 20 min.
19. Remove the supernatant.
20. Dissolve with 0.1 mL of water and filter the mixture through a 0.45 μm membrane filter prior to HPLC purification.
21. Purify the synthetic DNA via reverse-phase HPLC using the following recommended conditions:

    Gradient: 5–45 % Buffer B in Buffer A, 32 min

    Flow rate: 3.0 mL/min

    Detection: 254 nm
22. Collect the fraction containing the probe into a centrifuge tube and evaporate most of the acetonitrile using a SpeedVac evaporator.
23. Freeze the fraction in the tube by immersion in a Dewar bath using liquid nitrogen, and concentrate using a freeze dryer.
24. Check the molecular weight of the oligonucleotides using MALDI–TOF mass spectrometry.

3.8 Determination of the Probe's Concentration

25. Add 10 μL each of calf-intestine alkaline phosphatase, snake-venom phosphodiesterase, and P1 nuclease into an aliquot of the purified DNA (10 μL).
26. Incubate at 37 °C for 2 h.
27. Analyze digested solutions by HPLC using the following recommended conditions:

    Gradient: 3–10 % Buffer B in Buffer A, 20 min

    Flow rate: 3.0 mL/min

    Detection: 254 nm
28. Determine the concentration by comparing the HPLC peak areas with those of standard solution.

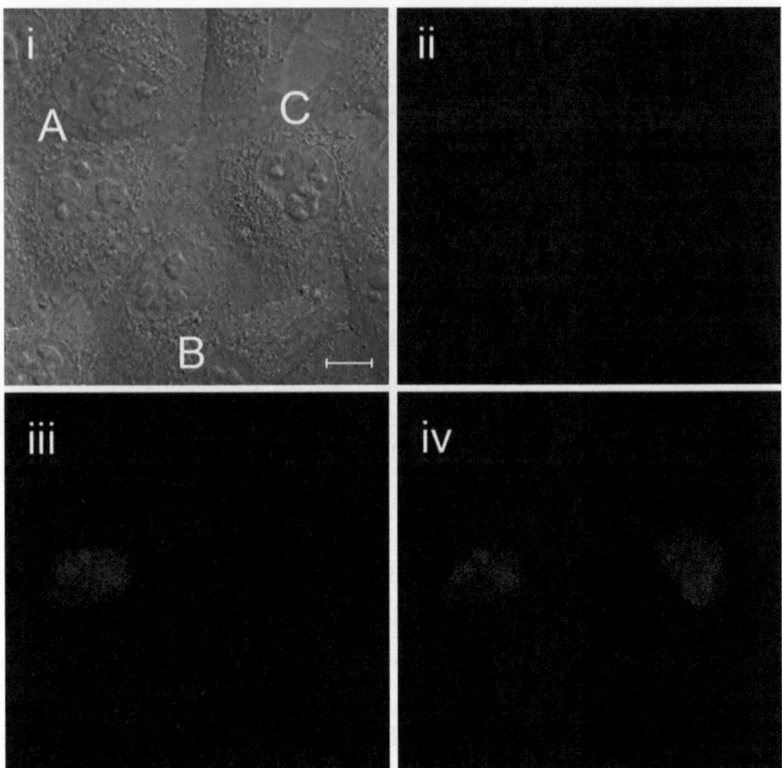

Fig. 4 Uncaging of probes in HeLa cells. (*i*) DIC image including the cell nuclei where $T_6^{(R)\text{-NB}}D_{514}T_6$ was microinjected (cells A and C); (*ii*) no laser irradiation; (*iii*) image acquired after exclusive irradiation of cell A with a 405 nm solid-state LED laser; and (*iv*) image acquired after exclusive irradiation of cell B with a 405 nm laser. Bar, 10 μm

3.9 Imaging and Uncaging (Fig. 4)

1. Culture HeLa cells (*see* **Note 11**).
2. Maintain cells at 37 °C, 5 % CO_2 in an incubation system during the observation.
3. Prepare a 20 μM filtered solution of the probe in water.
4. Inject the probe solution by glass needle microinjection using a pneumatic injector with glass needles and 3-D manipulators.
5. Acquire images with a motorized inverted microscope equipped with a 63× objective and confocal unit.
6. Perform uncaging with a 405 nm solid-state LED laser (40 mW, CW) using nine times laser radiation with full power (100 % setting in the software).
7. Excite the probes with the 514 nm Ar laser line for fluorescence imaging.
8. Analyze and process acquired images with the operating software Zen2008.

4 Notes

1. All glassware should be heat-treated at 120 °C for several hours and all reactions should be performed under nitrogen, to avoid moisture.
2. Perform all operations involving organic solvents and reagents in a well-ventilated fume hood, and wear gloves and protective glasses to prevent exposure to toxic chemicals and solvents.
3. Once the flask has cooled down, place an inert gas inlet into the septum of the flask, so that ambient air is not drawn into the dried flask.
4. At this stage, the procedure using 4,5-dimethoxy-α-methyl-2-nitrobenzyl alcohol (**1**) with high optical purity is introduced, but the caged probe prepared using racemic **1** without any optical resolution process shows almost the same efficiency of photoactivation. If using racemic **1**, the synthesis may be started from Subheading 3.3.
5. The starting material **1** can be purchased from chemical suppliers, but is expensive. This compound can be synthesized through reduction of 2-nitro-4,5-dimethoxyacetophenone, according to a protocol available in the literature [11].
6. Nucleoside **4** can be synthesized according to a protocol available in the literature [12].
7. Because the pale-yellow foam obtained as two diastereomeric isomers is unstable, it should be stored as (*R*)-**5** at −20 °C. Only the necessary quantity of (*R*)-**5** should be converted to the amidite (*R*)-**6** at the time of synthesis of the caged probes.
8. This protocol describes the automated synthesis, isolation, and characterization of oligonucleotides using (*R*)-**6**. In this protocol, the authors used an ABI 392 DNA/RNA synthesizer for the probe synthesis, but other automated DNA synthesizers are also available that would work in a fashion similar to that described in this protocol. For the coupling, 20 equivalents of amidites are used as 0.1 M solutions in acetonitrile. In the case of standard amidites, a coupling time of 180 s is sufficient to achieve quantitative yields. As (*R*)-**6** is less reactive than are standard phosphoramidites, the coupling time should be extended to at least 15 min. The isolation and purification of probes are conducted in this protocol using reverse-phase HPLC. The molecular mass of synthesized probes was checked by MALDI–TOF mass spectrometry (Ultraflex, Bruker Daltonics) in positive mode. Quantification of the concentration of each DNA was performed using HPLC analysis of fully digested DNA solutions.
9. To guarantee an efficient incorporation yield of phosphoramidite into oligodeoxynucleotides, it is necessary to prepare a fresh

and well-dried compound (*R*)-**6**, as the phosphoramidite (*R*)-**6** is labile and traces of moisture interfere with the coupling reaction.

10. Dye **9** can be synthesized according to a protocol available in the literature [12].

11. Culture HeLa cells in DMEM containing 10 % heat-inactivated fetal bovine serum, 50 U/mL penicillin, and 50 mg/mL streptomycin under a humidified atmosphere with 5 % CO_2 at 37 °C. For experimental use, culture cells (passages 5–9) in glass-based dishes. Prior to microscope observation, wash the culture medium and change it to phenol-red-free DMEM.

References

1. Tokunaga K, Tani T (2008) Monitoring mRNA export. Curr Protoc Cell Biol 22:1–20
2. Bancaud A, Huet S, Rabut G, Ellenberg J (2010) Fluorescence perturbation techniques to study mobility and molecular dynamics of proteins in live cells: FRAP, photoactivation, photoconversion, and FLIP. In: Goldman RD, Swedlow JR, Spector DL (eds) Live cell imaging: a laboratory manual, 2nd edn. Cold Spring Harbor Laboratory Press, New York
3. Lippincott-Schwartz J, Snapp E, Kenworthy A (2001) Studying protein dynamics in living cells. Nat Rev Mol Cell Biol 2:444–456
4. Cole NB, Smith CL, Sciaky N, Terasaki M, Edidin M, Lippincott-Schwartz J (1996) Diffusional mobility of Golgi proteins in membranes of living cells. Science 273:797–801
5. Dundr M, McNally JG, Cohen J, Misteli T (2002) Quantitation of GFP-fusion proteins in single living cells. J Struct Biol 140:92–99
6. Dantzig JA, Higuchi H, Goldman YE (1998) Studies of molecular motors using caged compounds. Methods Enzymol 291:307–348
7. Adams SR, Tsien RY (1993) Controlling cell chemistry with caged compounds. Annu Rev Physiol 55:755–784
8. Sando S, Kool ET (2002) Imaging of RNA in bacteria with self-ligating quenched probes. J Am Chem Soc 124:9686–9687
9. Mhlanga MM, Tyagi S (2006) Using tRNA-linked molecular beacons to image cytoplasmic mRNAs in live cells. Nat Protoc 1:1392–1398
10. Ikeda S, Kubota T, Wang DO, Yanagisawa H, Umemoto T, Okamoto A (2011) Design and synthesis of caged fluorescent nucleotides and application to live cell RNA imaging. Chembiochem 12:2871–2880
11. Dyer RG, Turnbull KD (1999) Hydrolytic stabilization of protected *p*-hydroxybenzyl halides designed as latent quinone methide precursors. J Org Chem 64:7988–7995
12. Ikeda S, Okamoto A (2008) Hybridization-sensitive on–off DNA probe: application of the exciton coupling effect to effective fluorescence quenching. Chem Asian J 3: 958–968

Index

Dmitry M. Kolpashchikov and Yulia V. Gerasimova (eds.), *Nucleic Acid Detection: Methods and Protocols*, Methods in Molecular Biology, vol. 1039, DOI 10.1007/978-1-62703-535-4, © Springer Science+Business Media New York 2013

MIX
Papier aus verantwortungsvollen Quellen
Paper from responsible sources
FSC® C105338

If you have any concerns about our products,
you can contact us on
ProductSafety@springernature.com

In case Publisher is established outside the EU,
the EU authorized representative is:
Springer Nature Customer Service Center GmbH
Europaplatz 3, 69115 Heidelberg, Germany

Printed by Libri Plureos GmbH
in Hamburg, Germany